PURINE METABOLISM IN MAN—II

Physiology, Pharmacology, and Clinical Aspects

ADVANCES IN EXPERIMENTAL MEDICINE AND BIOLOGY

Recent Volumes in this Series

Volume 72
FUNCTIONS AND METABOLISM OF PHOSPHOLIPIDS IN THE CENTRAL AND PERIPHERAL NERVOUS SYSTEMS
Edited by Giuseppe Porcellati, Luigi Amaducci, and Claudio Galli • 1976

Volume 73A
THE RETICULOENDOTHELIAL SYSTEM IN HEALTH AND DISEASE: Functions and Characteristics
Edited by Sherwood M. Reichard, Mario R. Escobar, and Herman Friedman • 1976

Volume 73B
THE RETICULOENDOTHELIAL SYSTEM IN HEALTH AND DISEASE: Immunologic and Pathologic Aspects
Edited by Herman Friedman, Mario R. Escobar, and Sherwood M. Reichard • 1976

Volume 74
IRON AND COPPER PROTEINS
Edited by Kerry T. Yasunobu, Howard F. Mower, and Osamu Hayaishi • 1976

Volume 75
OXYGEN TRANSPORT TO TISSUE–II
Edited by Jürgen Grote, Daniel Reneau, and Gerhard Thews • 1976

Volume 76A
PURINE METABOLISM IN MAN II: Regulation of Pathways and Enzyme Defects
Edited by Mathias M. Müller, Erich Kaiser, and J. Edwin Seegmiller • 1977

Volume 76B
PURINE METABOLISM IN MAN II: Physiology, Pharmacology, and Clinical Aspects
Edited by Mathias M. Müller, Erich Kaiser, and J. Edwin Seegmiller • 1977

Volume 77
RETINITIS PIGMENTOSA: Clinical Implications of Current Research
Edited by Maurice B. Landers, III, Myron L. Wolbarsht, John E. Dowling, and Alan M. Laties • 1977

Volume 78
TISSUE HYPOXIA AND ISCHEMIA
Edited by Martin Reivich, Ronald Cobern, Sukhamay Lahiri, and Britton Chance • 1977

Volume 79
ELASTIN AND ELASTIC TISSUE
Edited by Lawrence B. Sandberg, William R. Gray, and Carl Franzblau • 1977

PURINE METABOLISM IN MAN—II

Physiology, Pharmacology, and Clinical Aspects

Edited by
Mathias M. Müller and Erich Kaiser
University of Vienna
Vienna, Austria
and
J. Edwin Seegmiller
University of California, San Diego
La Jolla, California

PLENUM PRESS • NEW YORK AND LONDON

Library of Congress Cataloging in Publication Data

International Symposium on Purine Metabolism in Man, 2d, Baden, Austria, 1976.
Purine metabolism in man, II.

(Advances in experimental medicine and biology; v. 76)
Includes index.
CONTENTS: Pt. A. Regulation of pathways and enzyme defects.–Pt. B. Physiology, pharmacology, and clinical aspects.
1. Purine metabolism–Congresses. 2. Uric acid metabolism–Congresses. 3. Metabolism, Inborn errors of–Congresses. I. Müller, Mathias M. II. Kaiser, Erich, Dr. med. III. Seegmiller, J. E. IV. Title. V. Series. [DNLM: 1. Purines–Metabolism–Congresses. 2. Gout–Enzymology–Congresses. 3. Purine-pyrimidine metabolism, Inborn errors–Congresses. 4. Carbohydrates–Metabolism–Congresses. 5. Lipids–Metabolism–Congresses. W1 AD559 v. 76 1976 / QU58 I635 1976p]
QP801.P8I56 1976 612'.0157 76-62591
ISBN 0-306-39090-6 (vol. 76B)

Proceedings of the second half of the Second International Symposium on Purine Metabolism in Man, held in Baden, Vienna, Austria, June 20–26, 1976

A Division of Plenum Publishing Corporation
227 West 17th Street, New York, N.Y. 10011

Printed in the United States of America

PREFACE

The study of gouty arthritis has provided a common meeting ground for the research interests of both the basic scientist and the clinician. The interest of the chemist in gout began 1776 with the isolation of uric acid from a concretion of the urinary tract by the Swedish chemist SCHEELE. The same substance was subsequently extracted from a gouty tophus by the British chemist WOLLASTONE in 1797 and a half century later the cause of the deposits of sodium urate in such tophi was traced to a hyperuricemia in the serum of gouty patients by the British physician Alfred Baring GARROD who had also received training in the chemical laboratory and was therefore a fore-runner of many of today's clinician-investigators.

The recent surge of progress in understanding of some of the causes of gout in terms of specific enzyme defects marks the entrance of the biochemist into this field of investigation. The identification of the first primary defect of purine metabolism associated with over-production of uric acid, a severe or partial deficiency of the enzyme hypoxanthine-guanine phosphoribosyltransferase was achieved less than a decade ago. The knowledge of the mechanism of purine over-production that it generated led shortly to the identification of families carrying a dominantly (possibly X-linked) inherited increase in the activity of the enzyme phosphoribosylpyrophosphate synthetase as a cause of purine over-production. Yet this is only a start as these two types of enzyme defects account for less than five per cent of gouty patients.

The rapid pace at which new knowledge of aberrations of human purine metabolism is being acquired is adequate reason for holding the Second International Symposium on Purine Metabolism in Man (Baden, Austria, June 20 - 26, 1976) just three years after the first symposium was convened. It also marks the bicentennial aniversary of the discovery of uric acid by SCHEELE. The table of contents shows a further consolidation of our understanding of the mechanisms involved in the synthesis and degradation of purines and the aberrations produced in regulation of these processes by well characterized defects in purine metabolism. In addition are reports of newly discovered defects in enzymes of purine metabolism not previously presented at the last symposium. Homozygousity for deficiency of adenine phosphoribosyltransferase has now been identified in three children, two of whom presented with calculi of the urinary tract composed of 2,8-dihydroxyadenine thus setting at rest previous speculations based on studies of heterozygotes for this disorder.

On the basis of recent experiments the understanding of renal handling of urate has been further increased indicating a pre- and post-secretory reabsorption. The significance of protein-binding of urate is still open for discussion. However the knowledge of mechanisms regulating purine transport through membranes has improved by development of rapid micromethods.

A whole new area of considerable importance for the future is the association of an impaired function of the immune system in children with a gross deficiency of either of two sequential enzymes of purine interconversion, adenosine deaminase or purine nucleoside phosphorylase. Further investigation of the mechanism of this phenomenon gives promise of extending substantially our knowledge of the normal control of the immune response.

We wish to acknowledge the support of the Dean of the Faculty of Medicine of the University of Vienna, DDr. O. Kraupp and financial support from Dr. Madaus and Co. The contribution of the Organizing Committee and of the Scientific Committe in arranging the meeting and the details of the program is also gratefully acknowledged.

MATHIAS M. MÜLLER

ERICH KAISER

J. EDWIN SEEGMILLER

CONTENTS OF VOLUME 76 B

BIOCHEMISTRY OF PURINE TRANSPORT

BIOCHEMICAL PHARMACOLOGY

MECHANISM OF GOUTY INFLAMMATION

CLINICAL ASPECTS OF PURINE METABOLISM

HYPERURICEMIA AS A RISK FACTOR

CONTENTS OF VOLUME 76 A

De Novo Synthesis: Phosphoribosylpyrophosphate and Phosphoribosylpyrophosphate Synthetase

De Novo Synthesis: Phosphoribosylpyrophosphate Amidotransferase

Nucleotide Metabolism

Salvage Pathway

Catabolism

MUTATIONS AFFECTING PURINE METABOLISM IN MAN

General Aspects

Phosphoribosylpyrophosphate Synthetase

Adenine Phosphoribosyltransferase

Hypoxanthine-Guanine Phosphoribosyltransferase: Partial Deficiency

Purine Nucleoside Phosphorylase Deficiency

RELATIONSHIP BETWEEN CARBOHYDRATE, LIPID AND PURINE METABOLISM

Carbohydrate and Purine Metabolism

Lipid and Purine Metabolism

METHODOLOGY

RENAL HANDLING OF URIC ACID

Richard E. Rieselbach

University of Wisconsin Medical School at the Mount Sinai Medical Center
950 North 12th Street
Milwaukee, Wisconsin 53233

The pathogenesis of gout has intrigued physicians for centuries. A pathogenetic role for the kidney was first suggested by Garrod more than 125 years ago (1). During the ensuing century, many studies were performed attempting to measure urate excretion in patients with gout. However, the role of the normal kidney in maintaining urate homeostasis was not clearly defined. The first comprehensive study designed to investigate the mechanism of urate handling by the normal kidney, employing contemporary principles of renal physiology, was carried out by Berliner and colleagues in 1950 (2). They concluded that "the difference between the clearance of urate and the clearance of inulin must be due to either non-filterability of a large fraction of the plasma urate, to tubular reabsorption of a major fraction of the filtered urate, or to some combination of these two". It is remarkable that in the same year, Praetorius and Kirk described a patient with marked hypouricemia, who had a urate to inulin clearance ratio of 1.46 (3). These authors concluded that urate was either secreted or synthesized in the kidney. Nevertheless, the filtration-reabsorption theory of renal urate handling remained widely accepted until 1957, when Gutman and Yü again raised the question of tubular secretion of urate (4). Although direct evidence had not been obtained supporting a more complex system for renal urate handling, Gutman reasoned that the paradoxical retention of uric acid observed after the administration of pyrazinamide might be the result of a tubular secretory system responsive to pharmacologic inhibition. Thereafter, in 1959, Gutman and colleagues proceeded to publish data providing strong support for the presence of urate secretion (5). In 1961, Gutman assembled the evidence obtained from man and other mammals indicating that plasma urate is virtually completely filterable at the glomerulus, that it is reabsorbed by active tubular transport and that the kidney has the capacity for active tubular secretion of uric acid, as depicted in Fig. 1A (6). This mechanism for renal urate handling was generally accepted throughout the next decade until new data began to appear during the early 1970's from man and the chimpanzee suggesting that post-secretory reabsorption played an important role in renal urate handling (7). This raised the possibility of alternative mechanisms for the renal handling of urate, as noted in Fig. 1B-1E.

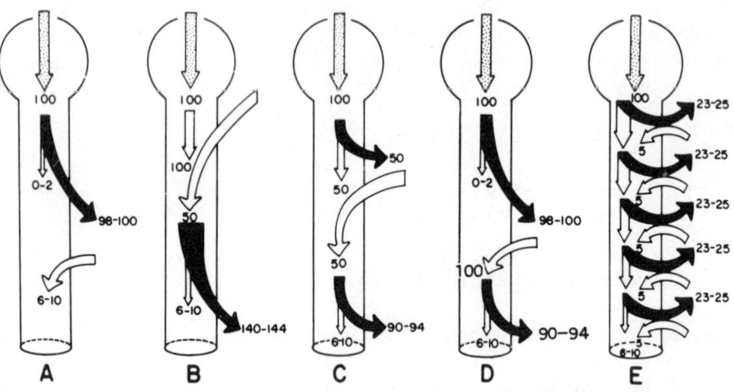

Figure 1. Five possible models for urate transport in man. The stippled arrows represent filtered urate, solid arrows represent urate reabsorption, and open arrows indicate either tubular secretion of urate or urate remaining in tubular fluid after reabsorption. Numerical values indicate hypothetical orders of magnitude of the transport processes. In "A" the traditional view is depicted, with most filtered urate reabsorbed and all secretion taking place subsequently. Alternatively, no pre-secretory reabsorption takes place in "B"; all reabsorption occurs distally to secretion. In "C" and "D", varying amounts of pre- and post-secretory reabsorption take place (see text). Finally, secretion and reabsorption are depicted as occurring simultaneously along a substantial length of the nephron ("E").

At the present time, no data are available which allow definitive characterization of the normal mechanism for renal urate handling in man. However, the model proposed in Fig. 1D appears to be a most attractive hypothesis, in that it is consistent with much of the recently published data in man and the chimpanzee. Thus, for the purpose of this presentation, this hypothesis will be examined in detail, some of the more compelling supporting evidence will be reviewed, and this hypothesis will be embellished with some admittedly speculative concepts as to the specific control systems which might be operative in maintaining urate homeostasis via the kidney.

This four component hypothesis involves 1) filtration of plasma urate at a rate approaching that of inulin, 2) virtual complete reabsorption of filtered urate by a high capacity, high affinity transport mechanism, 3) subsequent secretion of urate at a rate possibly approximating the original filtered load by a system susceptible to modulation by various hormones, metabolites and drugs, and 4) reabsorption of a major percentage of secreted urate by a high capacity, variable affinity transport mechanism of undetermined location, where reabsorptive transport is influenced by drugs, extracellular fluid volume, and to a limited extent, by urine volume. This, in health, the renal contribution to urate homeostasis is primarily mediated by the balance between the rate of urate secretion and its subsequent reabsorption.

I. Filtration of Plasma Urate

Although early studies demonstrated complete ultrafilterability of urate (8), the extent of urate filterability recently has come into question again. These recent studies have indicated that urate in serum or plasma is only about 80% dialyzable at 4°C. with most of the binding occurring in the albumin fraction (9). While binding decreases as physiologic temperature is approached, the extent of protein binding in normals at body temperature has not been clearly defined. Recent data derived from a variety of experimental conditions in one laboratory indicate that a substantial percentage of plasma urate (15-30%) may be protein-bound at physiologic temperature (10), while application of two different in vitro techniques in another laboratory reveal absence of significant urate binding to human serum proteins (11).

II. Reabsorption of Filtered Urate

On the basis of data published by Berliner et al in 1950 (2) and later by Lathem and Rodnan (12), it has been generally accepted that a substantial percentage of filtered urate is reabsorbed. In order to more fully characterize this component of urate transport, 6 normal subjects were studied in our laboratory in the presence of hyperuricemia (Table I). A mean plasma urate concentration of 13.5 mg% was attained by administration of pyrazinamide (in order to block urate secretion) and yeast RNA for several days; then clearance studies were performed under continued pyrazinamide (PZA) suppression of urate secretion. Although mean filtered urate was elevated to 17.4 mg/min, fractional urate excretion was only .51%. Thus, mean urate reabsorption was 17.3 mg/min, reaching 21.9 mg/min in one subject who attained a plasma urate level of 20 mg%. Similar studies on these patients during normouricemia (at a mean plasma urate level of 6.26 mg%) revealed no difference in the fractional excretion of urate, while filtered urate was 8.6 and reabsorbed urate 8.5 mg/min. Thus, during hyperuricemia, the rate of urate excretion factored by glomerular filtration rate was negligibly increased by just .03 mg/min. If filtered and secreted urate were reabsorbed cumulatively at a site distal to secretion (Fig. 1B or 1C), simulation of a normal reabsorptive load at the post-secretory site in the absence of secretion by doubling the filtered load of urate, as accomplished in this experimental setting, should have resulted in a normal rate of urate excretion, which approximates .5 mg/min/100 ml GFR. On the contrary, these data indicate an absence of any increase of urate excretion despite the presence of a marked increase in filtered load. Thus, the data would appear to indicate that filtered urate undergoes virtual complete reabsorption by a high capacity, high affinity transport system located proximal to the tubular locus of urate secretion. The reabsorptive mechanism for filtered urate is uniquely characterized; a mechanism of similar capacity and affinity could not be operative for the processing of secreted urate, as we shall subsequently consider; otherwise, uric acid excretion in man would be negligible. The nature of this pre-secretory reabsorptive mechanism in man would appear to limit the significance of urate protein binding as a factor influencing excretion of filtered urate. Since it would appear that virtually no filtered urate escapes reabsorption in normal man up to plasma levels exceeding 20 mg%, the exact percentage of urate escaping glomerular filtration due to protein binding is of little physiological consequence.

Table I. Uric Acid Excretion in 6 Normal Males One Hour After Administering 3 gm Pyrazinamide

	Normouricemia	Hyperuricemia	P<
GFR (ml./min.)	139 ± 8	129 ± 9	NS
Plasma urate (mg.%)	6.26 ± .64	13.5 ± .9	.005
Filt. urate (mg./min.)	8.63 ± .86	17.4 ± 1.5	.005
Reabsorbed urate (mg./min.)	8.57 ± .86	17.3 ± 3.7	.005
UV_{urate}/GFR (mg./100 ml.)	.04 ± .004	.07 ± .01	NS
FE urate (%)	.71 ± .07	.51 ± .06	NS

III. Tubular Secretion of Urate

In 1959, Yü and Gutman demonstrated a modest degree of net urate secretion in subjects with mild renal insufficiency after treatment with a uricosuric agent, sulfinpyrazone, in association with the infusion of mannitol and uric acid (5). Subsequently, that laboratory was able to demonstrate net urate secretion in the mongrel dog subjected to appropriate pharmacologic maneuvers (13). Thus, the 3 component concept of renal urate handling was proposed and generally accepted, and it was assumed that secreted urate might well approximate the quantity of urate excreted in bladder urine.

Pyrazinamide (PZA) suppression test as a means of characterizing urate transport. In the mid 1960's, the PZA suppression test was developed in an attempt to quantitate urate secretion in normal man (14). This test involved the performance of three 20-minute clearance periods, the mean of which was taken to be the rate of urate excretion as related to glomerular filtration rate (Fig. 2).

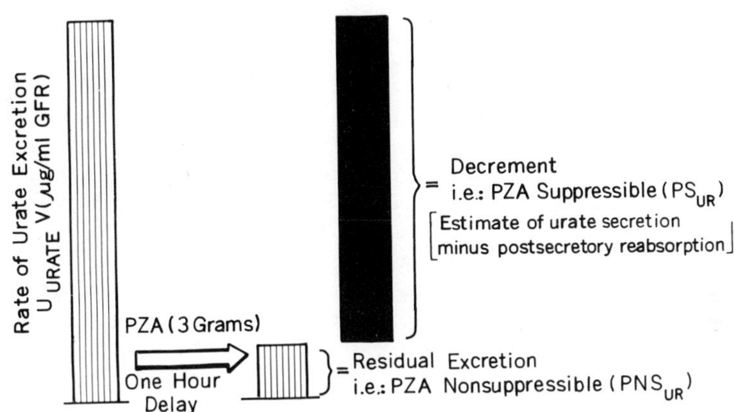

Figure 2. Pyrazinamide (PZA) suppression test. PZA suppressible urate excretion is the decrement (solid bar) in urate excretion factored by GFR following oral administration of PZA (see text).

One hour after administering 3 grams of PZA by mouth, three additional clearance periods were obtained. That period representing the lowest urate excretion as related to GFR was noted; this residual excretion was considered to represent either non-reabsorbed filtered urate and/or secreted urate not suppressed by PZA. The decrement in excretion (PZA suppressible urate excretion) was considered as a minimal estimate of urate secretion. Interpretation of data which evolved from the test was based upon two assumptions: 1) suppression of urate excretion by PZA occurs through inhibition of secretion rather than augmentation of the reabsorption of filtered urate and, 2) all tubular urate secretion occurs in those segments of the nephron distal to all reabsorptive sites.

During the early 1970's, the first assumption was questioned (15) until Weiner and Tinker published the first definitive studies on the renal pharmacology of pyrazinamide (16). In this, and a subsequent study by Fanelli and Weiner (17) it was demonstrated that a metabolite of pyrazinamide, pyrazinoic acid, is responsible for its anti-uricosuric effect. In the mongrel dog (and the chimpanzee) pyrazinoate was found to exert a bi-phasic action on renal urate handling. At low plasma concentrations a striking anti-uricosuric effect was observed, whereas at high plasma concentrations uric acid excretion was increased. Thus, if PZA were to exert any effect on reabsorptive urate transport, inhibition rather than enhancement would be expected. It appeared likely that the three gram dose of PZA utilized in man had an effect limited to inhibition of urate secretion.

The second assumption necessary for validity of the test as a quantitative estimate of urate secretion has not been confirmed. On the contrary, two major bodies of data have appeared indicating that urate entering the renal tubule via secretion may be predominantly reabsorbed at more distal sites in the nephron. First, Fanelli and colleagues (18) have observed that the rate of urate excretion in chimpanzees in some instances approached levels nearly 100% greater than the amount filtered following Mersalyl administration. This exceedingly high rate of tubular urate secretion in an animal handling urate similarly to man, unmasked by the inhibition of urate reabsorption after Mersalyl administration, is convincing evidence that the amount of urate excreted in the final urine is only a small fraction of that which is actually secreted by the nephron. Secondly, separate studies from the laboratories of Steele (19) and Diamond (20) have indicated that the uricosuric responses to agents such as Probenecid or intravenous chlorothiazide are greatly diminished in subjects pretreated with PZA. Assuming that PZA predominantly inhibits the tubular secretion of urate and does not effect the renal binding, transport or availability of the uricosuric agent (assumptions subsequently validated), these data suggest that a significant component of the uricosuric response to these compounds relates to a direct inhibitory effect upon post-secretory reabsorption. In addition, in hypouricemic patients with Wilson's Disease and Hodgkin's Disease, the markedly excessive excretion of urate in relation to plasma urate levels has been dramatically reduced by PZA treatment, to a level which might be expected in normal subjects at this reduced plasma urate level (21,22). By reducing the urate load to an apparently defective post-secretory reabsorptive mechanism in these patients, urate excretion is markedly diminished. Thus, it now appears likely that the PZA suppression test yields data which bear a relationship to the rate of urate secretion, but these data cannot be viewed as a quantitative representation of secretion.

As originally employed, the test was utilized in an attempt to examine the role of urate secretory transport in determining the rate of urate excretion (14). Secretion was estimated in 10 normal subjects by observing the decrement in urate excretion produced by a maximally anti-uricosuric dose of PZA (Fig. 3).

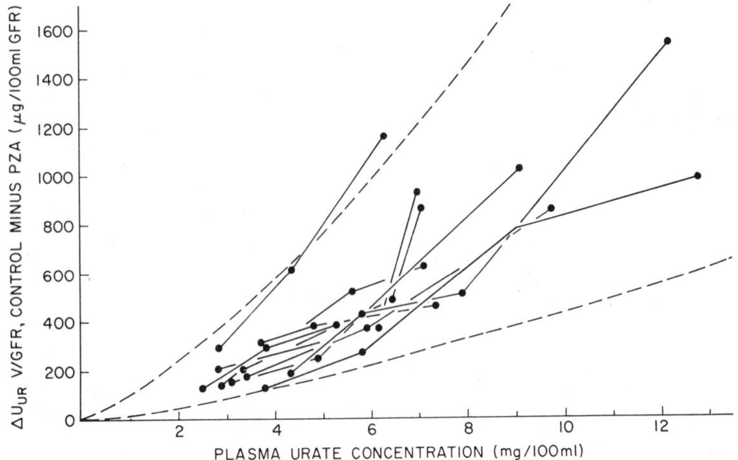

Figure 3. Pyrazinamide (PZA)-suppressible urate excretion in normal subjects. The maximal decrement in urate excretion after PZA administration (Δ $U_{ur}V/GFR$, control minus PZA) is plotted for 10 normal subjects, each studied at three or more different plasma urate levels. The plasma urate level was elevated by administration of yeast RNA and decreased by allopurinol. The dotted lines indicate the 95% regression tolerance limits. Adapted from data of Steele and Rieselbach (14).

Each subject was studied first in a baseline normouricemic state, next after decrease in the plasma urate following allopurinol, and finally after increasing plasma urate by yeast RNA loading. The estimated tubular secretion of urate (the decrement in urate excretion produced by PZA), was a direct function of the plasma urate concentration. These observations strongly suggested that 1) the kidney of normal man undergoes an augmented rate of secretory transport with increasing availability of urate to the secretory site, and 2) that this appeared to be the primary factor serving to increase urate excretion during hyperuricemia, thereby minimizing the hyperuricemic response to an increase in uric acid synthesis. In view of subsequently generated data relating to post-secretory reabsorption, as noted previously, these data from the mid-sixties in normal man may now be viewed in a different context (Fig. 4). A regression treatment of pyrazinamide suppressible urate excretion (PS_{ur}) on plasma urate concentration is depicted for the 32 studies on these 10 normal subjects. In view of the fact that tubular secretion of urate may well approximate glomerular filtration, yet urate excretion is normally less than 10% of filtered urate, a quantity approximating 90% of secreted urate may well be reabsorbed. This figure depicts that possibility

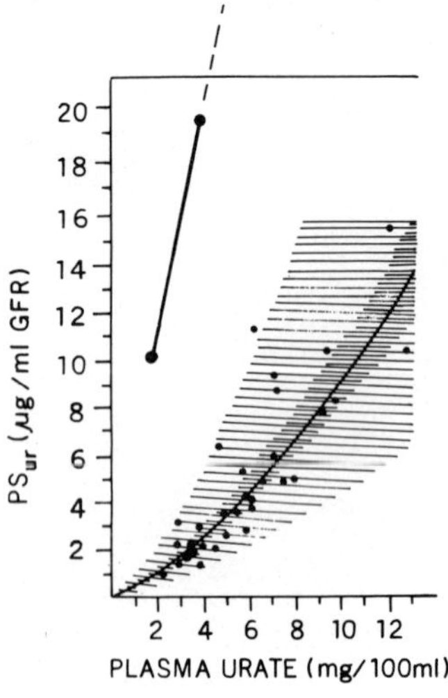

Figure 4. Regression treatment of data presented in Fig. 3, with 95% confidence limits for prediction of normalcy (lightly shaded area). Actual secretion might be 10 times greater than PS_{ur}, while still bearing the same qualitative relationship to plasma urate concentration, as depicted by large points for actual secretion at plasma urate concentrations of 2 and 4 mg% (see text).

by indicating an actual tubular secretion of urate of 10 times the level observed for the PZA suppressible component of urate excretion at a plasma urate concentration of 2 mg% and 4 mg%. If one were to employ the same ratio of secretion to PS_{ur} with increasing plasma urate levels, the relationship between secretion and the plasma urate concentration might be as indicated by the dashed line in Fig. 4. Thus, secretion of urate is clearly of much greater magnitude than previously suspected, probably occurs predominantly in the proximal tubule, and that urate secreted most likely undergoes substantial reabsorption in remaining nephron segments. The possibility of discrete modalities for urate transport being localized to specific segments of the proximal tubule is supported by recent observations utilizing scanning electron microscopy. These studies have delineated at least three distinctive morphological segments of the proximal tubule in the mouse, rat and monkey (23).

IV. Post-secretory Reabsorption of Urate

In order to regulate urate homeostasis in the precise manner to which we are accustomed, the normal human kidney must possess an extremely effective mechanism for processing the large quantity of secreted urate, which could be as great as 10 grams/day. In that data characterizing the post-secretory reabsorptive mechanism in man have been most difficult to obtain, we are left with several possible hypotheses. It might be appropriate to cite a well-established mechanism relating to sodium transport, in order to gain some insight into what will be proposed as an attractive hypothesis characterizing post-secretory urate reabsorption. As noted in Fig. 5A, urinary sodium excretion equals 140 μEq/min, if the fractional excretion of sodium is 1% in the presence of a GFR of 100 ml/min and a plasma sodium concentration of 140 μEq/ml. If the filtered load of sodium is increased by 50% due to a 50% increase in GFR (Fig. 5B), fractional excretion of sodium remains constant (assuming constancy of other variables influencing reabsorptive sodium transport), thereby limiting the increase in sodium excretion to 70 μEq/min, as oppossed to an increment of 7,000 μEq/min in filtered load. Thus, accelerated reabsorptive transport of sodium precisely buffers any increase in filtered load of sodium, thereby maintaining integrity of extracellular fluid volume via a mechanism termed glomerulo-tubular balance. The same mechanism serves to minimize any decrease in sodium excretion associated with a decrease in filtered load of sodium (Fig. 5C). The physiologic basis for this mechanism is not well understood, but may be related to alterations in renal hemodynamics associated with changes in GFR.

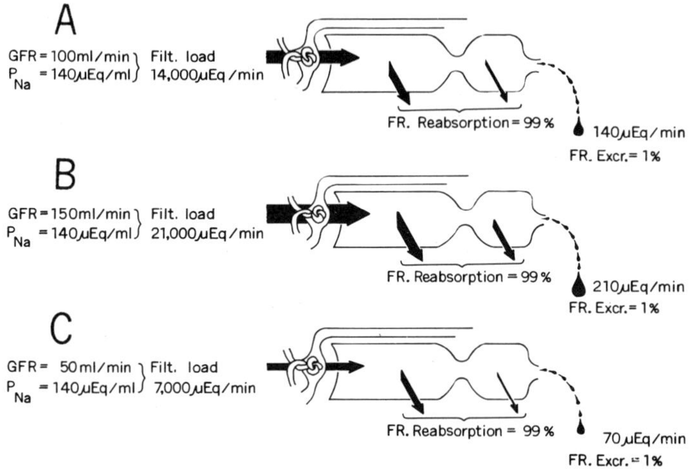

Figure 5. Glomerulo-tubular balance for sodium. In "B", if the filtered load of sodium is increased by 50% due to a 50% increase in GFR, the increase in actual sodium excretion is minimal if fractional sodium excretion remains constant. In "C", the same mechanism is depicted for a decrease in filtered load; the absolute decrease in sodium excretion is limited if fractional excretion remains the same (see text).

It appears reasonable to propose a mechanism of similar characteristics for post-secretory urate reabsorption, which might be designated "secretory-reabsorptive balance". As noted in Fig. 6A, the filtered load of urate at a GFR of 100 ml/min and a plasma urate concentration of .05 mg/ml is approximately 5 mg/min, and is depicted as undergoing virtual complete reabsorption. Given a filtration fraction of 20% (i.e.: renal plasma flow = 500 ml/min) the "peritubular load" of urate is 25 mg/min. If the net secretory transport of urate were 20% of this quantity, by virtue of diffusion into the tubular lumen of a portion of urate actively transported by a site located at the contraluminal membrane (24), the secretory load would be 5 mg/min, an amount equivalent to the filtered load. If fractional excretion of secreted urate were set at 10%, 4.5 mg/min of urate would be reabsorbed at the post-secretory site and .5 mg/min would be excreted in the final urine. If the plasma urate concen-

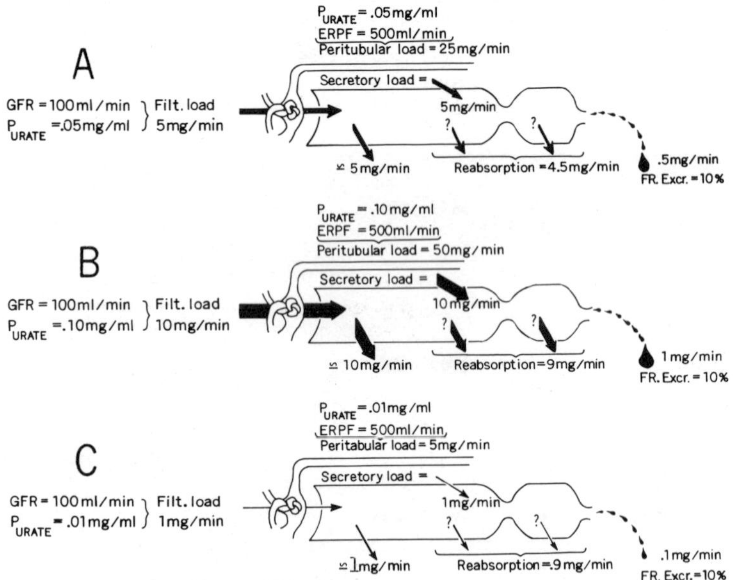

Figure 6. "Secretory-reabsorptive balance" for urate. In the presence of a constant fractional excretion of secreted urate, it is proposed that major fluctuations in secretory load are masked by alterations in absolute reabsorptive rate which allow maintenance of a constant fractional excretion. "B" depicts a doubling of "secretory load" attained as the result of a doubling of plasma urate concentration. The "secretory load" in "C" is reduced to 5 mg/min by virtue of a decrease to .01 mg/ml of plasma urate concentration (see text).

tration were to be doubled, this would result in a doubling of the secretory load, as noted in Fig. 6B. If fractional excretion remained the same at 10%, total excretion would increase to only 1 mg/min rather than to 5.5 mg/min, as would be the case if the entire increment in secretory load were to be excreted. The same mechanism would obtain in the case of a decrease in plasma urate to .01 mg/ml (Fig. 6C). Some uric acid would continue to be excreted in the final urine, i.e.: .1 mg/min.

This secretory-reabsorptive interaction, while appearing to be unduly complicated and possibly even counterproductive, makes good teleologic sense in view of the marginal solubility limits of urate in urine. Buffering of abrupt changes in secretory rate induced by an increase in secretory load decreases the possibility of uric acid stone formation by distributing any increase in excretion due to a transient increase in urate load over a longer period of time. The exact mechanism and site of post-secretory reabsorption remain to be clarified. Recent micropuncture data indicating substantial urate reabsorption in the loop of Henle of rats is of great interest (25). Also, variation in urine flow has been observed to exert a moderate effect on urate excretion in man, thereby suggesting a reabsorptive site in the distal tubule or collecting duct (26,27).

Factors Modulating Renal Urate Excretion

Inherent in the foregoing hypothesis is the concept that in health, renal urate excretion is predominantly modulated by only two of the four components of the suggested transport system, secretion and the level of post-secretory reabsorption. Alterations in urate excretion could be accomplished in three ways through alterations in these two processes:

(1) Altered urate secretion: As previously noted, urate excretion progressively increases with increasing concentration of plasma urate. Since filtered urate undergoes virtual complete reabsorption, increased excretion is accomplished via an increase in secretion, but with only a fraction of that increment actually excreted. Another means of increasing secretion could involve an increase in the other determinant of secretory load, effective renal plasma flow (Fig. 6). A possible example of this mechanism occurs in sickle cell anemia. Ten patients with sickle cell disease studied by Diamond et al had a PZA suppressible urate excretion, (PS_{ur}) two times that of controls in conjunction with a 25% increase in renal plasma flow (28). In this manner, these patients were able to maintain urate homeostasis in the presence of a substantial increase in urate production. It would follow that

secretion would be diminished in the presence of a decreased plasma urate concentration, or a decrease in renal plasma flow. In addition, a wide variety of hormones, metabolites and drugs decrease renal urate excretion, most likely through decreasing secretion (29).

(2) Altered post-secretory reabsorption: While the level of post-secretory reabsorption appears to be set at a constant percentage in order to minimize oscillations in excretion which would otherwise be produced by apparent substantial fluctuations in secretion due to influences previously noted, "secretory-reabsorptive balance" does appear to be susceptible to alteration by pharmacologic agents. Just as diuretic agents reset glomerulo-tubular balance for sodium by inhibiting reabsorptive transport, commonly used uricosuric agents appear to act similarly, presumably at the post-secretory site, with a resultant increase in fractional urate excretion. Since these agents gain entrance to tubular fluid via secretion (in that they are weak organic acids) by transport mechanisms presumably located in the same nephron segment as that for urate (and in some cases, possibly by the same transport mechanism), significant concentration of these drugs in the nephron is probably obtained only at the site of post-secretory reabsorption.

Again, as with sodium, alterations in extracellular fluid volume significantly effect fractional excretion of urate. This effect appears to be predominantly at the locus of post-secretory reabsorption, in that the uricosuric effect of volume expansion is greatly diminished by PZA suppression of secretion (30). The anti-uricosuric effect of chronic diuretic administration appears to be related to volume depletion, in that this effect may be ablated by replacement of urinary losses induced by diuretics (31). This mechanism undoubtedly plays a role in the hyperuricemia observed in other settings associated with ineffective arterial volume, such as congestive heart failure and the nephrotic syndrome, although associated increased levels of angiotensin II could produce secretory inhibition (32).

(3) Altered secretion combined with altered fractional excretion: A simultaneous increase in urate secretion and its fractional excretion would appear to be the most effective means of attaining an increase in excretion per nephron. In chronic renal failure, homeostasis can be maintained only if each residual nephron undergoes a progressive and substantial increase in excretory capacity as the nephron number decreases due to advancing disease (33). In this setting, normouricemia is maintained until the GFR falls below 30 ml/min by a progressive increase in urate excretion per nephron, which is due exclusively to an increase in PZA suppressible urate excretion, until far advanced renal failure supervenes (34).

In an attempt to characterize this adaptation in urate excretion associated with a reduction of nephron population, but in the absence of variables associated with renal disease, our laboratory has studied urate excretion in 15 normal subjects before, and 7-10 days after nephrectomy for organ donation (35). PZA suppressible urate excretion was determined in these 15 subjects, who had no apparent renal disease, hyperuricemia or history of clinical gout, by calculating the decrement in urate excretion per ml inulin clearance following a 3 gram dose of PZA. In order to assure a uniform degree of volume expansion prior to each study, one liter of .45 normal sodium chloride was administered intravenously over a 45-60 minute equilibration period, simultaneously with the inulin infusion. In addition, 250 ml of water was administered by mouth during each clearance period. The initial study was performed several days prior to unilateral nephrectomy, at a time of unrestricted dietary intake. Following nephrectomy, a period of at least 7 days was allowed to elapse to minimize any possible post-operative catabolic effect upon purine metabolism. All subjects were ambulatory at the time of the second study and had consumed an unrestricted diet for at least 3 days. All medications were discontinued at least 24 hours prior to study. A third study was performed in 10 of the subjects from 3-9 months following nephrectomy. Diet was again unrestricted and no medications were being administered. In 4 of the subjects, additional studies were performed following the first post-nephrectomy study, both at a time when plasma urate concentration had been reduced with allopurinol administration, and after subsequent induction of hyperuricemia by feeding of yeast RNA for a period of 4 days prior to study.

Table II. Urate Excretion in 15 Subjects - Studies Before and 2-3 wks. After Nephrectomy

	Before (2 kidneys)	After (1 kidney)	P<
P_{urate} (mg.%)	4.9 ± .2	5.2 ± .3	NS
C_{inulin} (ml./min.)	120 ± 7.2	75.5 ± 2.8	--
UV_{urate} (mg./min.)	.51 ± .04	.50 ± .03	NS
PS_{urate} (mcg./ml. GFR)	3.6 ± .3	7.0 ± .4*	.001
Fe_{Na} (%)	1.8 ± .2	2.5 ± .2	.01

*GFR corrected to reflect pre-nephrectomy GFR, ie: actual nephron population

As noted in Table II, mean plasma urate concentration and urate excretion (UV_{urate}) were not changed significantly by the removal of one-half of the normal nephron population. GFR in the remaining kidney increased more than 20% while PZA suppressible urate excretion almost doubled. The increase in glomerular filtration rate, along with an increase in fractional excretion of sodium, allowed sodium balance to be maintained by residual nephrons of the remaining kidney after contralateral nephrectomy. Pre- and post-nephrectomy PS_{ur} are depicted for each subject in Fig. 7. That the increase in urate excretion was due almost completely to an increase in the pyrazinamide suppressible component is confirmed by Fig. 8, which indicates virtual complete reabsorption of filtered urate both before and after nephrectomy. The diagonal line represents complete reabsorption of filtered urate. Irrespective of the presence of 1 or two kidneys, all points fall immediately below the line, indicating virtually complete urate reabsorption. Expressed as fractional reabsorption of filtered urate, the mean figure for 1 kidney of all donors (97.4%) was not significantly different than that of two kidneys (98.4%; p >.5).

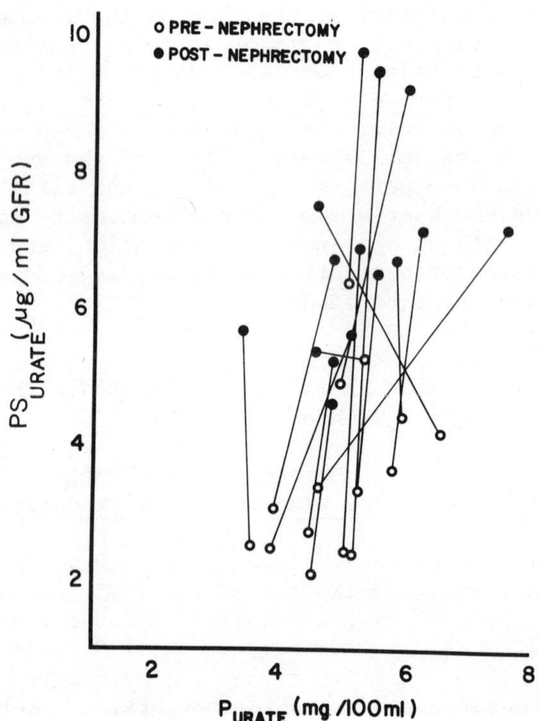

Figure 7. Individual data for pre- and post-nephrectomy PS_{ur} in 15 kidney donors. A marked increase occurred in 14 of the 15 subjects.

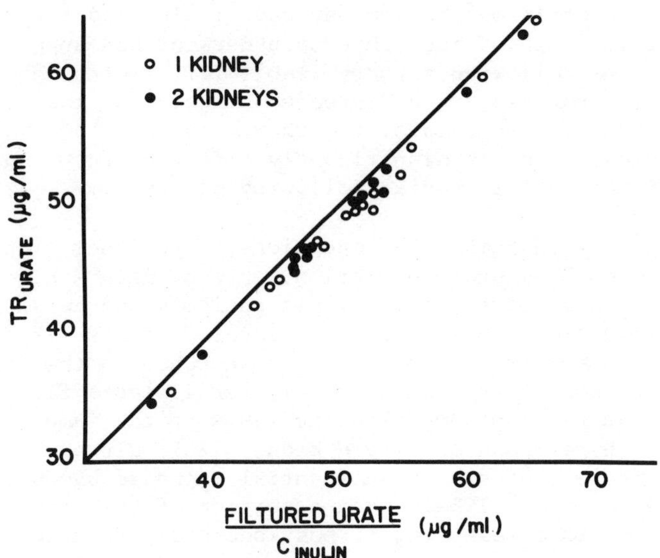

Figure 8. Absorption of filtered urate before and after nephrectomy. All points fall immediately below the diagonal line, indicating virtually complete reabsorption of filtered urate both before and after nephrecomy.

Table III. Follow-up Urate Excretion in 10 Subjects - Studies Performed at Mean of 4.6 Months After Nephrectomy

	First Post-Nephr. Study	Follow-up Post-Nephr. Study	P<
P_{urate} (mg.%)	5.2 ± .3	5.1 ± .3	NS
C_{inulin} (ml./min.)	71.9 ± 3.2	76.3 ± 4.3	NS
UV_{urate} (mg./min.)	.48 ± .03	.41 ± .03	NS
PS_{urate} (mcg./ml. GFR)	6.9 ± .4*	6.2 ± .4*	NS
Fe_{Na} (%)	2.7 ± .3	2.4 ± .2	NS

*GFR corrected to reflect pre-nephrectomy GFR, ie: actual nephron population.

So as to rule out the possibility that poorly defined factors related to the post-operative period, such as altered purine and/or protein metabolism may have influenced the results of these studies, ten of the subjects underwent PZA suppression tests 3-9 months following surgery (Table III). Although mean PS_{ur} and urate excretion rate decreased slightly at the time of the repeat study, these values, together with the other parameters measured, were not significantly different from those derived from the intial studies following nephrectomy.

Thus, following removal of one kidney, the other kidney approximately doubles urate excretion, thereby maintaining urate homeostasis. This increment is almost entirely due to an increase in PS_{ur}. Since plasma urate was not elevated, it is likely that any increase in secretion was due to an increase in the other determinant of secretory load (Fig. 6), renal plasma flow. It is well known that renal plasma flow increases to the same degree as GFR in the hypertrophied single kidney (36); although not measured in these subjects, a substantial increase undoubtedly occurred. Also, it is likely that a portion of the increment in PS_{ur} was due to a resetting of post-secretory fractional reabsorption, as possibly reflected by the increase in fractional excretion of sodium. Thus, it would appear that alterations in both control mechanisms acted synergistically, resulting in a marked increase in the PZA suppressible component of urate excretion, just as is seen in chronic renal failure. It is clear that the nephrons in the remaining normal kidney underwent an adaptation which allowed maintenance of extracellular fluid volume homeostasis by increasing the fractional excretion of sodium, (ie: altering glomerulo-tubular balance for sodium), in conjunction with an increased filtered load of sodium. By analogy, it would appear that these same nephrons could have sustained an alteration in "secretory-reabsorptive balance" for urate, in conjunction with an increase in secretory load.

An additional series of studies was performed on 4 of these transplant donors, in an attempt to define the relationship between PS_{ur} and plasma urate concentration in the presence of a pre-existing elevation in PS_{ur}. In these 4 subjects, post-nephrectomy PS_{ur} was measured following pharmacologic induction of hypo- and hyperuricemia (Fig. 9). Decrements and increments of PS_{ur} occurred promptly in response to changes in the plasma urate concentration. The subject noted by open triangles received 4 times the quantity of yeast RNA as the other 3 participants, resulting in a marked increase in PS_{ur} to 17.5 μg/ml with co-existent modest hyperuricemia at 8.8 mg%. Mean fractional reabsorption of filtered urate following RNA loading was 98%, not unlike that observed in normouricemic subjects with 2 kidneys (98.2%). Thus, in the presence of normouricemia and increased secretion probably secondary to increased renal plasma flow, alterations in the other component of secretory load (ie: plasma urate concentration) produce substantial changes in urate secretion.

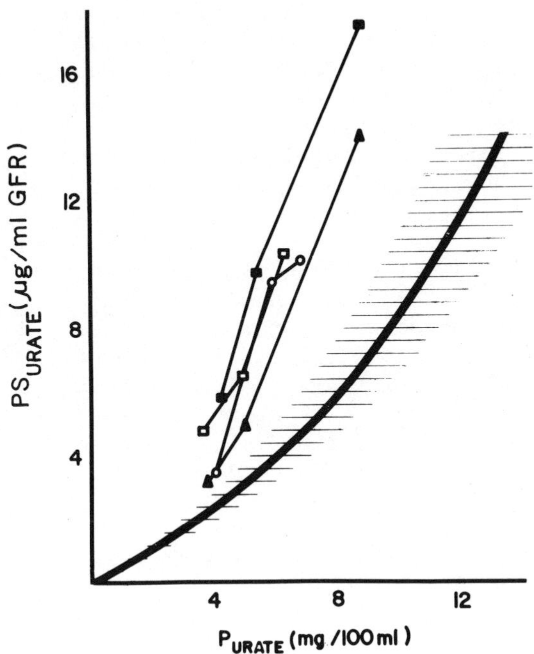

Figure 9. Relationship between PS_{ur} and plasma urate concentration in the presence of a pre-existing elevation in PS_{ur} in 4 subjects (see text). Shaded area represents 95% confidence limits of data in normals (see Fig. 3).

Intrinsic Abnormalities of the Control Mechanism

It would appear that some patients with gout have diminished secretion of urate as the basis for their hyperuricemia (37), with a recently described hereditary defect of urate secretion in gouty chickens appearing to support this pathogenetic theory in man (38). While these gouty patients maintain a "normal" level of urate excretion for a given purine intake, they do so at the cost of hyperuricemia. Alternatively, stated in another manner, these patients can attain a normal level of urate excretion only through a supra-normal stimulus to an apparently sluggish secretory mechanism; they have a sub-normal urate excretion when rendered normouricemic by allopurinol administration (37). Instead of the normal relationship for PS_{ur} vs. P_{ur} as noted in Fig. 3, their curve is shifted to the right, below the 95% confidence limits of the general population (37). Administration of uricosuric agents to these patients appears to result in the resetting of post-secretory fractional reabsorption at a lower level, thus facilitating excretion of a substantially greater percentage of secreted urate, in a manner similar to a diuretic agent depressing fractional reabsorption of sodium, with a resultant net increase of sodium excretion. Thus, under the influence of uricosurics, study state urate excretion remains at the same level as in hyperuricemia, but in the presence of a normal plasma urate concentration. Increased fractional excretion of secreted urate compensates for a sluggish rate of secretion, thus hyperuricemia and its symptomatic consequences are avoided.

Intrinsic renal lesions may also serve as the basis for hypouricemia. It would appear that three different varieties of deficient tubular reabsorption of urate have been described. One defect apparently involves diminished reabsorption of secreted urate, in that PZA administration eliminates excessive urate excretion in some patients with Wilson's Disease and Hodgkin's Disease (21,22). The patient described by Green, et al (39) who had an isolated urate reabsorptive defect, still excreted 29% of filtered urate after the administration of PZA. In the patient described by Sperling et al (40), who also had hypercalciuria and osteoporosis, urate excretion also was decreased only minimally after PZA. An increase in C_{urate} to 75 ml/min after probenecid treatment in this patient supported the idea that reabsorptive sites responsive to probenecid remained relatively intact. These data support the possibility of a pre-secretory reabsorptive defect in these two patients. A third seemingly distinct defect involves what appears to be nearly total inhibition of urate reabsorption in view of the extremely high C_{urate} observed. Of the 3 cases reported of this type (3,41,42), only Simkin's patient underwent pharmacologic studies (42). Inhibitors of urate secretion (including probenecid) reduced the C_{urate}/C_{inulin} to approximately 1.0. These experiments of nature, in conjunction with the other data in man which have been presented, serve to support the hypothesis involving a 4-component model of urate transport in the human kidney, as proposed in Fig. 1D. If normal urate transport were to take place in man as proposed in Fig. 1E, with simultaneous reabsorptive and secretory transport of urate within the same segment of the nephron, it is difficult to conceive how these seemingly three distinct patterns of deficient urate reabsorption could occur.

Thus, in conclusion, available data in man and the chimpanzee tend to support the model for renal urate handling proposed in Fig. 1D. While some aspects of the model as proposed herein are purely speculative, the available data appear to best fit this conceptual framework. To date, micropuncture studies in other species have yielded conflicting data; it is admittedly difficult to integrate some of that data into the system proposed herein. The question of simultaneous bidirectional urate transport is particularly troublesome. At least no significant intrarenal urate synthesis or uricolysis would appear to take place in the chimpanzee or man (43), whereas these considerations do add to the complexity of interpreting data from other species.

Future Experimental Approaches

As to the future, a clear understanding of the pathophysiologic mechanisms underlying disordered urate homeostasis awaits a more complete understanding of normal physiology. Over the past two decades, studies in man utilizing pharmacologic manipulations have been provocative. However, at present it would appear that this approach does not hold great promise of providing the additional data required to elucidate the details of the apparently intricate mechanism for renal urate handling in man. Nevertheless, one promising area meriting further exploration in man involves pharmacologic studies of patients with a post-secretory reabsorptive defect for urate. These patients provide an opportunity to study the secretory mechanism more directly, in that its characteristics are partially unmasked by deficient reabsorption of secreted urate.

Recent developments in microanalytical techniques for uric acid, and the associated micropuncture studies of renal urate handling in mammals provide some basis for optimism (25,44,45). If animal models which are relevant to human urate physiology can be effectively utilized, observations could be forthcoming which would provide both the basis for more definitive interpretation of presently available data and the stimulus for further studies in man. On the other hand, uncritical cross species interpretation of data from those animal models with questionable relevance to man is not likely to add to our understanding.

Another optimistic note for the future relates to the possible resurrection of the chicken as an experimental model. While uric acid excretion has been studied in the chicken for more than 50 years (46), application of recently derived concepts of renal urate handling to this experimental model (in the presence of allopurinol inhibition of intrarenal urate synthesis) has some attraction. The presence of an extremely active secretory transport mechanism for urate, a relatively minor component of post-secretory reabsorption and a portal system allowing direct pharmacologic access to the secretory mechanism could be exploited to yield an increased understanding of normal urate handling in man. Hopefully, in the years to come, data derived from these and other experimental models will provide the basis for an analysis of the normal mechanism for renal urate handling in man which includes substantially less speculation than put forth in this presentation.

REFERENCES

1. Garrod, A.B.: Observations on certain pathological conditions of blood and urine in gout, rheumatism in Bright's disease. Tr. M-Chir. Soc. 31:83, 1848.
2. Berliner, R.W., Hilton, J.G., Yü, T.F., and Kennedy, T.J., Jr.: The renal mechanism for urate excretion in man. J. Clin. Invest. 29:396, 1950.
3. Praetorius, E. and Kirk, J.E.: Hypouricemia: with evidence for tubular elimination of uric acid. J. Lab. Clin. Med. 35:865, 1950.
4. Gutman, A.B. and Yü, T.F.: Renal function in gout. With a commentary on the renal regulation of urate excretion, and the role of the kidney in the pathogenesis of gout. Am. J. Med. 23:600, 1957.
5. Yü, T.F. and Gutman, A.B.: A study of the paradoxical effects of salicylate in low, intermediate and high dosage on the renal mechanism for excretion of urate in man. J. Clin. Invest. 38:1298, 1959.
6. Gutman, A.B. and Yü, T.F.: A three-component system for regulation of renal excretion of uric acid in man. Trans. Assoc. Am. Physicians 74:353, 1961.
7. Rieselbach, R.E. and Steele, T.H.: Influence of the kidney upon urate homeostasis in health and disease. Am. J. Med. 56:665, 1974.
8. Yü, T.F. and Gutman, A.B.: Ultrafilterability of plasma urate in man. Proc. Soc. Exp. Biol. Med. 84:21, 1953.
9. Klinenberg, J.R. and Kippen, I.: The binding of urate to plasma proteins determined by means of equilibrium dialysis. J. Lab. Clin. Med. 75:503, 1970.
10. Steele, T.H.: Pyrazinamide suppressibility of urate excretion in health and disease: relationship to urate filtration. Proc. of Int. Symposium on Amino Acid Transport and Uric Acid. Innsbruck, June, 1975; in press.
11. Kovarsky, J., Holmes, E.W. and Kelley, W.N.: Absence of significant urate binding to human serum proteins. Clin. Res. 24:331A, 1976.
12. Lathem, W. and Rodnan, G.P.: Impairment of uric acid excretion in gout. J. Clin. Invest. 41:1955, 1962.
13. Yü, T.F., Berger, L., Kupfer, S., and Gutman, A.B.: Tubular secretion of urate in the dog. Am. J. Physiol. 199:1199, 1960.
14. Steele, T.H. and Rieselbach, R.E.: The renal mechanism for urate homeostasis in normal man. Am. J. Med. 43:868, 1967.
15. Holmes, E.W., Kelly, W.N., and Wyngaarden, J.D.: The kidney and uric acid excretion in man. Kidney Int. 2:115, 1972.

16. Weiner, I.M. and Tinker, J.P.: Pharmacology of pyrazinamide: metabolic and renal function studies related to the mechanism of drug-induced urate retention. J. Pharmacol. Exp. Ther. 180:411, 1972.
17. Fanelli, G.M., Jr. and Weiner, I.M.: Pyrazinoate excretion in the chimpanzee. Relation to urate disposition and the actions of uricosuric drugs. J. Clin. Invest. 52:1946, 1973.
18. Fanelli, G.M., Jr., Bohn, D., Riley, S., and Weiner, I.M.: Effects of mercurial diuretics on renal transport of urate in the chimpanzee. Am. J. Physiol. 224:985, 1973.
19. Steele, T.H. and Boner, G.: Origins of the uricosuric response. J. Clin. Invest. 52:1368, 1973.
20. Diamond, H.S. and Paolino, J.S.: Evidence for a post-secretory reabsorptive site for uric acid in man. J. Clin. Invest. 52:1491, 1973.
21. Wilson, D.M. and Goldstein, H.P.: Renal urate excretion in patients with Wilson's disease. Kidney Int. 4:331, 1973.
22. Bennett, J.S., Bond, J., Singer, I., and Gottlieb, A.: Hypouricemia in Hodgkin's disease. Ann. Intern. Med. 76:751, 1972.
23. C. Craig Tisher, Anatomy of The Kidney. In: The Kidney, ed. B.M. Brenner and F.C. Rector, Jr., W.B. Saunders Co., Philadelphia, 1976.
24. Zmuda, M.J. and Quebbemann, A.J.: Localization of renal tubular uric acid transport defect in gouty chickens. Am. J. Physiol. 229:820, 1975.
25. Gregor, R., Lang, F., Deetjen, P., and Knox, F.G.: Sites of urate transport in the rat nephron. Proc. 2nd Int. Symposium on Purine Metabolism in Man, in press.
26. Meisel, D. and Diamond, H.: Effects of vasopressin on uric acid excretion: evidence for distal nephron reabsorption of urate in man. Clin. Sci. Molec. Med. 51:1, 1976.
27. Engle, J.E. and Steele, T.H.: Variation of urate excretion with urine flow in normal man. Nephron 16:50, 1976.
28. Diamond, H.S., Meisel, A., Sharon, E., Holden, D., and Cacatian, A.: Hyperuricosuria and increased tubular secretion of urate in sickle cell anemia. Am. J. Med. 59:796, 1975.
29. Emmerson, B.T. and Ravenscroft, P.J.: Abnormal renal urate homeostasis in systemic disorders. Nephron 14:62, 1975.
30. Manual, M.A. and Steele, T.H.: Pyrazinamide suppression of the uricosuric response to sodium chloride infusion. J. Lab. Clin. Med. 83:417, 1974.
31. Steele, T.H. and Oppenheimer, S.: Factors effecting urate excretion following diuretic administration in man. Am. J. Med. 47:564, 1969.
32. Ferris, T.F. and Gorden, P.: Effect of angiotensin and norepinephrine upon urate clearance in man. Am. J. Med. 44:359, 1968.
33. Bricker, N.S., Klahr, S., Lubowitz, H., and Rieselbach, R.E.: Renal function in chronic renal disease. Medicine 44:263, 1965.

34. Steele, T.H. and Rieselbach, R.E.: The contribution of residual nephrons within the chronically diseased kidney to urate homeostasis in man. Am. J. Med. 43:876, 1967.
35. Shelp, W.D. and Rieselbach, R.E.: Increased bidirectional urate transport per nephron following unilateral nephrectomy. Am. Soc. Neph. Abstracts, p. 60, 1968.
36. Donadio, J.V., Farmer, C.D., Hung, J.C., Tauye, W.N., Hallenback, G.H., and Shorter, R.G.: Renal function in donors and recipients of renal allotransplantation. Ann. Intern. Med. 66:105, 1967.
37. Rieselbach, R.E., Sorensen, L.B., Shelp, W.D., and Steele, T.H.: Diminished renal urate secretion per nephron as a basis for primary gout. Ann. Intern. Med. 73:359, 1970.
38. Austic, R.E., Cole, R.K.: Impaired renal clearance of uric acid in chickens having hyperuricemia and articular gout. Am. J. Physiol. 223:525, 1969.
39. Greene, M.L., Marcus, R., Aurbach, G.D., Kazam, E.S., and Seegmiller, J.E.: Hypouricemia due to isolated renal tubular defect. Dalmation dog mutation in man. Am. J. Med. 53: 361, 1972.
40. Sperling, O., Weinberger, A., Oliver, I., Lieberman, U.A., and DeVries, A.: Familial hypouricemia, hypercalciuria and osteoporosis: a new syndrome. Isr. J. Med. Sci. 9:1114, 1973.
41. Khachadurian, M.D. and Arslanian, M.J.: Hypouricemia due to renal uricosuria. Ann. Intern. Med. 78:547, 1973.
42. Simkin, P.A., Skeith, M.D., and Healey, L.A.: Suppression of uric acid secretion in a patient with renal hypouricemia. Isr. J. Med. Sci. 9:1113, 1973.
43. Weiner, I.M. and Fanelli, G.M.: Renal urate excretion in animal models. Nephron 14:33, 1975.
44. Gregor, R., Lang, F., and Deetjen, P.: Handling of uric acid by the rat kidney. Pfleugers. Arch. 324:279, 1971.
45. Roch-Ramel, F. and Weiner, I.M.: Excretion of urate by the kidneys in Cebus monkeys. A micropuncture study. Am. J. Physiol. 224:1369, 1973.
46. Mayrs, E.D.: Secretion as a factor in elimination by the bird's kidney. J. Physiol. 58:276, 1924.

THE EFFECT OF URINE FLOW RATE ON URATE CLEARANCE

B.T. EMMERSON, P.J. RAVENSCROFT and G. WILLIAMS

University of Queensland

Princess Alexandra Hospital, Brisbane, Australia 4102

The concept of renal clearance implies that the relationship between the excretion of a substance in the urine and its plasma concentration are such that the clearance is independent of the rate of urine flow, above a certain minimum value. Personal observation of consecutive hourly clearances of urate and creatinine often revealed more variation in the urate clearance than in the creatinine clearance, and this seemed greatest when there was fluctuation in the urine flow rate. At such times, one was left with the impression that an increase in the urine flow rate was followed by a delayed increase in the urate clearance. Brøchner-Mortenson (1937) had noted that urate excretion, and the urate clearance, fell when the urine flow rate was reduced below 1 ml/minute but detected no augmentation of the urate clearance with further increases in the urine flow rate above this. For these reasons, it was decided to look specifically at the effect of variation in urine flow rate as induced by a water diuresis upon urate excretion and urate clearance. There were also reports of changes in the urate clearance with alteration in the urine pH and because of this, and to eliminate any possible effects due to variations in the pH of the urine, the urine was maintained alkaline throughout the present study by the oral administration of sodium bicarbonate. The possibility that an alkaline urine might also promote the excretion of uric acid by means of non-ionic diffusion was also a factor.

EXPERIMENTAL

The effect of an increasing water diuresis was therefore studied in a group of healthy control subjects by measuring consecutive hourly renal clearances of urate, inulin, creatinine

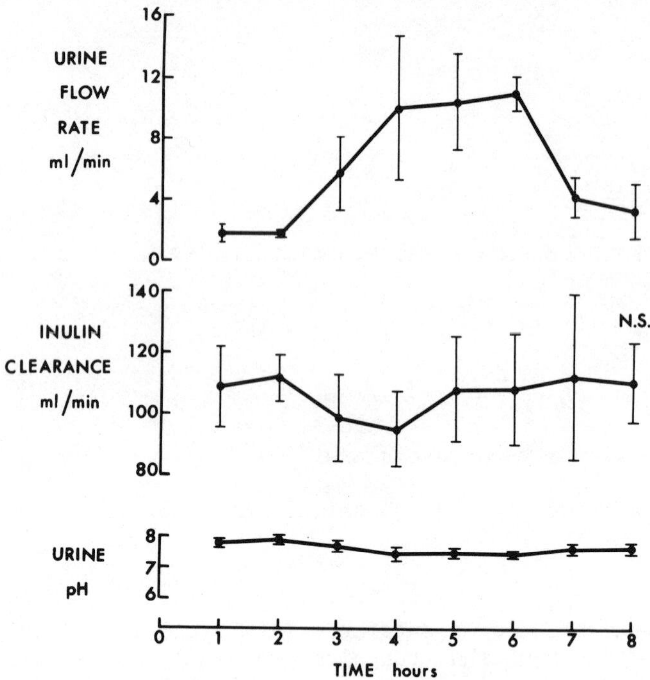

Fig. 1. The effect of a water load between the second and fifth hours on urine flow rate and inulin clearance.

and urea. The other parameters to be measured were chosen because it was generally accepted that variation in urine flow rate did not affect the renal clearance of inulin or creatinine, although an increased urine flow rate might be expected to promote the excretion of the readily diffusible urea. Eight consecutive renal clearances, each of one hour's duration, were therefore measured on the one day, with urine volumes which varied from less than 2 ml/minute at the beginning of the study to greater than 10 ml/minute four hours later, finally falling to about 3 ml/minute. Urine was collected by voluntary voiding in these healthy young males. A purine free breakfast was taken, before the first collection and a purine free lunch some three hours later. The consecutive clearances began at 9.00 a.m. and water was given as 250 ml at 8.00 a.m., 50 ml at 9.00 a.m. and 10.00 a.m., 1000 ml at 11.00 a.m. and 500 ml each at 12 md, 1.00 p.m. and 2.00 p.m. This method of water administration produced a significant diuresis (Figure 1) so that it was possible to assess the various clearances at both low urine volumes (less than 2 ml/minute) and high urine volumes (greater than 10 ml/minute). One could also study the rate of change of clearances with rate of change in the urine

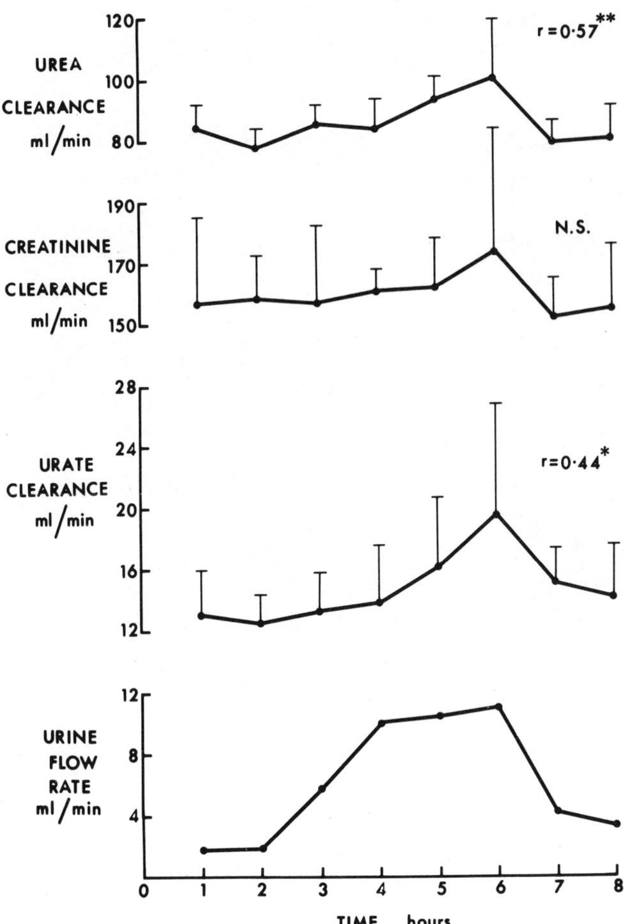

Fig. 2. The effect of increasing urine flow rate on the clearances of urea, creatinine and urate.

flow rate. It will be noted that there was no significant change in the glomerular filtration rate (as indicated by the inulin clearance) with the change in urine flow rate. The urine pH was relatively steady at between 7.4 and 7.9 during the study. This increase in urine flow rate was associated with a significant rise in the mean urate and urea clearances but not in the mean creatinine clearance (Figure 2). The correlation coefficients were 0.44 ($p < 0.05$) between urine flow rate and urate clearance, 0.28 (N.S.) between urine flow rate and creatinine clearance and 0.57 ($p < 0.01$) between urine flow rate and the urea clearance. Although the percentage change in urate clearance with the

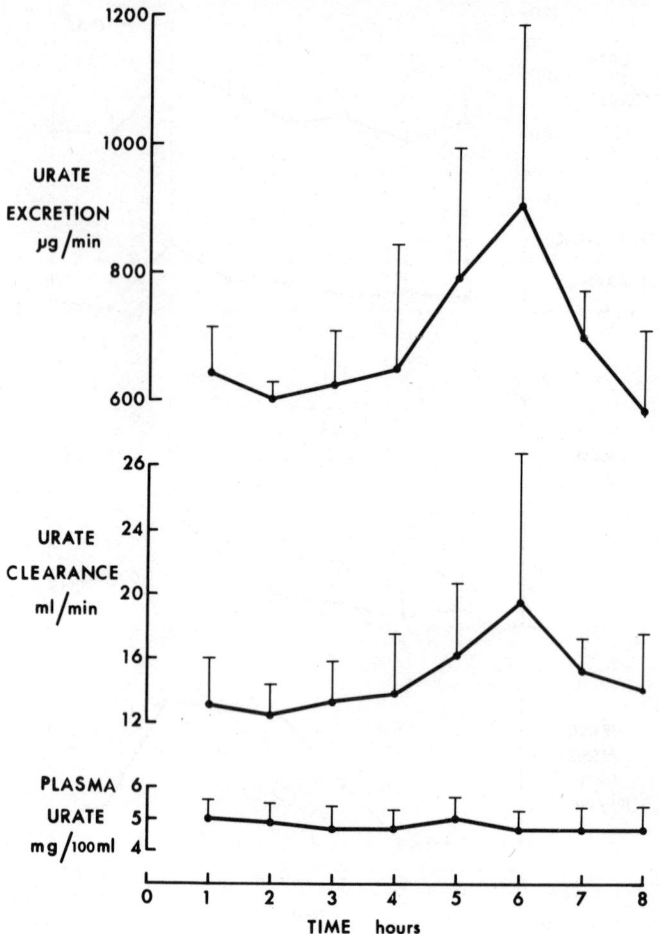

Fig. 3. The effect of the water load on urate clearance and its components.

diuresis was greater than that for the urea clearance, there was more variation between subjects and hence a higher standard deviation with regard to the urate clearance than for the urea clearance. When variations in inulin clearance were eliminated by calculating partial correlation coefficients, a highly significant correlation between urine flow rate and urate clearance was demonstrated ($r = 0.54$, $\rho < 0.01$). Partial correlation coefficients between urine flow rate and urate clearance were not significant when either urea or creatinine clearance was held constant.

When the various components of the urate clearance, namely UV urate and the P urate, were studied (Figure 3), it was clear that there was relatively little change in the serum urate due to the altered urine flow rate and that the major change occurred in the urate excreted in the urine (UV urate), thereby suggesting that the increased urate clearance of an alkaline diuresis is due to the increased renal excretion of urate.

The results were also subjected to covariance analysis and this confirmed the highly significant association between urine flow rate and the urea clearance. It also failed to show any significant variation in the slope of the regression lines of urea clearance on urine flow rate for individual subjects (Figure 4).

With regard to the urate clearance, on the other hand, although there was a significant over-all association between the urine flow rate and the urate clearance ($p < 0.01$), there were significant differences between individuals in the slopes of the regression lines of urate clearance on urine flow rate (Figure 5).

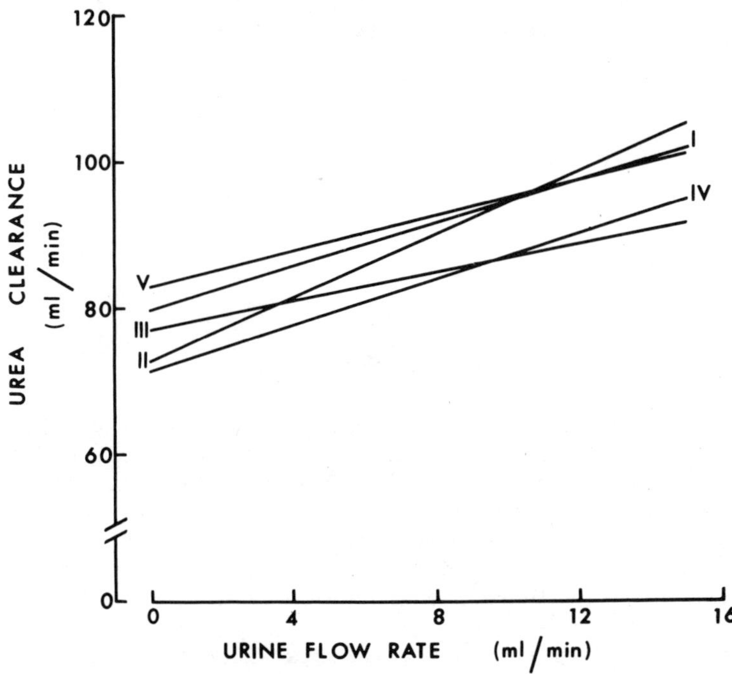

Fig. 4. The regression of urea clearance on urine flow rate for the five individuals studied.

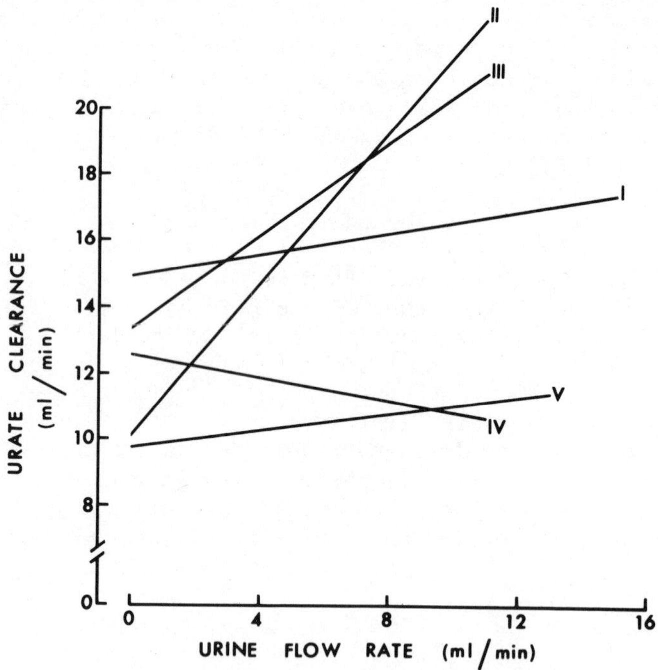

Fig. 5. The regression of urate clearance on urine flow rate for the five individuals studied.

DISCUSSION

This study was planned primarily to determine whether an alkaline diuresis promoted urate excretion, rather than to study the mechanism of its production. The positive water balance at any time would not have exceeded 1 litre, but this would have induced some degree of plasma volume expansion, which in turn would have promoted reduced tubular reabsorption of urate (Steele, 1969) and hence an increased excretion of uric acid in urine.

It is unlikely that any alkalosis which might have developed from the sodium bicarbonate administration would have contributed to the increased urate clearance because the urine was alkaline at all values of the urine flow rate and, if it has any effect at all, an alkalosis might tend to reduce the urate clearance (Meyer et alii, 1975). Although a positive correlation has been described between urate and creatinine clearance in healthy subjects (Ramsay et alii, 1975), this is also unlikely to be the whole explanation for the increase in urate clearance with a water diuresis because the rise in urate clearance was significant whereas that for the creatinine clearance was not.

In summary, it would appear that, in the presence of an alkaline urine, the rate of urine flow appears to have an effect upon the urate clearance, which is not as great as the effect of flow rate upon the urea clearance, and which varies considerably in magnitude from person to person. Perhaps this variation reflects the wider variability between individuals in the renal handling of urate,when compared with other parameters of renal function, which has already been observed in many other situations. Although further study of this problem is needed, it would appear that the induction of an alkaline diuresis may promote significant urate excretion in some individuals.

REFERENCES

1. BRØCHNER-MORTENSEN, K. Acta Med. Scand., Supp 84:167 (1937).

2. STEELE, T.H. J. Lab. Clin. Med. 74:288 (1969).

3. MEYER, W.J., GILL, J.R. and BARTTER, F.C. Ann. Intern. Med. 83:56 (1975).

4. RAMSAY, L., LEVINE, D., SHELTON, J., BRANCH, R. and AUTY, R. Ann. Intern. Med. 83:903 (1975).

EFFECTS OF VITAMINS ON THE RENAL HANDLING OF URIC ACID

Irving H. Fox, H.B. Stein and S.L. Gershon

Purine Research Laboratory, University of Toronto

Rheumatic Disease Unit, Wellesley Hospital, Toronto

The use of vitamins in megadoses has been popular in our modern society for the treatment of many problems. Although these compounds are often of dubious therapeutic value except for the specific management of vitamin deficiency, it has often been assumed that they are innocous in a large dose. Diseases associated with the excessive ingestion of certain vitamins are well known and new adverse effects are continually being described.

The existence of vitamins which are weak organic acids suggested the possibility that they could interfere with the renal tubular handling of uric acid in man. Indeed, hyperuricemia and the precipitation of acute gout in patients receiving nicotinic acid for type II hyperlipoproteinemia suggested such a potential effect (1-3). In addition the formation of renal calculi, a hazard associated with ascorbic acid ingestion (4,5), implied that a renal effect on uric acid handling might also occur.

We have evaluated the effect of nicotinic acid (6) and ascorbic acid (7) on the renal handling of uric acid in man. The effect of oral nicotinic acid was compared with the effect of nicotinamide or pyrazinamide. In 9 patients studied, one gram of nicotinic acid caused a mean decrease of 62% in the $C_{urate}:C_{creatinine}$ during the first 4 hours after its administration (Figure 1). A quantitatively similar but more sustained diminution of $C_{urate}:C_{creatinine}$ was observed after 3 g oral pyrazinamide in 3 patients. Four to 6 hours after oral ingestion, 1 g nicotinamide caused a minimal diminution in $C_{urate}:C_{creatinine}$ in 3 patients.

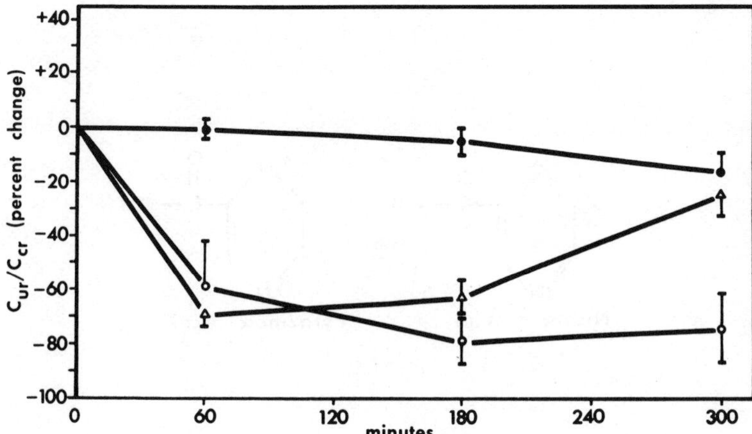

Fig. 1. Effect of nicotinic acid, nicotinamide, and pyrazinamide on the renal handling of uric acid. The Curate:Ccreatinine was expressed as the percentage change from the control values 1 to 6 hours after drug ingestion. Each point represents the mean value plus or minus the standard error of the mean. Δ—Δ nicotinic acid, 1 g; o—o pyrazinamide, 3 g; •—• nicotinamide, 1 g.

Nicotinic acid could potentially diminish the clearance of uric acid by a decrease in the glomerular filtration rate or by an alteration in the renal tubular handling of uric acid. Since the creatinine clearance was not substantially decreased, the mechanism of the nicotinic acid effect on the renal tubule was studied by demonstrating its interaction with sulfinpyrazone, iopanic acid and acetylsalicylic acid. Nicotinic acid inhibited the uricosuria following sulfinpyrazone or iopanoic acid. Acetysalicylic acid did not inhibit the diminution of percent Curate:Ccreatinine associated with oral nicotinic acid.

Nicotinamide is a structural analog of pyrazinamide and nicotinic acid is a structural analog of pyrazinoic acid (Figure 2). Pyrazinamide inhibits the renal clearance of uric acid by virtue of its conversion to pyrazinoic acid, the active metabolite. The structural similarities suggest that nicotinic acid may affect the renal tubular handling of uric acid by a mechanism similar to pyrazinoic acid whatever that might be. However, there are two major differences between the effects of nicotinic acid and pyrazinamide: (a) nicotinic acid has a more transient effect, and (b) acetylsalicylic acid inhibits the decreased renal clearance of uric acid related to pyrazinamide but not the decrease related to

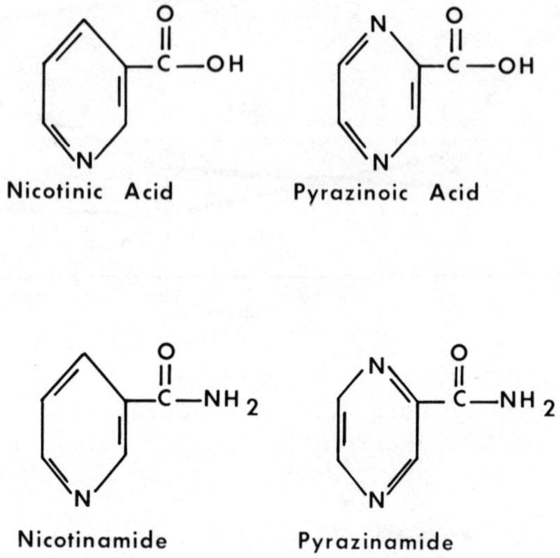

Nicotinic Acid

Pyrazinoic Acid

Nicotinamide

Pyrazinamide

Fig. 2. Structure of nicotinic acid, nicotinamide and analogs.

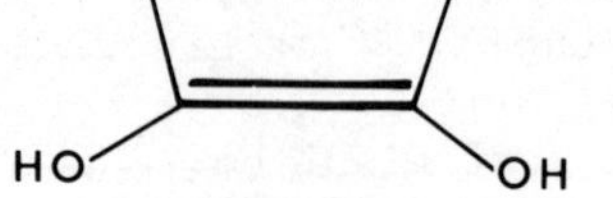

Fig. 3. Structure of ascorbic acid.

nicotinic acid. Nicotinic acid is handled in the canine kidney by a bidirectional tubular mechanism similar to uric acid. The analogous transport system for these two organic acids suggests a physiological basis by which nicotinic acid could interfere with the renal excretion of uric acid.

Studies were next performed with the organic acid vitamin, ascorbic acid (Figure 3). This vitamin is handled in the kidney by glomerular filtration and tubular reabsorption. The effect of this compound on the renal excretion of uric acid was assessed (Figure 4). In 9 subjects 4.0 g of ascorbic acid caused a maximum increase to 102 $\pm$ 14% in Curate:Ccreatinine 2 to 6 hours after its administration. The peak effect varied between period 3 and period 4 in different patients. In contrast 0.5 and 2.0 g of ascorbic acid

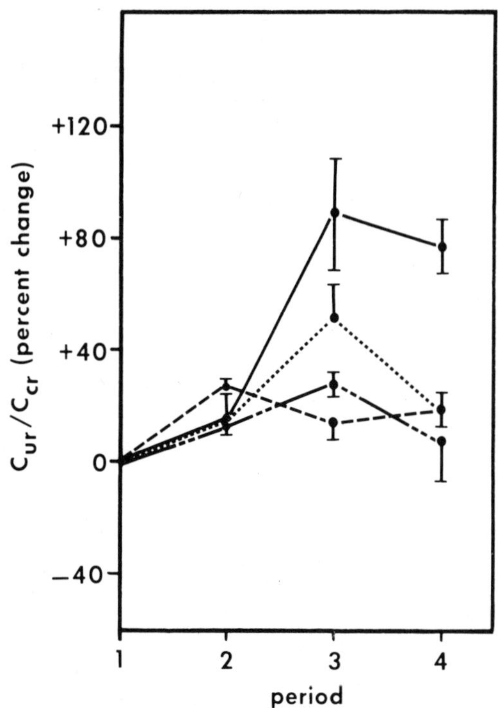

Fig. 4. The effect of ascorbic acid on the renal handling of uric acid. Curate:Ccreatinine is expressed as a percentage change plus or minus the standard error from the control value (period 1). Each period represents a 2 hour collection. Ascorbic acid: •—•, 4 g; •...•, 2 g; (•--•), 0.5 g;(•—-•), 0g.

increased Curate:Ccreatinine to 28 $\pm$ 3% and 52 $\pm$ 12% respectively. When no ascorbic acid was given, the Curate:Ccreatinine increased to 31 $\pm$ 5% of the control value. A statistically significant increase ($P < .01$) above the trial without medication only occurred with 4.0 g ascorbic acid.

Ascorbic acid could potentially increase the renal clearance of uric acid by a decrease in the binding of uric acid to plasma protein, by an increase in the glomerular filtration rate or by an alteration in the renal tubular handling of uric acid. Urate binding to plasma protein in vitro was not decreased by ascorbic acid in concentrations from 0 to 20 mg/dl. No increase in the mean creatinine clearance from the control period was found after the ingestion of 4.0 g of ascorbic acid. In 3 of 4 patients acetylsalicylic acid competely inhibited ascorbic acid induced uricosuria, while in 1 patient this was not observed. Pyrazinamide reversed the uricosuric effect of ascorbic acid. The modification of the uricosuric effect of ascorbic acid by acetylsalicylic acid and pyrazinamide implied that related renal sites of activity may exist for these compounds.

Our studies have also shown that long term therapy with ascorbic acid 8 g per day can lower the serum uric acid (Table 1). The diminution of the serum uric acid ranged from 1.2 to 3.1 mg/dl while the mean increase of uric acid clearance ranged from 55 to 74%.

These observations suggest that ascorbic acid is uricosuric and nicotinic acid is antiuricosuric by renal tubular mechanisms. Therefore, these vitamins could interfere with the clinical evaluation of disorders of uric acid metabolism by altering the serum and urine uric acid. By acutely changing the serum uric acid they could potentially precipitate acute gout. Finally, they illustrate the multipotential pharmacologic actions of megadoses of simple vitamins.

TABLE 1

EFFECT OF LONG TERM THERAPY WITH ASCORBIC ACID

8 g/day

	Serum uric acid (mg/dl)		Cur:Ccr (mean % increase)	Duration of therapy (Days)
	Control	Lowest value		
1	6.1	4.5	74	7
2	9.8	6.7	55	6
3	4.1	2.9	61	3

REFERENCES

1. Christensen, A., Anchor, W.P., Berge, K.G., et al. 1961. Nicotinic acid treatment of hypercholesterolemia. J.A.M.A. 177:546-550.

2. Berge, K.G., Anchor, W.P., Christensen, N.A., et al. 1961. Hypercholesterolemia and nicotinic acid: a long term study. Am. J. Med. 31:25-36.

3. Parsons, W.B. 1961. Studies of nicotinic acid use in hypercholesterolemia. Arch. Intern. Med. 107:653-667.

4. Vitamin C - were the trials well controlled and are large doses safe? 1971. Med. Lett. Drugs Ther. 13:46-48.

5. Briggs, M.H. 1973. Side effects of vitamin C. Lancet 2:1439.

6. Gershon, S.L. and Fox, I.H. 1974. Pharmacologic effects of nicotinic acid on human purine metabolism. J. Lab. Clin. Med. 84:179-186.

7. Stein, H.B., Hasan, A. and Fox, I.H. 1976. Ascorbic acid-induced uricosuria. A consequence of megavitamin therapy. Ann. of Int. Med. 84:385-388.

The Uricosuric Action of Amino Acids in Man

F. Matzkies, G. Berg

Abteilung für Stoffwechsel und Ernährung (Vorsteher: Prof. Dr. Dr. h. c. G. Berg) in der Medizinischen Klinik mit Poliklinik der Universität Erlangen-Nürnberg 8520 Erlangen, Krankenhausstr. 12

During infusion of fructose, sorbitol and xylitol, but not during glucose or maltose, we found a considerable augmentation of uric acid levels in healthy young males (3-7).

This increase of uric acid was dose-dependant. For fructose and sorbitol a dosage of 1,5 g/kg/h is needed, whereas xylitol augments uric acid after a dosage of only 0,25 g/kg/h BW (3-7).

Now, we investigated the uric acid metabolism during infusion of a carbohydrate mixture containing fructose, glucose and xylitol at a rate of 2:1:1 (LGX = Triofusin E-1000, Pfrimmer). In addition the metabolism of uric acid was examined after administration of this carbohydrate mixture given in combination with amino acids (Aminofusin L-10 % carbohydratfree, Pfrimmer) or fat emulsion (Lipofundin S-10, Braun-Melsungen).

Methods

6 healthy young men received a long-term infusion of LGX at a dosage of 0,5 g/kg/h (1).

8 men were given the same carbohydrate mixture at a rate of 0,6 g/kg/h together with amino acids with a rate of 0,1 g/kg/h (2).

8 men received an intravenous infusion of the carbohydrate mixture (0,5 g/kg/h) together with amino acids (0,1 g/kg/h) and fat

at a rate of 0,1 g/kg/h.

In 8 men only amino acids at a rate of 0,1 g/kg/h were infused. The infusion time in all experiments was 12 hours. Blood samples for determination of uric acid by uricase-method were taken at the beginning, after 6 hours and at the end of the infusion time.

During the infusion period the urine was collected and the renal excretion rate of uric acid determined.

Results (Tables 1, 2,)

During the 12 hour carbohydrate infusion the uric acid level rose from 6 mg/dl to 6, 4 mg/dl.

In contrast, in all experiments with amino acids the level of uric acid showed a significant fall.

Table 1: Uric acid during infusion of carbohydrates, amino acids and fat

Solution and dosage	Uric acid level mg/100 ml 0	6 h	12 h
LGX 0, 5 g/kg/h fructose, glucose and xylitol 2:1:1	6, 0 ± 0, 8	6, 6 ± 0, 6	6, 4 (6, 0-8, 4)
LGX 0, 6 g/kg/h + Amino acids 0, 1 g/kg/h	6, 8 ± 0, 9	5, 8 ± 0, 9 s	4, 9 ± 0, 8 s
LGX 0, 5 g/kg/h + Amino acids 0, 1 g/kg/h + fat emulsion 0, 1 g/kg/h	5, 5 ± 0, 5	5, 6 ± 0, 8	5, 1 ± 0, 9 s
Amino acids 0, 1 g/kg/h	6, 2 ± 1, 5	5, 6 ± 1, 2 s	5, 0 ± 1, 3 s

After LGX and amino acids we observed a decrease from 6, 8 mg/dl to 4, 9 mg/dl. During infusion of carbohydrates, amino acids and fat the uric acid level decreased from 5, 5 to 5, 1 mg/dl. After amino acid alone the uric acid level showed a significant fall from 6, 2 to 5, 0 mg/dl

During carbohydrate infusion the elimination rate of urate was 40 ± 6 mg/h.

Table 2: Renal uric acid elimination during infusion of carbohydrates, amino acids and fat.

Solution and dosage		Elimination rate mg/h	mg/12 h
LGX (Triofusin E-1000) 0,5 g/kg/h		40 ± 6	507 ± 76
LGX + Amino acids	0,6 g/kg/h 0,1 g/kg/h	68 (59-77)	845 (730-961)
LGX + Amino acids + fat	0,5 g/kg/h 0,1 g/kg/h 0,1 g/kg/h	56 (28-142)	667 (335-1720)
Amino acids (Aminofusin 10%) 0,1 g/kg/h		35 ± 9	423 ± 112

After carbohydrate and amino acid the renal elimination was 68 mg/h and after infusion of carbohydrates, amino acids and fat we found an elimination rate of 56 mg/h.

During infusion of amino acids alone, the uric acid elimination was 35 mg/h. Results are listed in table 2.

Discussion

The augmentation of uric acid level during infusion of fructose and xylitol is well known (3-7).

In our long-term infusion experiments with fructose (0,25 g/kg/h) and xylitol (0,125 g/kg/h) over a period of 12 hours we found an increase of uric acid. This phenomen is caused by an augmented breakdown of ATP and an elevated lactate level in blood (4-7).

It was therefore very surprising to see, that uric acid levels in ours experiment with amino acids together with the carbohydrate mixture showed a significant fall (4). This fall was due to an increased renal excretion of 68 mg/h.

In the study with carbohydrates, amino acids and fat there was a decrease of uric acid. Because of the lower levels at the beginning, the renal excretion was some what lower.

After amino acid infusion no elevation of uric acid is known. We found a significant decrease after a 12 hour infusion of a total dose of 1,2 g/kg/12 h. The renal excretion rate was 35 mg/h.

This value is a little bit lower than after carbohydrate infusion. But taking into account that there is no evidence of an increased de novo synthesis of uric acid this elimination rate is high. In the untreated control group (n=13) the uric acid elimination was 30(22-43) mg/h.

Conclusion

During infusion of amino acids at a rate of 0,1 g/kg/h BW over a period of 12 hours a significant fall of uric acid excists due to an augmented renal excretion. The fall of uric acid can still be seen when amino acids are infused together with a fat emulsion or carbohydrate mixture.

References

1. Berg, G., F. Matzkies, H. Heid, W. Fekl, M. Conolly: Wirkungen einer Kohlenhydratkombinationslösung auf den Stoffwechsel bei Langzeitinfusion. Z. Ernährungswiss. 14 (1975), 64.

2. Berg, G., F. Matzkies, H. Heid. W. Fekl: Wirkungen einer Kohlenhydratkombinationslösung auf den Stoffwechsel bei gleichzeitiger Applikation von Aminosäuren. Z. Ernährungswiss. 14 (1975), 163.

3. Berg, G., F. Matzkies, H. Heid: Zur Wirkung hochdosierter Sorbitinfusionen auf den Kohlenhydrat-, Fett- und Harnsäurestoffwechsel, den Säure-Basen-Haushalt und die Elektrolytkonzentrationen im Serum und Urin bei gesunden Männern. Dtsch. med. Wschr. 99 (1974), 2352.

4. Matzkies, F.: Untersuchungen zur Pharmakokinetik von Kohlenhydraten als Grundlage ihrer Anwendung zur parenteralen Ernährung. Z. Ernährungswiss. 14 (1975), 184.

5. Matzkies, F.: Harnsäurestoffwechsel nach Kohlenhydratzufuhr. In: V. Becker: Gastroenterologie und Stoffwechsel. Witzstrock-Verlag Baden-Baden Brüssel 1974 s. 286.

6. Matzkies, F., G. Berg: Fruktoseinduzierte Hyperurikämie bei Gesunden sowie Patienten mit Gicht und sekundärer Hyperurikämie. Med. Welt 25 (1974), 414.

7. Matzkies, F., H. Heid, W. Fekl, G. Berg: Wirkungen einer Kohlenhydratkombinationslösung auf den Stoffwechsel bei hochdosierter Kurzzeitinfusion. Z. Ernährungswiss. 14 (1975), 53.

URATE EXCRETION IN NORMAL AND GOUTY MEN

Peter A. Simkin

Division of Rheumatology, Department of Medicine

University of Washington, Seattle, Washington

Serum levels of urate are easily measured and have been systematically studied in many different settings. We now know about urate values in populations all over the globe, in a great variety of human diseases, in divers under the sea, in men in outer space, and practically everywhere in between. These studies have been intriguing in the differences they have revealed within and between groups. We know almost nothing, however, of what these differences mean. The reason for our ignorance is simple: our present tools are far too cumbersome to apply to these vast problems. Inulin clearances, restricted diets, purine loads, 24-hour urine collections and isotopic turnover studies are expensive and cumbersome for both patients and investigators. The complexity of such studies discourages us from asking large scale questions about the renal handling of urate in different hyperuricemic states. In search of more effective tools, I have reviewed and reevaluated classic data on urate excretion by normal and gouty men. The aim of this analysis was to develop simple expressions which might be helpful in evaluating groups and individuals with hyperuricemia.

Data were reviewed from six classic papers[1-6] in which uric acid excretion was examined at varying concentrations of plasma urate in both normal and gouty men. The initial calculations are based on a simple plot of UV uric acid against plasma urate. The UV urate (in mg/min) was chosen rather than urate clearance because in a two-way plot we are primarily interested in the slope of the resultant function. When UV is plotted against P, the slope of the curve at any given point (regardless of the overall shape of the curve) will be UV/P or the clearance of urate in ml/min. Any curve other than a straight line tells us directly how the clearance changes at different plasma levels. When, on the other hand, clearance is plotted on the vertical axis, the slope of the resultant curve is in ml^2/mgmin and is impossible for us to meaningfully interpret.

The UV urate has also been factored by the glomerular filtration rate (inulin clearance) for each data point and only subjects with GFR greater than 70 ml/min are included. Using UV urate per 100 ml of GFR offers the advantage of expressing urate excretion in terms of a standard unit of functional renal mass and is a device favored by a number of authors in this field including Dr. Rieselbach[7].

Serum urate levels were increased by oral purine loads or decreased by pretreatment with allopurinol. All data between 2 and 13 mg% were included.

From the six papers reviewed, there were 84 observations on normal individuals with a mean GFR of 115 ± 14 ml min and 132 points from gouty patients with a mean GFR of 104 ± 19 ml/min.

In gouty individuals, the data fit a plot of log UV urate against P with a correlation coefficient of 0.87. When a regression was run on the same data plotted in a linear fashion the correlation coefficient was only 0.74.

The exponential character of this curve has been well appreciated in the past, and reflects the fact that increments in plasma urate at high concentrations have a more dramatic effect on UV urate than do similar increments when the urate concentration is low.

The data from normal individuals also best fit a plot of log UV urate against P with a correlation coefficient in this instance of 0.82.

When both normal and gouty data are plotted together, there is considerable overlap with many observations from gouty individuals falling within the normal range.

The most striking feature of this figure is that the regression lines for normal and gouty data are parallel with respective slopes of 0.084 ± 0.006 and 0.083 ± 0.004. The parallel nature of these slopes permits two interesting generalizations. First, for any given plasma concentration of urate the average gouty individual excretes 41% less urate than does the normal individual. Second, to achieve any given level of urinary urate excretion the gouty individual must have a mean serum urate value 1.8 mg% higher than that of the normal subject. Some previous authors have emphasized that the gouty kidney responds to a urate load in a normal fashion. The parallel nature of these two regression lines illustrate that that is precisely the case. For both gouty and normal individuals the increment in urate excretion is similar. It is clear from these data, however, that the average gouty kidney labors under a substantial handicap in its attempt to handle the urate presented to it by the plasma.

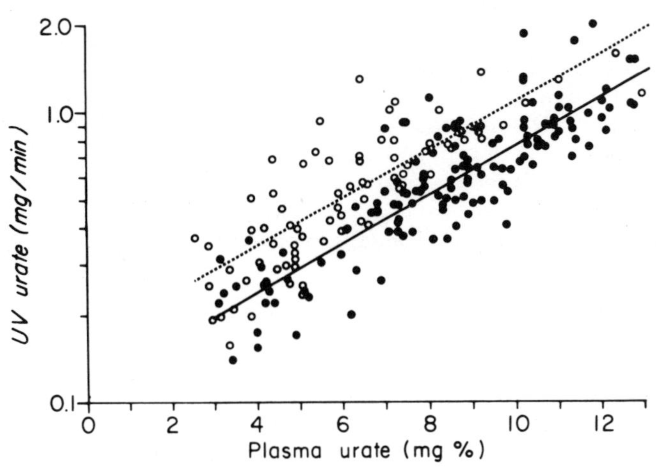

Figure 1

UV URATE EXCRETION AT VARYING PLASMA URATE CONCENTRATIONS IN NORMAL (O) AND GOUTY (●) MEN.

An additional feature of the parallel nature of these slopes is that we may employ the basic regression equation to remove the effect of differing plasma urate concentrations on the UV urate. For both the normal and the gouty subjects, the data are grouped equally on both sides of the regression line. In the equation

$$(1)\quad y = ax+b$$

we can substitute and solve for b

$$(2)\quad b = \log UV - 0.084P$$

without distorting the relationship of the normal and the gouty points. This device permits the direct comparison of UV urate data obtained under varying plasma urate loads.

One of the rare attempts to follow up on epidemiologic observations is the paper by Drs. Healey and Bayani-Sioson[8] on urate excretion in normal Filipino and Caucasian men studied before and after an oral purine load. In this study the UV urate was determined from 24-hour urine collections and the GFR from creatinine clearance. When these data were analyzed by equation (2) six observations on Filipinos and none on Caucasians fell more than two standard deviations below the Caucasian mean. This analysis supports the authors' contention that some but not all Filipinos have a significant defect in their ability to excrete urate. The conclusion is supported at least as well by baseline data as by those after a purine load thus indicating that purine loading is unnecessary to identify defective excretion of urate.

The 24-hour urine collections employed in this study of Filipinos are cumbersome and have well recognized problems of inaccurate timing and incomplete collection of the specimen. Fortunately, it is possible to determine the UV urate on a random urine sample. The UV urate per unit GFR may be written:

$$(3)\quad \frac{U_u\,\text{mg/100 ml} \times V_u\,\text{ml/min}}{C_c\,\text{ml/min}}$$

where U_u and V_u are the urate concentration and flow rate of urine and C_c is the clearance of creatinine. Substituting U_cV_c/P_c for C_c we obtain:

$$(4)\quad \frac{U_u\,\text{mg/100 ml} \times V_u\,\text{ml/min} \times P_c\,\text{mg/100 ml}}{U_c\,\text{mg/100 ml} \times V_c\,\text{ml/min}}$$

This formula may then be cancelled as follows:

$$(5)\qquad \frac{U_u\, \cancel{mg/100\ ml} \times \cancel{V_u}\ \cancel{ml}/min \times P_c\, mg/100\ \cancel{ml}}{U_c\, \cancel{mg/100\ ml} \times \cancel{V_c}\ ml/min}$$

to yield

$$(6)\qquad \frac{U_u \times P_c\ (mg/min)}{U_c\ \ (100\ ml/min)}$$

The advantage of this derivation is that it is convenient and cannot be affected by errors in the timing or collection of urine specimens. In preliminary experiments we have found that UV urate values obtained from random mid-morning specimens are entirely comparable to those obtained with 24-hour urine specimens in the past. It is obvious that the numerical value for UV urate will be very similar to the urinary urate to creatinine ratio that has been usefully employed by others in screening hyperuricemic patients. Since the plasma creatinine is often 1 the numerical value of the ratio will be little altered by including the value of plasma creatinine. The plasma creatinine is easily obtained, however, and its addition means that we can directly obtain a value which properly represents UV urate.

Through the use of equation (6), it should be easy to evaluate the effect of many different variables on UV urate. By adding equation (2) we can usefully compare UV urate values obtained at widely different plasma urate concentrations. I hope that these simple formulations will prove to be of value in evaluating hyperuricemic states and the many environmental and pharmacologic agents which effect the renal elimination of urate.

REFERENCES

1. Seegmiller, J.E., Grayzel, A.I., Howell, R.R., Plato, C. J. Clin. Invest. 1962, 41, 1094.
2. Nugent, C.A., Tyler, F.H. J. Clin. Invest. 1959, 38,1890.
3. Yü,T.F.,Berger, L., Gutman, A.B. Amer. J. Med. 1962,33,829.
4. Steele, T.H., Rieselbach,R.E. Amer. J. Med. 1967, 43, 868.
5. Gutman, A.B., Yü,T.F.,Berger, L. Amer. J. Med. 1969, 47, 575.
6. Rieselbach, R.E., Sorensen, L.B., Shelp, W.D., Steele, T.H. Ann. Intern. Med. 1970, 73, 359.
7. Rieselbach, R.E., Steele, T.H. Amer. J. Med. 1974, 56, 665.
8. Healey, L.A., Bayani-Sioson, P.S. Arthritis Rheum. 1971, 14, 721.

THE EFFECT OF ACID LOADING ON RENAL EXCRETION OF URIC ACID AND AMMONIUM IN GOUT

T.Gibson, S.F.Hannan, P.J.Hatfield, H.A.Simmonds
J.S.Cameron, C.S.Potter and C.M.Crute

Arthritis Research Unit and Department of Medicine
Guy's Hospital Medical School, London SE1 9RT

A defect of ammonium production associated with a relatively acid urine has been described in patients with both gout (1) and idiopathic uric acid stone formation (2).

Gutman and Yu (1965) advocated an hypothesis which ascribed the concurrence of hyperuricaemia and decreased ammonium excretion to an inability of the kidneys to utilise glutamine for ammonia synthesis and increased conversion of glutamine in the liver to uric acid.

If this postulate were generally applicable it should be possible to demonstrate an inverse relationship between ammonium excretion and uric acid production. Attempts to resolve this question have produced conflicti answers. Plante, Durivage and Lemieux, (3) were able to demonstrate a reciprocal relationship between uric acid and ammonium excretion in gout patients on a high purine diet. This finding was not confirmed by Swales, Kopstein and Wrong (4).

In an attempt to re-examine the possible relationshi between ammonium and uric acid excretion, the current study explores the effect of ammonium chloride on these variables in both gouty patients and healthy volunteers undertaking controlled dietary regimes.

SUBJECTS AND METHOD

Eight patients with gout and an equal number of healthy controls participated in the study which was conducted on an out-patient basis. The controls were matched with the gout patients for age and creatinine clearance.

Urine osmolality was estimated after 15 hours of fluid deprivation and then each subject underwent a four day low purine diet containing less than 300 mg of purine nitrogen daily. On days three and four, 24 hour urine specimens were collected. On the fourth day, each sample voided was collected separately in order to establish individual variations of urine pH throughout a 24 hour period.

During the following 16 days, each participant ate a high purine diet containing an average of 410 mg of purine nitrogen daily. Separate 24 hour urine collections were made from the third to the sixteenth day of the high purine diet. On the sixth to ninth days inclusive ammonium chloride was given in a dose of 100 mg per Kg body weight. Three control subjects and two gouty patients repeated the whole 20 day study while eating only low purine foods.

All daily urine specimens were collected under toluene and paraffin and determinations of pH, total acid, ammonium, titratable acid, uric acid, urea and creatinine were made on each specimen. Venous blood samples were obtained on at least two separate days of each phase of the experiment· low purine; high purine; ammonium chloride plus high purine diet and recovery period while still eating a high purine diet. Measurements of uric acid, urea, creatinine and bicarbonate were performed on the blood. Uric acid was measured by a uricase method (5), total acid by the method of Jorgensen (6) and titratable acid and ammonium by the technique of Chan (7). Creatinine and urea were estimated by standard methods adapted to the autoanalyser.

RESULTS

Although patients with gout were matched with the control subjects in respect of age and creatinine clearance, urine concentrating ability after 15 hour fluid deprivation was significantly less in the gout patients (Table 1). The urinary pH of the gouty patients

TABLE 1
DETAILS OF GOUT AND CONTROL SUBJECTS

	GOUT	CONTROL
Mean age (range)	48 (31-68)	40 (30-59)
Sex	7M 1F	6M 2F
Creatinine Clearance ± SD ($ml/min/1{\cdot}73\ m^2$)	86 ± 8·0	101 ± 22*
Urine concentration ± SD (mOsm/l)	723 ± 120	942 ± 136+

*t = 1·86: N.S. +t = 3·13: $p < 0{\cdot}01$

throughout a single day tended to be lower than that of of the controls (Fig. 1) and the percentage diurnal variation ± SD of the gout subjects was 17·0 ± 8·1%, a value that was significantly lower than that of the controls (27·2 ± 8·5%) (t = 2·45; $p < 0{\cdot}05$).

On a low purine diet the mean ± SD blood uric acid of the gout patients was 6·46 ± 1·18 mg/100 ml and that of the controls 3·46 ± 0·74 mg/100 ml. Substitution of a high purine diet induced an increase of blood uric acid to 7·32 ± 1·53 mg/100 ml in the gout patients and to 3·86 ± 0·84 mg/100 ml in the controls. Striking increments of 2·0 and 3·0 mg/100 ml were seen in two gouty individuals but the mean increment for the gout patients as a whole was not significant (t = 1·25).

For convenience, the variables under investigation have been expressed as the mean of successive daily collections for each phase of the experiment. Thus the values for the low purine and high purine phases are each represented by the mean of two urine collections; the high purine plus ammonium chloride phase by the mean of four collections and the recovery phase by the mean of seven collections. Data from the five subjects who repeated the study while confined to a low purine diet

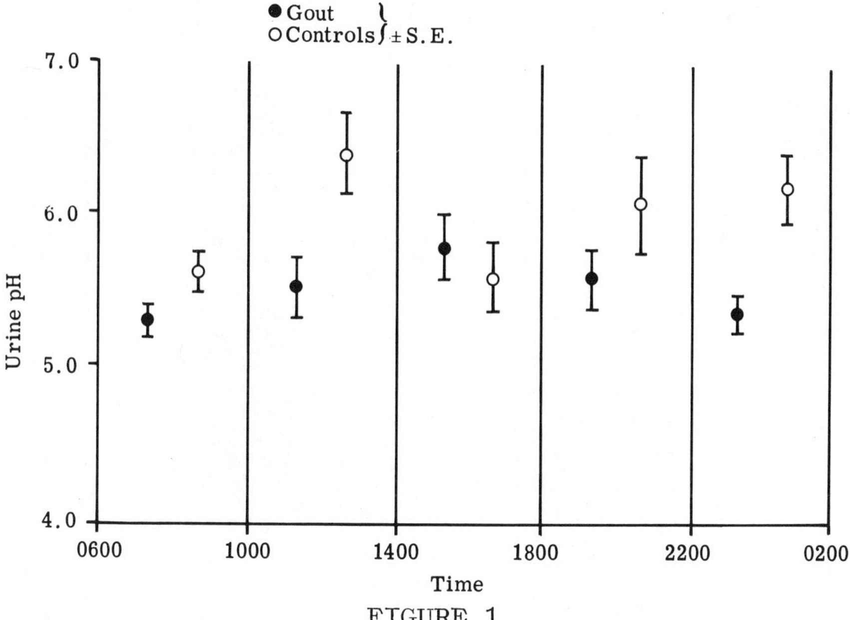

FIGURE 1

DIURNAL VARIATION OF URINE pH IN GOUT AND CONTROLS

were examined by comparing the means of three collection phases.

The results of the low purine experiment are outlined in Table 2. Ingestion of ammonium chloride was associated with modest decrements of uric acid excretion in all five subjects, the mean values being 67mg/24h. for the gouty and 31mg/24h. for the controls. In both groups there was a return to previous levels during the recovery phase.

A similar but more striking decrease of urine uric acid excretion was observed in every subject receiving a high purine diet. The mean values for each phase of the study are documented in Table 3. There were significant differences between uric acid excretion before and during ammonium chloride in both the gouty ($t = 3{\cdot}99$; $p<0{\cdot}01$) and controls ($t = 4{\cdot}03$; $p<0{\cdot}01$). The mean decrement for the gout patients was 105 mg/24h and for the controls 171 mg/24h and there was no significant difference between these values ($t = 0{\cdot}372$). A clear reciprocal relationship between the fall in uric acid excretion and the expected increase of ammonium excretion was demonstrable (Fig. 2).

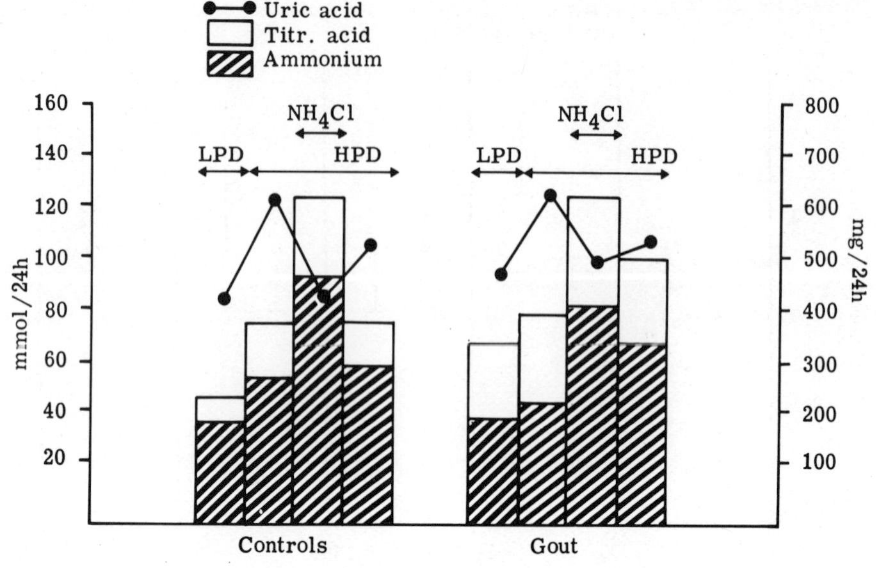

FIGURE 2

RESPONSE OF GOUT AND CONTROL SUBJECTS TO AMMONIUM CHLORIDE

In the controls as a whole, the changes in excretion of ammonium, titratable acid and uric acid were brisk and values returned to normal in an average of 2·5 days. By contrast, the same changes in the gout subjects were sluggish, the largest mean decrement in uric acid excretion being on the day following ammonium chloride administration. Recovery of acid excretion took an averag of 5·6 days.

The mean urine pH of the gout patients was lower and the mean titratable acid higher in all four phases of the study. The percentages of hydrogen ion excreted as ammonium in the gout patients was consistently less than that of the controls (Table 3). During high purine feeding the respective percentages ±SD were 56 ± 5·4% for the gouty and 71 ± 12·7% for the controls ($t = 3{\cdot}15$; $p < 0{\cdot}01$). In response to ammonium chloride, the percentage excreted as ammonium rose to 65 ± 5·2% in the gouty subjects whereas that of the controls remained constant. Nevertheless, a significant difference between the respective percentages persisted during this period ($t = 2{\cdot}47$: $p < 0{\cdot}05$) and in the recovery phase ($t = 2{\cdot}48$; $p < 0{\cdot}05$). The percentage of total acid excreted as ammonium in the gout and control on a high purine diet correlated with urine concentrating ability ($r = 0{\cdot}653$; $p < 0{\cdot}02$) and with creatinine clearance

Table 2

MEAN URINARY EXCRETION OF AMMONIUM, TITRABLE ACID AND URIC ACID IN TWO GOUT AND THREE CONTROL SUBJECTS CONFINED TO A LOW PURINE DIET

GOUT	AMMONIUM±SD (mmol/24h)	TITR.ACID±SD (mmol/24h)	URINE pH	URIC ACID±SD (mg/24h)
Low Purine	30·5 ± 11·1	24·1 ± 3·6	5·4	522 ± 120
Amm. Chloride	75·6 ± 1·4	51·1 ± 8·7	4·9	455 ± 66
Recovery	65·3	44 ± 14·3	4·9	486 ± 68
CONTROL				
Low Purine	43·4 ± 27·2	11·3 ± 11·0	6·6	424 ± 112
Amm. Chloride	83·5 ± 39	27 ± 13·9	5·3	393 ± 129
Recovery	55·3 ± 31·5	17·3 ± 7·7	6·0	455 ± 142

($r = 0{\cdot}602$; $p<0{\cdot}02$) but not with age ($r = 0{\cdot}339$).

The striking decrements of uric acid excretion following ammonium chloride were not associated with any change of blood uric acid levels. Modest increments of urea excretion were observed but creatinine excretion remained constant (Table 3).

DISCUSSION

The results of our studies confirm the relatively low urine pH and the impaired diurnal variation in many gouty subjects. It has been suggested that an acidic urine with disturbance of the diurnal rhythm may be an early manifestation of renal impairment secondary to gout and may precede deterioration of glomerular function (8).

The relevance of reduced ammonium excretion to low urine pH can be reasonably assumed from our data

Table 3

GOUT	AMMONIUM ± SD (mmol/24h)	TITR.ACID ± SD (mmol/24h)	URINE pH	URIC ACID ± SD (mg/24h)	CREATININE ± SD (mmol/24h)	UREA ± SD (mmol/24h)	BLOOD URIC ACID ± SD (mg/100 ml)
LOW PURINE	36·8 ± 11·0 (57%)	28·8 ± 11·8 (43%)	5·3	457 ± 115	12·7 ± 3·5	317 ± 157	6·46 ± 1·18
HIGH PURINE	42·7 ± 12·8 (56%)	35·2 ± 15·5 (44%)	5·3	606 ± 152	12·9 ± 4·7	353 ± 170	7·32 ± 1·53
AMM. CHLORIDE	79·9 ± 27·5 (65%)	43 ± 15·8 (35%)	4·8	501 ± 182	12·9 ± 4·0	373 ± 184	7·39 ± 1·63
RECOVERY	66·2 ± 16·6 (68%)	32·7 ± 11·3 (32%)	5·2	536 ± 143	13·0 ± 4·1	321 ± 156	6·67 ± 1·0
CONTROL							
LOW PURINE	35·5 ± 18·3 (77%)	13·3 ± 9·3 (23%)	6·2	414 ± 86	11·7 ± 3·4	278 ± 112	3·46 ± 0·74
HIGH PURINE	51·5 ± 17·2 (71%)	21·1 ± 12·8 (29%)	5·8	606 ± 156	12·8 ± 4·0	374 ± 157	3·86 ± 0·84
AMM. CHLORIDE	91·1 ± 33·8 (72%)	33·5 ± 10·8 (28%)	5·1	435 ± 136	12·8 ± 2·6	406 ± 131	3·78 ± 0·47
RECOVERY	56·8 ± 17·9 (75%)	18·8 ± 6·3 (25%)	5·9	524 ± 126	12·2 ± 2·6	349 ± 85	3·88 ± 0·43

MEAN URINARY EXCRETION OF AMMONIUM, TITRATABLE ACID, CREATININE AND MEAN BLOOD URIC ACID LEVELS. PERCENTAGE FIGURES INDICATE PROPORTION OF HYDROGEN ION EXCRETED AS AMMONIUM OR TITRATABLE ACID.

and from the observations of others. However, the hypothesis advanced by Gutman and Yu (1) to explain defective ammonium excretion in gout has not found wide acceptance. The finding has been ascribed by others to generalised renal impairment secondary to hyperuricaemia (9). A similar explanation has been advocated for reduced ammonium excretion in uric acid stone formers (10) and our finding of a significant correlation between other parameters of renal function and the percentage of hydrogen ion excreted as ammonium is also in accord with this concept.

Attempts to examine more closely the postulated relationship between defective ammonium excretion and hyperuricaemia have failed to produce consistent results. The absence of dietary supervision and inadequate periods of study may explain some of the disparities. The importance of dietary influence on the renal handling of acid loads has been previously emphasised (3,11). In the current study dietary purine intake was controlled and sufficient time allowed for equilibration of diets. Furthermore, prolonged periods of collection ensured that the relatively slow response and recovery of gout patients to ammonium chloride have been fully observed.

Our results demonstrate an unequivocal reduction of urine uric acid excretion in response to an ammonium chloride load. This was apparent in gouty patients and controls while they were eating both low and high purine diets. On a low purine diet the decrement in uric acid excretion was small but on a high purine diet the phenomenon was greatly accentuated. There was no significant difference between the reduction of urate excretion in gouty and control subjects. The concomitant rise and fall of ammonium excretion showed a reciprocal relationship with the changes of urine uric acid. These findings were similar to those of Plante and others (3) except that in the previous study the data was considered to demonstrate a reciprocal relationship only in gout subjects.

The above observations are in contrast to those of Swales and others (4) and Vogler and Drane (12) who failed to demonstrate any relationship between ammonium and uric acid excretion. Their studies differed in several essentials from our own, not least in the absence of a period of high purine intake and adequate recovery periods. On the other hand, small decrements of uric acid excretion have been noted after ammonium chloride administration to fasting patients (4) and to subjects not on dietary restrictions (13).

The fact that blood uric acid levels did not rise in conjunction with the fall in uric acid excretion suggests that the mechanism was not simple renal retention of urate. The most obvious interpretation of our findings is one which implies a reduction of uric acid production. Plante and others (3) were unable to reproduce a reciprocal effect when hydrochloric acid was substituted for ammonium chloride and they suggested that ammonium chloride, rather than acidosis per se, may influence glutamine synthesis in such a way that uric acid production is impaired. Our data do not allow us to comment further on this proposal.

REFERENCES

1. Gutman, A.B., and Yu, Ts'ai-Fan, "Urinary Ammonium Excretion in Primary Gout", J.Clin.Invest, 44, 1474 1965

2. Henneman, P.H., Wallach, S., and Dempsey, E.F., "The metabolic defect responsible for uric acid stone formation", J. Clin. Invest. 41, 537, 1962

3. Plante, G.E., Durivage, J., and Lemieux, G., "Renal Excretion of hydrogen in primary gout", Metabolism, 17, 377, 1968

4. Swales, J.D., Kopstein, J. and Wrong, O.M., "Renal excretion of ammonia and urate production; examination of Gutman-Yu Hypothesis". Metabolism, 21; 541 1972

5. Simmonds, H.A.,"A method of estimation of uric acid in urine and other body fluids", Clin. Chim. Acta. 15, 375, 1967

6. Jorgensen, K., "Titrimetric determination of the net excretion of acid/base in urine", Scand.J. Clin. Lab. Invest. 9, 287, 1957

7. Chan, J.C.M., "The rapid determination of urinary titratable acid and ammonium and evaluation of freezing as a method of preservation", Clin. Biochem. 5, 94, 1972

8. Pak-Poy, R.K., "Urinary PH in gout", Aust. Ann. Med. 14, 35, 1965

9. Klinenberg, J.R., Gonick, H.C., Dornfield, L.,

"Renal function abnormalities in patients with asymptomatic hyperuricaemia", Arth. Rheum. (suppl.) 18, 725, 1975

10. Metcalfe-Gibson, A., McCallum, F.M., Morrison, R.B. I., and Wrong, O.M., "Urinary excretion of hydrogen ion in patients with uric acid calculi", Clin. Sci. 28,325, 1965

11. Falls, W.F., "Comparison of urinary acidification and ammonium excretion in normal and gouty subjects" Metabolism, 21, 433, 1974

12. Vogler, W.R., and Drane, J.W., "Effect of allopurinol on urinary ammonia excretion in patients with gout", Metabolism, 18, 519, 1969

13. Thompson, G.R., "The effect of diazoxide, potassium chloride and ammonium chloride on serum and urinary uric acid", Arthritis Rheum. 8, 830, 1965

EVIDENCE FOR TWO SECRETORY MECHANISMS FOR ORGANIC ACID TRANSPORT IN MAN

Herbert S. Diamond and Allen D. Meisel

Downstate Medical Center

Brooklyn, New York

Pyrazinamide administration markedly impairs the uricosuric response to probenecid in man and in the chimpanzee (1,2). This effect of pyrazinamide on probenecid uricosuria has been attributed to inhibition of urate secretion by pyrazinamide and used as evidence for post secretory reabsorption of urate (1). However, other forms of drug interaction between probenecid and pyrazinamide have not been excluded. Recent evidence from studies in the chimpanzee and man suggest that urate and hippurate are secreted by different transport mechanisms (2,3). The present study was designed to further investigate the mechanism of pyrazinamide inhibition of the uricosuric response to probenecid in man and to compare the interactions between pyrazinamide and probenecid with those between PAH and probenecid.

Seventy-two clearance studies were performed on 18 subjects with normal GFR admitted to a Clinical Research Center and maintained on a normal protein, constant purine, isocaloric diet. All medications affecting urate excretion were discontinued at least 4 days prior to the study, and three days were allowed for dietary adjustment. Sodium bicarbonate (30 mg/kg daily in divided dose) was administered orally in order to minimize non-ionic passive reabsorption of foreign organic acids in the distal tubule.

Intravenous administration of PAH in doses sufficient to give plasma PAH levels of 30-80 mg/dl resulted in an increase in urate clearance from 8.1 ml/min in the control intervals to 11.5 ml/min after PAH. Urate excretion increased from 466 ug/min to 591 ug/min. (Fig. 1)

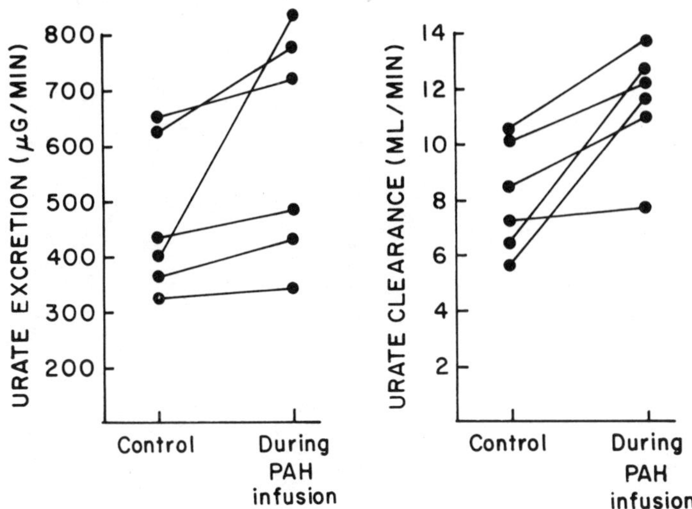

Fig. 1. The effect of PAH on urate excretion and urate clearance.

Administration of probenecid (1 gm intravenously during continuous PAH infusion in 5 subjects resulted in a decrease in both clearance of PAH from a control PAH clearance of 289 ml/min to a post probenecid PAH clearance of 161 ml/min and in tubular secretion of PAH from a control tubular maximum for PAH of 103 mg/min to a post probenecid tubular maximum for PAH of 55 mg/min (Fig. 2). In each subject studied, probenecid administration was associated with approximately a 50% decrease in tubular secretion of PAH.

Administration of probenecid 1 g intravenously resulted in a mean increase in urate clearance of 38 ml/min. The uricosuric response to probenecid was diminished to 13 ml/min during continuous PAH infusion (Fig. 3). The diminished uricosuric response to probenecid during PAH infusion was associated with a marked decrease in probenecid excretion from a control excretion of 20.1 ug/min to 9.2 ug/min during PAH infusion.

Pyrazinamide administration alone resulted in approximately

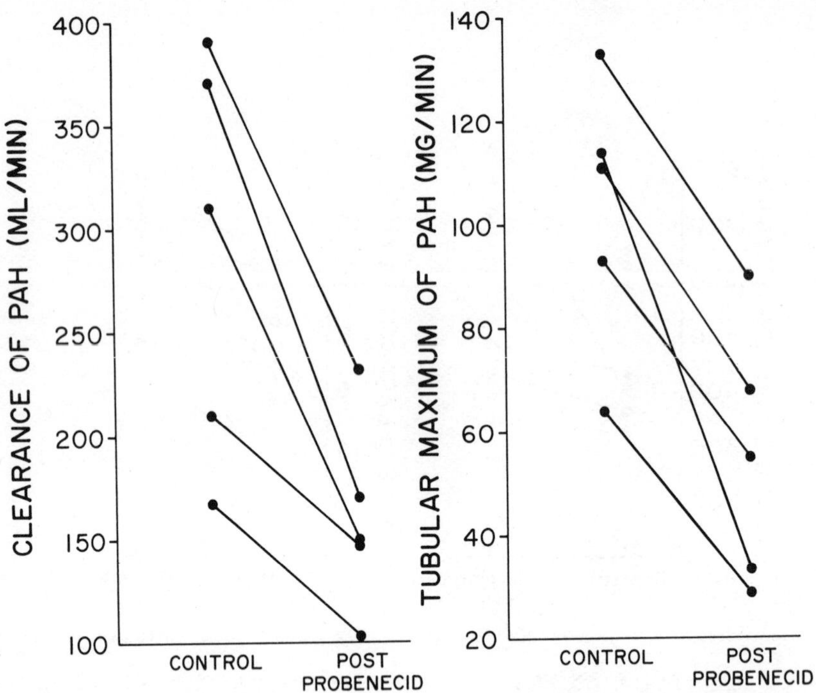

Fig. 2. The effect of probenecid on PAH clearance and on tubular maximum of PAH.

an 85% decrease in urate clearance. Pyrazinamide resulted in a marked attenuation of the uricosuric response to intravenous probenecid. Urate clearance after probenecid alone increased by 45 ml/min; after probenecid plus pyrazinamide urate clearance was unchanged from control clearance (Fig. 4). However, pyrazinamide did not inhibit probenecid excretion.

During PAH infusion, uricosuric response to probenecid paralleled probenecid excretion, whereas after pyrazinamide, the uricosuric response to probenecid was independent of probenecid excretion. (Fig. 5).

Pyrazinamide had no effect on PAH clearance or tubular secretion of PAH. Control PAH clearance was 589 ml/min. PAH clearance after pyrazinamide was 551 ml/min.

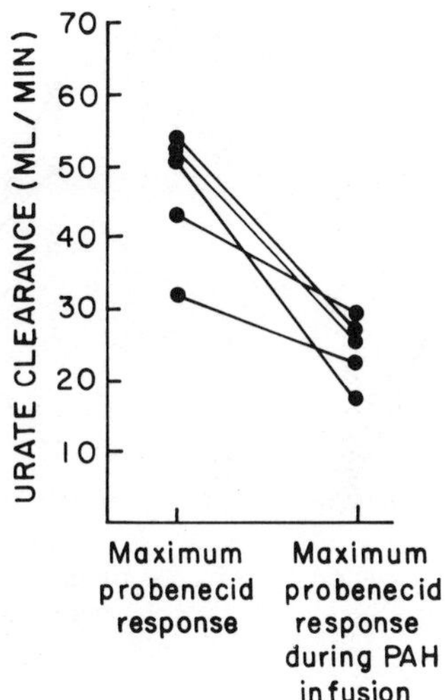

Fig. 3. The effect of PAH on probenecid induced uricosuria. PAH infusion diminished the uricosuric response to probenecid.

Probenecid excretion was similar at endogenous urate levels and after urate loading by RNA feeding and the uricosuric response to probenecid was intact or enhanced after RNA feeding.

In the present study, PAH administration inhibited probenecid excretion and probenecid administration resulted in reduced PAH excretion. In contrast, probenecid excretion was not effected by pyrazinamide administration or urate loading. As previously reported, PAH was weakly uricosuric and pyrazinamide did not alter PAH excretion. PAH had no effect on the urate retaining action of pyrazinamide. Thus, these results are consistent with the previous suggestion that in man and the chimpanzee, the organic acid secretory carrier responsible for most urate secretion and inhibited by pyrazinoic acid is distinct from the carrier accounting for secretion of probenecid and PAH.

In serum, probenecid is 99% bound to serum albumen and, therefore, enters the urine largely by tubular secretion (4). The decrease in uricosuric response to probenecid observed during PAH infusion was proportionate to the reduction in probenecid excretion, suggesting that PAH reduced probenecid uricosuria by competing with probenecid for excretion.

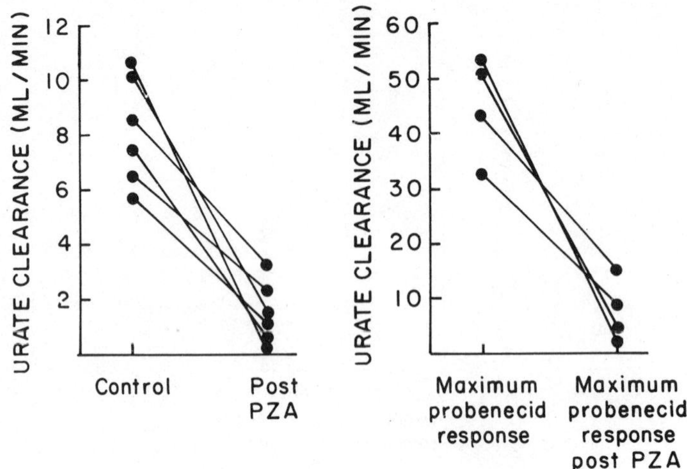

Fig. 4. The effect of pyrazinamide on urate clearance. Pyrazinamide inhibits urate clearance and attenuates the uricosuric response to intravenous probenecid.

Pyrazinamide administration markedly reduced urate excretion both under control conditions and during probenecid induced uricosuria. Since pyrazinamide had no effect on probenecid excretion, inhibition of probenecid uricosuria by pyrazinamide is most likely due to inhibition of urate secretion, thus reducing the concentration of urate available for reabsorption at the reabsorptive site inhibited by probenecid. Alternatively, pyrazinamide might interfere with probenecid binding to or effect upon a urate transport carrier located on the tubular luminal membrane.

The attenuation of the uricosuric response to probenecid resulting from inhibition of probenecid secretion by PAH suggests that probenecid must be secreted in order to exert its uricosuric effect. In man, the concentration of probenecid at the luminal membrane in the earliest segment of the proximal tubule may be low. Thus, probenecid may be a poor inhibitor of any urate reabsorption occurring in the earliest segment of the proximal nephron where intraluminal urate available for urate reabsorption might be largely derived from filtration. Probenecid might be a more effective inhibitor of urate reabsorption in more distal segments of the nephron where a larger fraction of intraluminal urate was derived from secretion. This possibility is consistent with the marked reduction in probenecid uricosuria observed during inhibition of urate secretion by pyrazinamide (1).

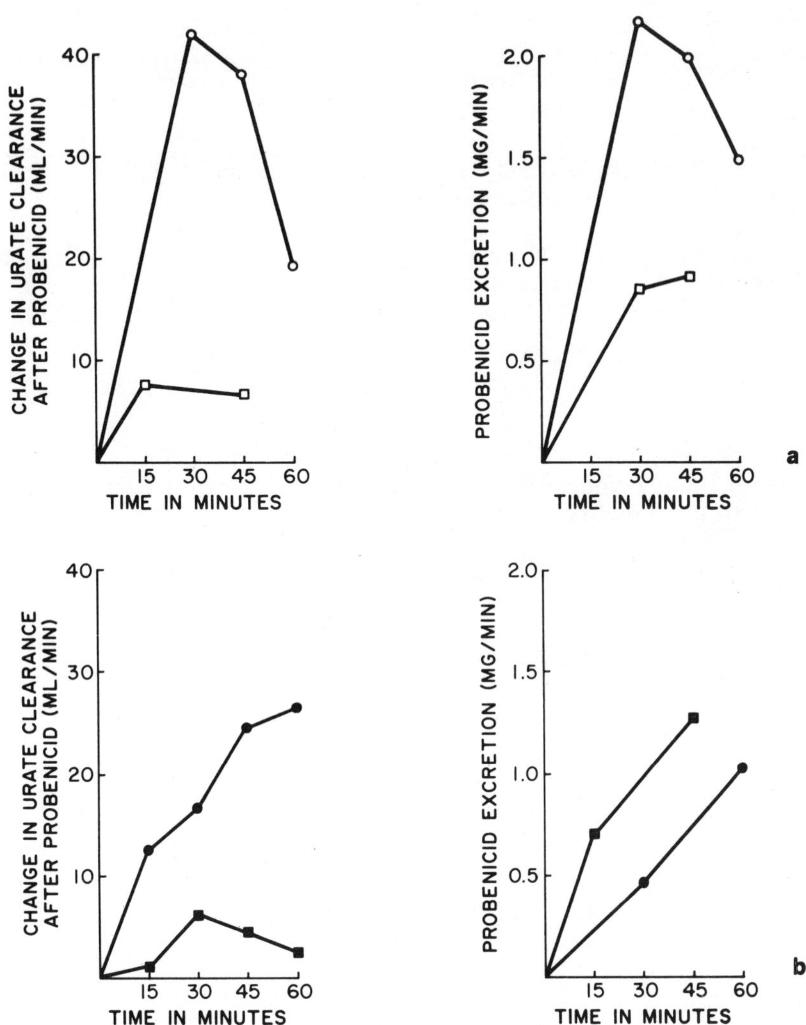

Fig. 5. The effect of PAH and PZA on probenecid induced uricosuria and probenecid excretion.
a) Both probenecid excretion and probenecid induced uricosuria (control, o - o) are reduced during PAH infusion (post PAH □ - □)
b) Probenecid induced uricosuria (control, ● - ●) is markedly inhibited after PZA (post PZA, ■ - ■) but probenecid excretion (control, ● - ●) is not diminished (post PZA, ■ - ■).

ACKNOWLEDGEMENTS

We are indebted to the staff of the Clinical Research Center for their invaluable aid in these studies. This work is supported by grant RR 318 from the General Clinical Research Center Program of the Division of Research Resources, and by a grant from the Arthritis Foundation.

Dr. Herbert Diamond is supported, in part, by an Irma T. Hirschl Career Scientist Award.

Dr. Allen Meisel is a Clinical Research Associate of the General Clinical Research Center Program of the Division of Research Resources and is supported, in part, by grant RR 318.

REFERENCES

1. Diamond, H.S. and J.S. Paolino. Evidence for a post secretory reabsorptive site for uric acid in man. J. Clin. Invest. 52:1491-1499, 1973.

2. Fanelli, G.M., D. Bohn and S. Reilly. Renal urate transport in the chimpanzee. Amer. J. Physiol. 220:613-620, 1971.

3. Boner, G. and T.H. Steele. Relationship of urate and p amino hippurate secretion in man. Amer. J. Physiol. 225:100-104, 1973.

4. Koch-Weser, J. and E.M. Sellars. Binding of drugs to serum albumen. N. Eng. J. Med. 294:311-316, 1976.

CLASSIFICATION OF URICOSURIC STATES BASED UPON RESPONSE TO PHARMACOLOGIC INHIBITORS OF URATE TRANSPORT

Herbert S. Diamond and Allen D. Meisel

Downstate Medical Center

Brooklyn, New York

Uricosuria occurring in several clinical states appears to be associated with characteristic patterns of response to uricosuric and urate retaining drugs.

To investigate the possibility that these different patterns of response to pharmacologic inhibitors of urate transport might reflect different mechanisms of uricosuria, the effects on urate excretion of administration of probenecid and pyrazinamide were observed in normal subjects under control conditions and after induction of uricosuria.

Seventy-five clearance studies were carried out in 20 volunteer subjects. All subjects were admitted to a Clinical Research Center and maintained on a normal protein, constant purine, isocaloric diet except as noted in specific study protocols. All medications known to affect serum or urine urate concentrations were discontinued at least one week prior to the studies. Thereafter, no drugs were administered except as indicated in the study protocols. Three days were allowed for dietary adjustment.

Uricosuria was induced by 3 methods (Table 1). In 7 subjects exogenous urate loading by feeding yeast RNA 4 g every 6 hours for a minimum of 3 days increased urate clearance by 6.4 ml/min. In 11 subjects, volume loading by rapid infusion of 2L of 3% saline increased urate clearance by 4.4 ml/min. In 23 subjects, partial inhibition of urate reabsorption was induced by administration of single doses of 4 different uricosuric drugs as probenecid (500 mg) in 9 subjects to increase urate clearance by 5.3 ml/min; diatrizoate 50 ml of a 50 percent solution intravenously in 7 subjects to

Table 1

INDUCTION OF URICOSURIA

MECHANISM	CONTROL URATE CLEARANCE (ml/min)	URATE CLEARANCE AFTER URICOSURIC AGENT (ml/min)
RNA 16 gm/day	6.8 ± 1.4	13.2 ± 2.5
Saline 2 liters	6.5 ± 0.9	10.9 ± 1.2
Uricosuric drugs		
Diatrizoate 50 cc	6.4 ± 1.2	14.6 ± 2.7
Probenecid 500 mg	5.1 ± 0.8	10.4 ± 1.1
Sulfinpyrazone 600 mg	4.1 ± 0.2	7.1 ± 1.3
ASA 3.6 - 5.4 gm	5.9 ± 2.1	11.7 ± 3.1

increase urate clearance 8.2 ml/min; sulfinpyrazone 200 mg in 2 subjects to increase urate clearance 3.0 ml/min and aspirin 100 mg/kg orally in 4 divided doses in 5 subjects to increase urate clearance 5.7 ml/min. There were no significant differences in the magnitude of the uricosuria induced by these methods.

The effect of probenecid administration during induced uricosuria was assessed by comparing the uricosuric response to probenecid alone with the uricosuric response to the same dose of probenecid administered during induced uricosuria. The uricosuric response to probenecid 1 g intravenously was unimpaired during uricosuria induced by RNA feeding or saline infusion (Figure 1). Peak urate clearance after probenecid plus RNA or saline infusion exceeded peak urate clearance in the same subject after probenecid alone. In contrast, the uricosuric response to probenecid was attenuated after any of the 4 uricosuric drugs tested (Figure 2). Peak urate clearance after probenecid alone was similar to peak urate clearance after probenecid plus sulfinpyrazone, diatrizoate or low dose probenecid itself. Peak urate clearance after probenecid plus aspirin was less than after probenecid alone.

The two uricosuric states with intact response to probenecid differed in response to pyrazinamide although pyrazinamide inhibited the uricosuric response in both states. After pyrazinamide

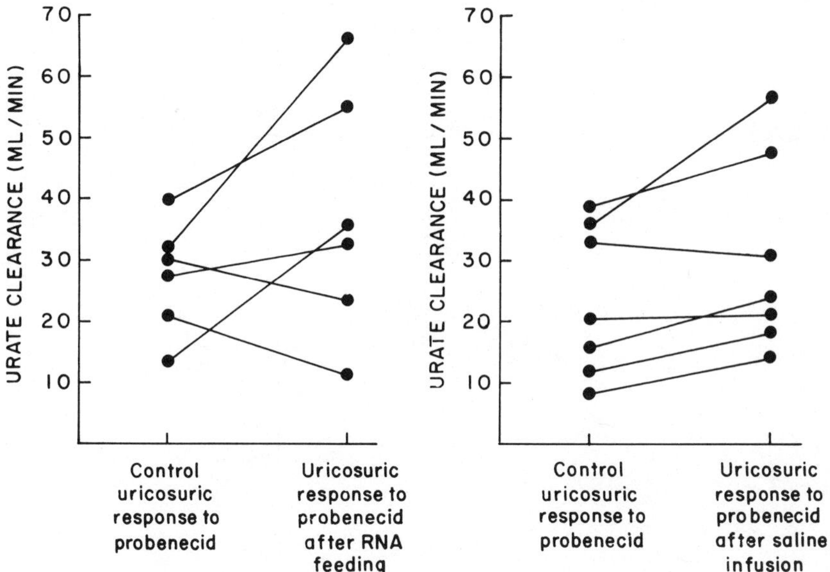

Fig. 1. Effect of RNA feeding and saline infusion on uricosuric response to probenecid.

administration, RNA feeding had no effect on fractional excretion of urate (Figure 3). In contrast after pyrazinamide administration fractional excretion of urate was increased by saline infusion.

Pyrazinamide also inhibited the uricosuric response to sulfinpyrazone, diatrizoate and probenecid, but had no effect on uricosuria induced by aspirin (Figure 4).

Thus, different mechanisms of induction of uricosuria were associated with distinct patterns of response of urate excretion to administration of pyrazinamide and probenecid. The uricosuria which followed exogenous urate loading by RNA feeding was associated with an intact uricosuric response to probenecid. RNA feeding was not uricosuric after pyrazinamide. The uricosuria which followed volume loading by saline infusion was also associated with an intact uricosuric response to probenecid, but saline infusion remained uricosuric after pyrazinamide administration. Pyrazinamide inhibited the uricosuria induced by probenecid, diatrizoate and sulfinpyrazone but did not inhibit uricosuria induced by aspirin.

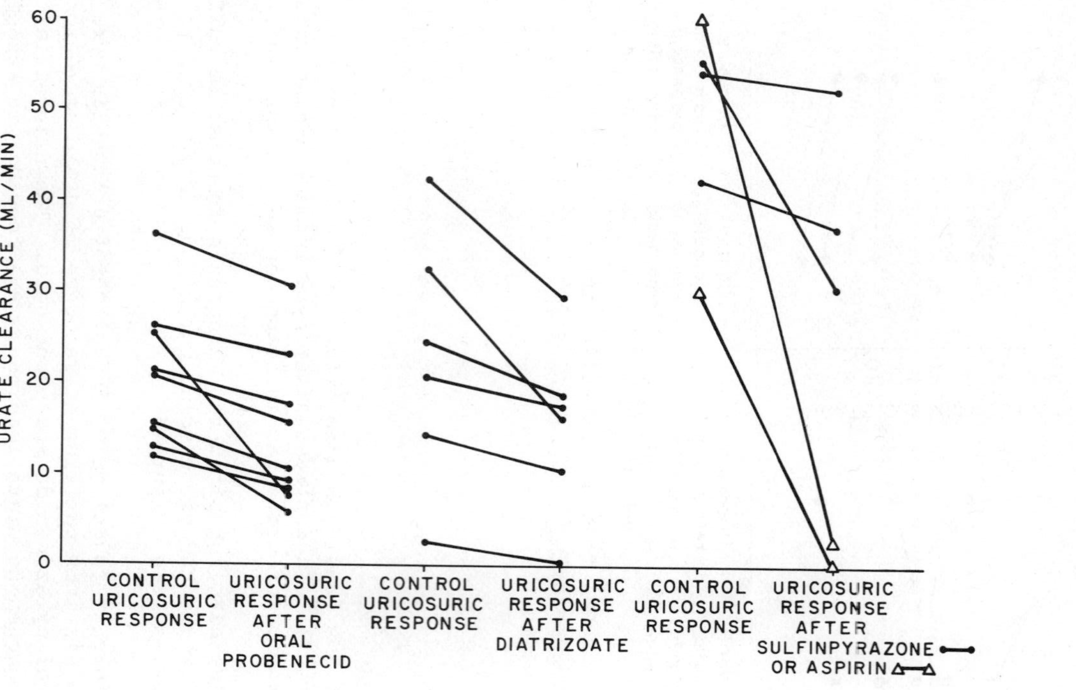

Fig. 2 Control uricosuric response to probenecid and uricosuric response after administration of probenecid, diatrizoate, sulfinpyrazone or asprin.

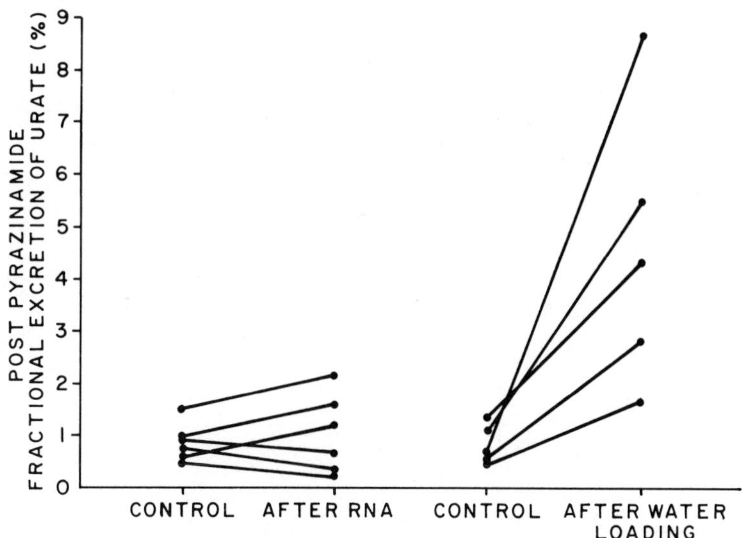

Fig. 3. Control urate excretion after pyrazinamide and uricosuric response to RNA and water loading after pyrazinamide.

The responses to probenecid and pyrazinamide in the present study are consistent with prior observations of the effects of pharmacologic inhibitors of urate transport and with responses observed in clinical uricosuric states. Uricosuria in patients with endogenous urate overproduction due to sickle cell anemia is associated with an intact response to probenecid and is completely suppressed by pyrazinamide administration, a pattern similar to that seen after exogenous urate loading by RNA feeding (1). Uricosuria associated with vasopressin induced volume expansion is only partially inhibited by pyrazinamide as is uricosuria following hypertonic saline infusion (2). The uricosuric response to probenecid in patients with multiple tubular reabsorption defects due to Fanconi syndrome is diminished and uricosuria in these patients

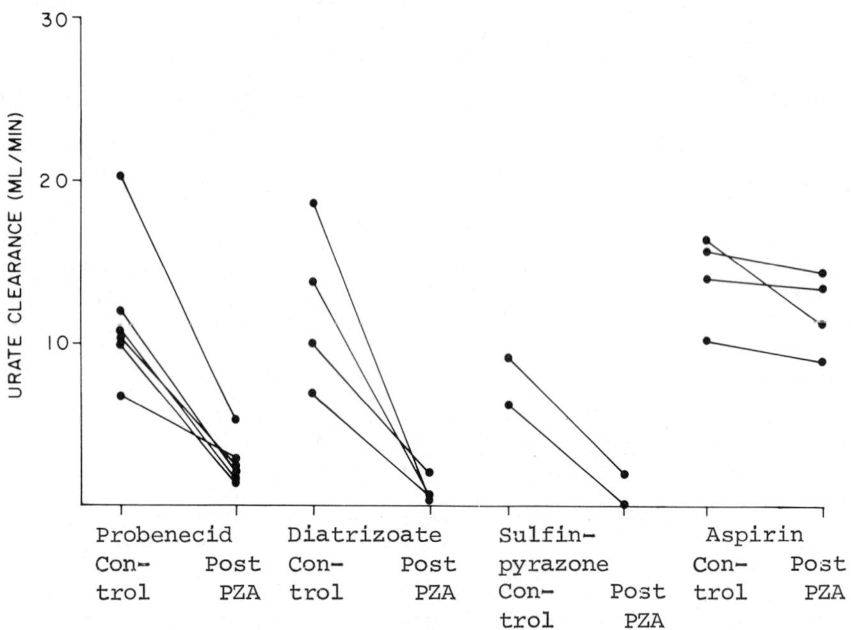

Fig. 4. Effect of pyrazinamide on the uricosuric response to probenecid, diatrizoate, sulfinpyrazone and aspirin.

is inhibited by pyrazinamide (3). This is similar to the response seen after partial inhibition of net urate reabsorption by uricosuric drugs.

The pattern of response to pharmacologic inhibitors of urate transport observed in this study and in clinical uricosuric states, might, in part, reflect different mechanisms of uricosuria. The methods used in the present study and prior studies in man are inadequate for definitive delineation of specific defects in tubular transport of urate in clinical and induced uricosuric states. Thus, a probable mechanism for each of these patterns can only be tentatively assigned.

An increase in endogenous urate load as in sickle cell anemia, or exogenous urate load as after RNA feeding results in increased plasma urate concentrations, and should result in an increase in

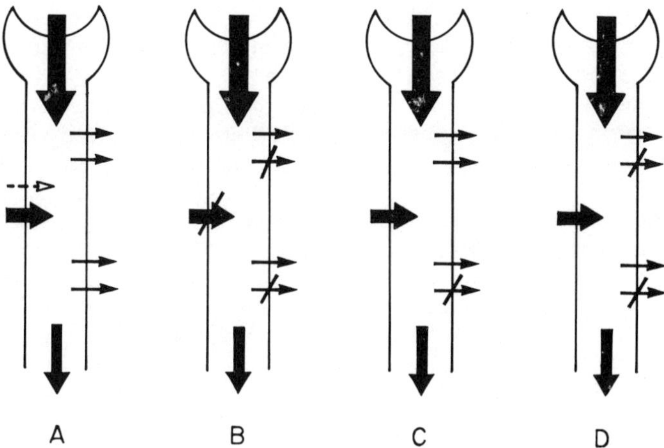

Fig. 5. Theoretical models for 4 possible mechanisms of uricosuria. Each mechanism may be associated with a different pattern of response to pharmacologic manipulation of urate transport. Dotted arrows indicate augmented transport, a slash through an arrow indicates diminished transport. "A" represents increased secretion, "B" diminished reabsorption partially offset by diminished secretion, "C" diminished reabsorption of secreted urate, and D diminished reabsorption of filtered (and secreted) urate.

both filtered load of urate and urate secretion (Figure 5A). Since filtered urate is almost completely reabsorbed, uricosuria in this situation would be the result of enhanced urate secretion without a compensating increase in reabsorption of secreted urate. Increased PAH clearance in patients with sickle cell anemia is consistent with enhanced secretion (1).

Volume loading is associated with inhibition of reabsorption of multiple solutes and has been shown to inhibit urate reabsorption in the proximal tubule in the rat (4). Thus, volume loading probably leads to inhibition of reabsorption of both filtered and secreted urate (Figure 5D). This is consistent with the observation of partial inhibition of urate excretion by pyrazinamide during uricosuria induced by volume loading (4). The intact response to probenecid might be explained if probenecid itself must be secreted to be uricosuric. Inhibition of urate reabsorption in the early proximal tubule during volume loading would result in a higher concentration of urate at more distal, probenecid affected sites.

Drug induced uricosuria has generally been attributed to inhibition of urate reabsorption. Uricosuria in disease states such as Fanconi syndrome, Wilson's disease and idiopathic hypercalcuria is associated with other defects in tubular reabsorption and is likely to represent a reabsorptive defect. The diminished response to probenecid observed here during drug induced uricosuria may be explained by drug interaction. However, this would not explain the similar diminished probenecid response seen in the Fanconi syndrome and Wilson's disease (3,6). The diminished probenecid response is consistent with diminished urate reabsorption including diminished reabsorption of secreted urate (Figure 5C). Aspirin is believed to inhibit both urate secretion and reabsorption (Figure 5B). This would explain the failure of pyrazinamide to affect aspirin induced uricosuria. The uricosuria in idiopathic hypercalcuria which is also not suppressed by pyrazinamide could be accounted for by a similar mechanism (7,8).

Acknowledgements

We are indebted to the Staff of the Clinical Research Center for their invaluable aid in these studies. This work is supported by grant RR 318 from the General Clinical Research Center Program of the Division of Research Resources, and by a grant from the Arthritis Foundation.

Dr. Herbert Diamond is supported, in part, by an Irma T. Hirschl Career Scientist Award.

Dr. Allen Meisel is a Clinical Research Associate of the General Clinical Research Center Program of the Division of Research Resources and is supported, in part, by Grant RR318.

References

1. Diamond, HS, Meisel, A, Sharon, E, Holden, D, and Cacatian, A: Hyperuricosuria and increased tubular secretion of urate in sickle cell anemia. Amer J Med 59:796, 1975.

2. Mees, EJD, VanAssendelft, PB, and Nieuwenhuis, MG: Elevation of uric acid clearance caused by inappropriate antidiuretic hormone secretion. Acta Medica Scandinavia 189:69-72, 1971.

3. Diamond, HS, Meisel, AD: Renal tubular transport of urate in Fanconi Syndrome. Adv Exp Med Biol: in press.

4. Weinman, EJ, Eknoyan, G and Suki, WN: The influence of the extracellular fluid volume on the tubular transport of uric acid. J Clin Invest 55:283-291, 1974.

5. Manuel, MA, Steele, TH: Pyrazinamide suppression of the uricosuric response to sodium chloride. J Lab Clin Med 83:417-427, 1974.

6. Wilson, DM, Goldstein, NP: Evidence for urate reabsorptive defects in patients with Wilson's disease. Adv Exp Med & Biol 41B:729, 1973.

7. Greene, ML, Marcus, R, Aurbach, GD, Kazam, ES, Seegmiller, JE: Hypouricemia due to isolated renal tubular defect: dalmation dog mutation in man. Amer J Med 53:361, 1972.

8. Sperling, O, Weinberger, A, Oliver, I, Liberman, UA and deVries, A: Hypouricemia, hypercalciuria and decreased bone density: a new syndrome. Ann Int Med 80:482, 1974.

FAMILIAL RENAL HYPOURICEMIA DUE TO ISOLATED TUBULAR DEFECT

D. Benjamin, O. Sperling, A. Weinberger, J. Pinkhas and A. de Vries

Department of Medicine D and Rogoff-Wellcome Medical Research Institute, Tel-Aviv University Medical School, Beilinson Medical Center, Petah Tikva, Israel

Renal hypouricemia, that is hypouricemia due to a renal abnormality reflected in increased uric acid clearance, is a rare condition, either inborn or acquired. When genetically determined, it usually appears in association with other tubular defects, such as in the Fanconi (1) and Hartnup syndromes (2) and Wilson's disease (3). Recently, we reported on a familial renal hypouricemia, associated with idiopathic hypercalciuria and decreased bone density (4). Renal hypouricemia due to an isolated renal tubular defect has been reported in a few subjects (5,6), but to our knowledge only once as a familial syndrome (7). We presently describe another family with renal hypouricemia, in whom the renal tubular defect relates to the handling of uric acid exclusively. The propositus, a 27-year-old Iraqian-born Jewish female, was followed at the Hematology Clinic since 1969, when a diagnosis of toxoplasmosis was established. All blood chemistry and hematological tests were normal, except for serum uric acid which ranged from 1.1 to 1.9 mg%. Urinary uric acid excretion on a regular diet ranged from 815-1008 mg/24 h, as determined colorimetrically (8). Serum ceruloplasmin level and urinary copper excretion were within the normal range. There was no glycosuria and the urinary pattern of amino acids was normal. Creatinine clearance was 110 ml/min and the tubular reabsorption of phosphate was 88% (normal range > 85%). Urinary calcium excretion, on a diet containing 900 mg of calcium per day, was 112-160 mg/24 h. Morning urinary pH was 5.2-5.8.

For the study of renal urate handling the patient was maintained on an essentially purine free diet. Uric acid in serum and urine was determined by an enzymatic spectrophotometric method (9).

The pyrazinamide (PZA) suppression test was performed according to Steele and Rieselbach (10). The proband's serum uric acid concentration during the two weeks' period of observation ranged from 1.43 to 1.51 mg%, and her urinary uric acid excretion from 620 to 662 mg/24 h. The average 24 h and 25 min uric acid clearance were 30 and 32.1 ml/min, respectively.

Administration of PZA (Table 1) decreased uric acid clearance from an average base-line value of 32.1 ml/min to a minimum of 23.3 ml/min, the $C_{uric\ acid}:C_{creatinine}$ ratio decreasing from 36.4% to 20.4%. Following probenecid administration (2.0 gm orally, Table 2), the uric acid clearance increased from a base-line value of 28.4 ml/min to a maximal value of 49.2 ml/min, and accordingly, the $C_{uric\ acid}:C_{creatinine}$ ratio from 31.9% to 51.8%.

The proband is an offspring of a non-consanguineous marriage. Two of her sisters were found to have renal hypouricemia: in one serum uric acid was 1.0 mg%, urinary uric acid 750 mg/24 h, and uric acid clearance 52 ml/min; in the other serum uric acid was found to be 1.2 mg%, urinary uric acid excretion 513 mg/24 h and uric acid clearance 29 ml/min. One other studied sister, and the proband's two brothers and children were all normouricemic and normouricosuric. No additional metabolic or renal abnormality was detected in the proband's two affected sisters: they had no glucosuria, phosphaturia or calciuria, and the urinary amino acid profiles were normal. These data on the family do not allow assessment of the mode of inheritance of the renal tubular defect.

In recent years a four-compartment model has been suggested for the renal handling of urate (11). Filtered urate is reabsorbed in the proximal tubule by a high capacity-high affinity system. Urate is secreted in the distal tubule, but some of the urate secreted is reabsorbed by a distal postsecretory mechanism which has a lesser affinity as compared to the proximal reabsorptive system.

Three different varieties of deficient tubular reabsorption of urate have been suggested to explain renal hypouricemia (11): a postsecretory defect in Wilson's disease (12) and in Hodgkin's disease (13), a proximal presecretory defect in the patients studied by Greene et al. (7) and Sperling et al. (4), and a total inhibition of urate reabsorption in the patient studied by Simkin (14) and in two other reported patients (5,6). The classification of renal hypouricemia into these three varieties was made by the measurement of the urate clearance and its response to two drugs: PZA and probenecid, the former decreasing and the latter increasing uric acid excretion (10,15). In normal subjects administration of PZA causes almost total suppression of uric acid excretion (10); in defective postsecretory reabsorption it eliminates the excessive excretion of urate (12,13); in defective proximal reabsorption it has only a

TABLE 1

EFFECT OF PYRAZINAMIDE ON URIC ACID EXCRETION IN THE PROBAND

Period min	$C_{uric\ acid}$ (ml/min)	Excreted urate (mg/min)	$C_{uric\ acid}:C_{creatinine}$ %
Control			
-60 to -26	24.7	0.354	35.2
-26 to 0	39.5	0.565	37.6
Pyrazinamide (3.0 gm orally)			
0 to 27	27.4	0.392	24.0
27 to 52	24.0	0.336	21.4
52 to 77	26.1	0.372	21.9
77 to 102	31.7	0.451	32.0
102 to 123	23.3	0.327	20.4
123 to 150	24.1	0.377	25.4

TABLE 2

EFFECT OF PROBENECID ON URIC ACID EXCRETION IN THE PROBAND

Period min	$C_{uric\ acid}$ (ml/min)	Excreted urate (mg/min)	$C_{uric\ acid}:C_{creatinine}$ %
Control			
-28 to 0	28.4	0.430	31.9
Probenecid (2.0 gm orally)			
0 to 29	34.1	0.515	36.3
29 to 50	47.1	0.712	35.9
50 to 77	49.2	0.746	51.8
77 to 100	37.7	0.565	51.1

minimal effect (4,7) and in total loss of reabsorption capacity it reduces the $C_{urate}:C_{inulin}$ ratio to 1.0 (14). In the proband, presently studied, the renal hypouricemia appears to be of the variety characterized by a defective proximal (presecretory) reabsorption. This is suggested by the response of the urate clearance to PZA and probenecid. Administration of PZA decreased uric acid excretion by only 29%. Moreover, probenecid, which in normal subjects blocks tubular reabsorption of uric acid to a large extent and thus increases the uric acid clearance by manyfold (15,16), caused in the proband only a 73% increase in uric acid clearance. Still, in view if the disputed interpretation of the PZA suppression test (17) we prefer to connotate the condition in this family as "renal hypouricemia with attenuated response of uric acid clearance to probenecid and PZA", rather than relate the abnormality to a defective tubular uric acid reabsorption, exclusively.

In respect to the response to PZA and probenecid administration the presently studied family behaved similar to the family with renal hypouricemia, reported by us previously (4). However, in that family the renal defect in uric acid handling appeared in combination with idiopathic hypercalciuria and decreased bone density. In the presently reported family, the renal hypouricemia reflects an isolated renal tubular defect, affecting uric acid handling only. In this respect this family is similar to that reported by Greene et al. (7); it is therefore the second such family to be reported.

REFERENCES

1. Wallis, L.A. and Engle, R.L. Am. J. Med., 22:13-23, 1957.
2. Baron, D.N., Dent, C.E., Harris, H., Hart, E.W. and Jepson, J.B. Lancet, 271:421-428, 1956.
3. Bishop, C., Zimdahl, W.T. and Talbott, J.H. Proc. Soc. Exp. Biol. Med., 86:440-441, 1954.
4. Sperling, O., Weinberger, A., Oliver, I., Liberman, U.A. and de Vries, A. Ann. Int. Med., 80:482-487, 1974.
5. Khachadurian, A.K. and Arslanian, M.J. Ann. Intern. Med., 78:547-550, 1973.
6. Praetorius, E. and Kirk, J.E. J. Lab. Clin. Med., 35:865-868, 1950.
7. Greene, M.L., Marcus, R., Aurbach, G.D., Kazam, E.S. and Seegmiller, J.E. Am. J. Med., 53:361-367, 1972.
8. Eichhorn, F., Zelmanovsky, S., Lew, E., Rutenberg, A. and Fanias, B. J. Clin. Path., 14:450-452, 1961.
9. Liddle, L., Seegmiller, J.E. and Laster, L. J. Lab. Clin. Med. 54:903-913, 1959.
10. Steele, T.H. and Rieselbach, R.E. Am. J. Med., 43:868-875, 1967.

11. Rieselbach, R.E. and Steele, T.H. Am. J. Med., 56:665-675, 1974.
12. Wilson, D.M. and Goldstein, N.P. Kidney Int., 4:331-336, 1973.
13. Bennett, J.S., Bond, J. and Singer, I. Ann. Intern. Med., 76:751-756, 1972.
14. Simkin, P.A., Skeith, M.D. and Heaky, L.A. Adv. Exp. Med. Biol., 41B:723-728, 1974.
15. Sirota, J.H., Yu, T.F. and Gutman, A.B. J. Clin.Invest. 31:692-700, 1952.
16. Gutman, A.B. and Yu, T.F. Am. J. Med., 23:600-622, 1957.
17. Holmes, E.W. and Kelley, W.N. Adv. Exp. Med. Biol., 41B: 739-744, 1974.

ISOLATION AND CHARACTERIZATION OF URICINE

Bernardo Pinto and Elisabeth Rocha

Laboratorio de Exploraciónes Metabólicas and Servicio de Urología, Ciudad Sanitaria, c/ Aragón nº 420, Barcelona-13, Catalunya, Spain

The red-yellowish color of the uric acid stones is in contradiction with the white color of the pure uric acid. However, no reasonable explanation has been given. At the first Symposium on Uric acid metabolism, held at Batyam, Israel, I presented the first report on the isolation of a pigment(s) from uric acid stones. Such finding did really explain, on physicochemical terms, the hitherto unexplained color of these stones. At the same time Dr. Kleeberg, from Haifa, found, in the uric acid stones, a similar pigment that could explain the special color of these stones. Although Dr. Kleeberg's finding was reported about one year later. However he has shown us evidence supporting the simultaneity of both findings

Although the color of the uric acid stones seems to be due to the presence of various specific pigments. One of them is permanently present in all the uric acid stones thus far analyzed (73 stones) while the other pigments are erratically found. Furthermore they appear to be either precursors or degradation products of a small molecule whose color changes from yellow-reddish to black-greenish, depending on the pH. It has a tremendous affinity towards uric acid. Based on this property we have named, uricine.

The present report attempts to describe the isolation identification and to propose a preliminary structure of uricine.

Experimental Procedure

Uricine was isolated from uric acid stones in five steps. Two hundred grams of powdered uric acid stones were extracted with 2M NaOH (7 l.) for 48 hours. The sediment was separated by centrifugation and re-extracted the same manner. The supernatant was brought to pH 1 with concentrated HCl, while a color change from yellow reddish to green did occur.The extract was concentrated by flash evaporation at 60º C. It is important to carry out this process in separated batches of 200 ml and to eliminate the white precipitate by filtration (cotton).

The extract amounting about 500 ml was deposited on the top of a 44 x 3.5 cm Dowex - 50 column equilibrated with 10 mM HCl.Elution was performed with 2 M HCl until the 290 nm absorption was zero and then with 0.1 M NaOH by monitoring either by fluorescence emission at 420 nm (excitation at 310 nm) or the 240 nm absorption.

Fractions containing the pigment were evaporated to dryness.The sodium bicarbonate was eliminated by extracting the residue with concentrated HCl.The sodium chloride formed it was separated by centrifugation followed by evaporation to dryness of the supernatant. The NaCl residue was separated from the pigment by taking advantage of the different solubility speed in water of the pigment versus the sodium chloride. The pigment was separated by filtration through Millipore filters. This step was repeated as many times as required until the x - ray diffraction pattern did not show any trace of NaCl.

Physicochemical properties of Uricine

The uricine possesses a major absorbing peak a 240 nm and two minor peaks a 520 and 650 nm depending on the alkaline (520 nm) or acid (650 nm) pH values of the solutions. Infra red spectra showed three major bending bands at 3,450, 1,660 and 1,090 cm^{-1}, one stretching band (1,260 cm^{-1}) and one wag band (795 cm^{-1}). The mass spectrography fragmentation pattern was defined by the 282, 277, 265, 229, 207, 133, 104, 97, 84, 64, 48, 44 and 28 M/e peaks. The highest peak appeared at 36 M/e but its hight depended of the drying process. It indicates is due to Cl. The nuclear magnetic resonance spectra had two main groups of bands corresponding to 1 - 3 and 7 - 8 P.P.M. The fluorescence spectra showed two maxima at 360 (acid) and 420 (alkaline), when the excitation wave length was set at 310 nm. The x - ray diffraction pattern have several lines of which the most intensive had d and 2 θ values of 23.12 and 3.847 (Table 1).

No melting point was found for uricine. The maximum electrophoretic mobility was at pH 5.5 with two isoelectric points at pHs 9.5 and 2.2. On paper chromatography it did not migrate in the system ethanol:ammonia:water (80:4:16) while on thin layer chromatography had a Rf of 0.85 in the system methanol - formic acid 90% - water

(160:30:10). Uricine was more soluble in alkaline solutions than in neutral or acid solutions. The greatest solubility was obtained at pH 12 (102 mg/ml). It was also more soluble in methanol (15.6 mg/ml) than in water (9.8 mg/ml). It was also more soluble in concentrated HCl (3.7 mg/ml) than in 10 mM HCl (0.8 mg/ml) solutions It was insoluble in benzene.

Table 1

X - Ray Diffraction Data of Uricine

2Θ	dÅ	I
13.00	6.810	SSW
15.09	5.870	M
17.06	5.198	W
20.96	4.208	WW
23.12	3.847	SSS
34.28	2.616	MS

S = Strong
W = Weak
M = Medium

The elementary microchemical analysis showed that uricine consisted of 55.85% carbon, 10.14 % nitrogen, 23.16% oxygen and 5.78% hydrogen.The proposed elementary composition was $C_{13} N_2 O_4 H_{17}$.

Intensely red pigment

Occasionally an intensely red pigment was found .This pigment had 105 molecular weight, the peak of the alkaline fluorescence emission appeared at 400 nm and the visible spectra had two absorbing peaks at 475 nm (acid pH) and 440 nm (alkaline pH). Furthermore, the water solubility was much higher than uricine.

This pigment was detected in some of the bladder uric acid stones. Erratically, was detected another pigment with similar properties as uricine, but higher molecular weight (385 m.w.).

Discussion

The isolation of uricine from the uric acid stones explain the hitherto unexplained typical color of these calculi.

The infrared spectra showed the presence of a strong wag band at 795 cm^{-1} that indicates the presence of a 5 atom ring. Coupled with the nitrogen detection in the elementary michroanalysis and the wide 3,450 cm^{-1} bending band suggest the presence of NH groups together with OH functions. This suggestion is supported by the mass spectrography fragmentation with 28, 44, 64, 84, 97 and 133 M/e peaks. All of this, indicates the presence of pyrrol rings among its structure. The higher solubility in concentrated acid together with the fluorescence emission and the presence of two nitrogens per mol-

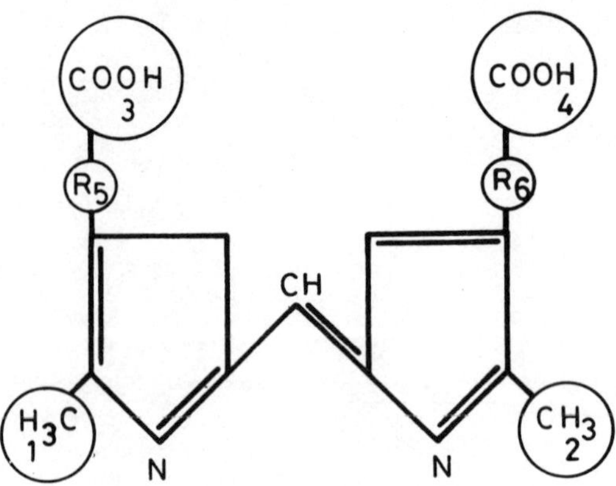

Fig. 1 - Proposed uricine structure. Functional - groups nº 1 - 2 are of unknown position. It is unknown the R_5 - R_6 length.

ecule clearly indicates a two pyrrol structure joined through a methyl bridge. The nuclear magnetic resonance indicates the presence of proton-interchangeable methyl groups and the infrared spectra shows the presence of carboxylic groups. The elementary composition support the existence of two carboxylic groups (Fig. 1). The color shift, depending on the pH changes, agrees with this structure.

The presence in the uric acid stones of other pigments with similar properties may suggest the interrelation with uricine either as precursors or degradation products.

Acknowledgement

We are indebted to Dr. Sole - Balcells, head of this Division, for his continue support and patronage.

URICINE - URIC ACID INTERACTIONS

Bernardo Pinto and Elisabeth Rocha

Laboratorio de Exploraciones Metabólicas and Servicio de Urologia, Ciudad Sanitaria, c/ Aragón nº 420, Barcelona-13, Catalunya, Spain.

We have shown in the previous paper that the special type of color of the uric acid stones (red - yellowish) is due to the presence of a pigment of low molecular weight, named uricine.

The uricine color is affected by the pH changes, varying from green, at pH 1, to red - yellowish. Uricine is permanently present in all uric acid stones thus far analyzed. Its structure appears to be constitued by two pyrrol rings joined through a methyl bridge.

The purpose of this paper is to show evidence on the interaction of uricine with uric acid.

Experimental Procedure

This study had two steps, phases or purpose:

1) To describe some of the properties of the uric acid - uricine interaction, when the uric acid was kept in solution at low concentrations.

2) To study the effect of the uricine - uric acid interaction, when the uric acid reached concentrations beyond the solubility product. This step could be considered as the investigation of the effect of uricine on uric acid precipitation.

1 - Uric acid - uricine at low uric acid concentration

It attempted to answer various questions, as the optimum binding pH, effect of uricine and uric acid concentrations on the reaction, time equilibrium, base binding specificity and aggregates formation.

The binding optimum pH was investigated in an assay system containing in one ml final volume: 100 μmoles of buffer at different pHs, from 1 to 9, 0.42 μmoles of uricine 0.30 μmoles of uric acid and approximately 20 x

10^3 cpm of ^{14}C - 2 - uric acid. Incubations were carried out at room temperature and then, the free and bound portions were separated, either by electrophoresis or filtration through 0.22 μ (diameter pore) Millipore filters. Radioactivity was counted in the system PPO-POPOP -toluene. The optimum binding occurred at pH 4.0, while at 4.8 an 80 % of the binding was found. The effect of uricine concentration was investigated in a similar system in which in an 1 ml final volume contained 100 μmoles of sodium acetate buffer (pHs 4.0 and or 4.8), 0.3 μmoles of uric acid, approximately 20 x 10^3 cpm of ^{14}C - 2 - uric acid and increasing amounts of uricine from 0.028 up to 0.56 μmoles. Free and bound portions were obtained as described.

The effect of <u>uric acid concentration</u> was carried out as above was described but adding to each incubation mixture 0.60 μmoles of pigment and increasing amounts of uric acid from 0.02 up to 0.30 μmoles / ml.

The amount of uricine - uric acid complexes formed is affected by increasing concentrations of uric acid, while the effect of increasing concentrations of uricine was limited by the concentration of uric acid.

The <u>binding equilibrium</u> constant found, was 2.2 x 10^{-10} M. The time equilibrium was investigated by incubating the same amount of buffer and final volume in the presence of 0.30 μmoles of uricine and 0.20 μmoles of uric acid. The incubation took place at room temperature and the reaction was stopped at 5, 15, 30, 60 minutes 2, 3, 4, 6, 8, 10, 12, 16, 20 and 24 hours. Free and bound portions were similarly obtained. The binding equilibrium was obtained after 4 hours incubation.

The binding equilibrium <u>specificity</u> of uricine towards the purine bases was studied by using sephadex columns at constant base concentration in the eluent. Each sephadex G - 100 column of 34 x 1.8 cm size was equilibrated with a solution containing per ml: 0.1 μmoles of each one of the different purine bases named: uric acid, guanine, adenine, xanthine and hypoxanthine, approximately 15 x 10^3 cpm of the ^{14}C respective base and 100 μmoles of sodium acetate buffer (pH 4.8). Each sample contained 0.72 μmoles of uricine. After incubation at room temperature for 24 hours, the sample was placed on the top of the sephadex column. In some of the columns, in which the eluent was uric acid, the uricine was replaced in the samples by 100 μmoles of uric acid.

One ml fractions were collected and in each one the radioactivity and the fluorescence were determined. The elution profile showed a peak in which the radioactivity and fluorescence were matched together followed by a depression of the radioactivity together with appearance of a second fluorescence peak (free uricine). If the uric acid towards itself is termed 100 the uricine percentage binding towards the different bases was 122 for uric acid, 50 - xanthine, 44.44 - guanine, 27.77 - hypoxanthine, and 5.55-adenine.

The formation of aggregates was investigated by using sucrose centrifugation gradients ranging from 0 to 2 % sucrose concentrations. The solutions were buffered with 10 mM sodium acetate (pH 4.8). Gradients nº 1 and 2 had 0.1 mM uric acid, while gradient nº 3 - 5 had not. One ml volume samples incubated at room temperature for 24 hours and each numbered sample was layered on each respective numbered gradient. Samples nº 1 and 4 contained 0.1 μmoles of uric acid and 2 x 10^6 cpm of ^{14}C - 2 - uric acid samples nº 2 and 5 had the same amount of radioactive and no - radioactive uric acid, plus 3.25 μmoles of uricine. Sample nº 3 had 3.25 μmoles of uricine, as only component, besides the buffer.

Ultracentrifugation was performed in a Spinco - L 5 , by using a S.W. 41 rotor, set at 23º C, 40,000 rpm, for 48 hr. without brake. After the run 25 fractions were collected.

These gradients showed that uric acid did not move when it was alone. However it moved further down when uric acid was present in the buffer system and even if moved much further when uricine was present in the samples (Fig. 1).

2 - Uric acid - uricine interaction at high uric acid concentration

Incubation mixtures were carried out as described but adding increasing amounts of uric acid from 0.3 to 4 μmoles/ml in the presence of 0.42 μmoles of uricine. Precipitates were separated by filtration through 0.22 μ (pore diameter) Millipore filters.

The uric acid retained on the filters followed an exponential kinetics type of curve, that was steeper in the presence of uricine.

The precipitation constants were calculated with the following equation

$$y = e^{k \cdot x}$$

y - is the amount of purine bases retained on the filters x - is the amount of purine bases added to the incubation mixtures, k - is the precipitation constant and e - is the Nepperian number.

The precipitation constants were respectively 0.23 x 10^{-5} M for adenine, 0.63 x 10^{-5} M - guanine, 0.69 x 10^{-5} M - hypoxanthine, 0.60 x 10^{-5} M - xanthine, 10.75 x 10^{-5} M for uric acid without uricine and 11.51 x 10^{-5} M for uric acid in the presence of uricine.

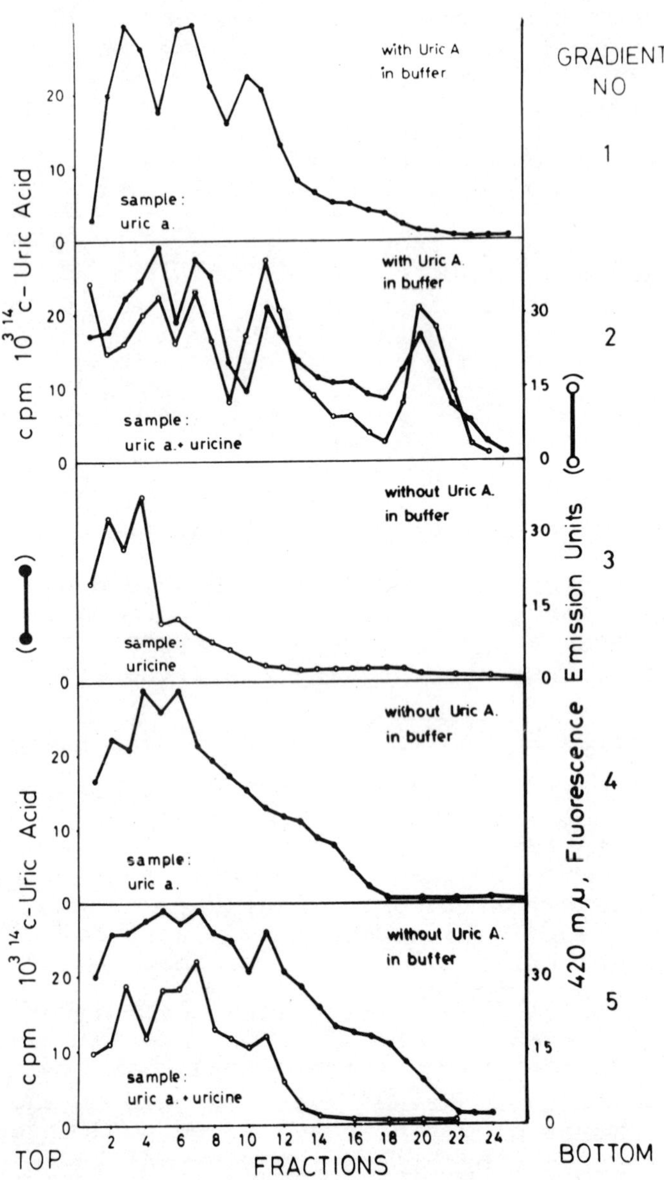

Fig. 1 - Uric acid and uric acid - uricine aggregates formation.

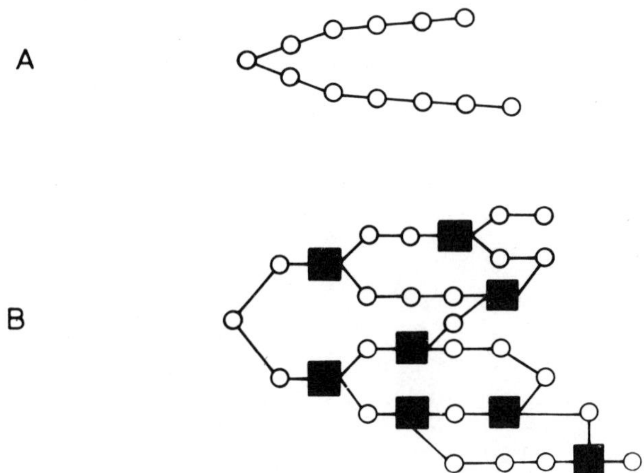

Fig. 2 - Proposed mechanism of action of uricine - (black).A- Uric acid interaction without uricine.- B- Uricine - uric a. interaction.

Discussion

Uricine binds to uric acid and less strongly to the other purine bases. The uric acid - uricine interaction is affected by the pH (optimum at pH 4.0) and the uric acid concentration, that was the rate limiting step of the reaction.

Uric acid appears to bind to itself as shown by the sephadex column and the centrifugation gradients. The uric acid interacts with itself forming aggregates of different size such an interaction follows an exponential type of kinetics. This behaviour could be explained by acting as substrate and activator of itself. The uricine - uric acid interaction seems to accelerate or stimulate such a behaviour (Fig. 2).

Acknowledgement

We are indebted to Dr. Sole - Balcells, head of this Division, without his support this work could not been done.

URINARY EXCRETION OF URICINE

Bernardo Pinto, Elisabeth Rocha and Francisco J. Ruiz Marcellan

Laboratorio de Exploraciónes Metabólicas and Servicio de Urologia, Ciudad Sanitaria, c/ Aragón nº 420, Barcelona-13, Catalunya, Spain

In the previous papers we have described the isolation characterization and interaction with uric acid of a pigment existing in the uric acid stones. These findings explain the typical color of uric acid stones. Due to the specific and high affinity of this pigment toward its binding with the uric acid was named, uricine.

Uricine appears to facilitate or increase the formation of uric acid aggregates and therefore the uric acid precipitation.

This paper attempt to shed light on the possible origin of uricine as well as is urinary excretion in a group of normals, uric acid, gouty and calcium stone formers with hyperuricuria.

Experimental Design

Uricine Origin

Incubation mixtures of 10 mM phosphate buffer (pH 7.5) contained 1.5 g in 10 ml final volume of homogenized human kidney pieces and 20 mg of bilirubin plus 10^6 cpm of ^3H - bilirubin (E_1 and E_2). Incubations were carried out at 37º C for 1 hr. In blanks the bilirubin was left out. Pigments were extracted with methanol. Uricine was separated from bilirubin by thin layer chromatography in the system methanol - formic acid 90% - water (160:30:10). Uricine was identified by infrared, U.V.-visible, fluorescence and mass spectra.

This experiment showed that bilirubin was partially transformed in uricine.

Uricine extraction from urine

Aliquots of 10 ml from the 24 hours urine samples were brought up to pH 1 with concentrated HCl, then, they were deposited on 13 x 1 cm size. Dowex-50 columns, equilibrated with 10 mM HCl. Columns were eluted with 2 M HCl until the 290 nm absorption was negligible, then, the column was washed with 10 mM NaOH until the pH of the effluent was 10 or higher. The whole alkaline column washed was evaporated to dryness, at 60º C, and the residue was dissolved in 10 ml of water. Uricine quantification was performed by comparison the fluorescence emission with a standard curve, when the excitation and emission wave - lengths were respectively set at 310 and 420 nm. The cuvettes for the fluorescence readings contained 3.4 ml of 0.1 M NaOH and aliquots of 20 up to 100 µl from each alkaline column washed.

The potential capacity of uricine, in each subject, to bind uric acid was calculated with the following equation:

$$(U - Ur) = U \cdot Ur \cdot k \cdot \frac{Pb}{100}$$

U - Ur is the number of µmoles of the bound uric acid - uricine complex. U and Ur are respectively the number of µmoles of uric acid and uricine per liter of urine. K is the equilibrium binding constant (2.2×10^{-10} M), Pb is the binding percentage, when the optimum binding at pH 4.0 is designated 100.

Table 1

Uricine excretion in control subjects

Subject Nº	Uric Acid mg/24 h	Uricine mg/24 h	Potential Binding Uric Acid - Uricine 10^{-3}M
1	320	690	0.04
2	180	618	0.07
3	195	381	0.13
4	280	261	0.02
5	235	225	0.02
6	255	120	0.02
7	310	85	0.02
8	270	82	0.01
9	360	82	0.01
10	240	75	0.02
11	315	50	0.04

Uricine excretion in the control group of subjects varied from 50 - 690 mg/24 h while the mean uric acid - uricine potential binding was 0.036×10^{-3} M (Table 1).

About half of the uric acid stone formers excreted larger amounts of uricine than the control group (Table 2). However all the uric acid stone formers (except nº 26) had higher uric acid - uricine potential binding values, than the control group.

Table 2
Urinary Uricine excretion in recurrent uric acid stone formers

Patient Nº	Uric Acid mg/24 h	Uricine mg/24 h	Potential Binding Uric Acid Uricine 10^{-3}M
12	507	2,220	3.25
13	485	1,360	0.22
14	532	1,247	0.38
15	953	1,140	0.42
16	1,104	1,111	1.71
17	625	982	1.87
18	980	925	2.94
19	160	920	0.89
20	520	833	0.68
21	780	781	1.75
22	1,104	727	0.93
23	780	685	1.27
24	434	654	0.28
25	816	590	0.31
26	325	508	0.01
27	1,046	494	0.81
28	571	458	1.30
29	890	456	0.37
30	1,094	342	0.34
31	930	289	0.14

Table 3
Uricine excretion in calcium recurrent stone formers. Data from the urine.

Patient Nº	Uric Acid mg/24 h	Uricine mg/24 h	Potential Binding Uric Acid Uricine 10^{-3}M
32	630	472	0.27
33	1,080	424	0.03
34	1,111	380	0.08
35	1,131	326	0.01
36	1,254	276	0.01
37	660	220	0.04

All the recurrent calcium stone formers (in this study) were uric acid hyperexcreters, nevertheless, the 24 hr. uricine excretion was within the normal range (table 3). The uric acid - uricine potential binding was normal in 5 subjects out of 6.

The uric acid - uricine potential binding was increased, 4 gouty patients out of 5. However the mean value was much lower than in the uric acid stone formers (Table 4)

Table 4
Uricine excretion in gouty patients without clinical history of stone formation. Data from the urine.

Patient Nº	Uric Acid mg/24 h	Uricine mg/24 h	Potential Binding Uric Acid Uricine $10^{-3}M$
38	867	422	0.17
39	1240	380	0.22
40	835	319	0.03
41	847	306	0.29
42	725	181	0.24

Discussion

This paper describes the preliminary experiments that show the uricine - bilirubin interrelationship. Bilirubin appears to be transformed into uricine by the homogenized kidney. Although the bilirubin appears to be metabolically related to uricine, could be that bilirubin were not the true substrate and it may be a slow rate type of reaction.

The uricine - uric acid potential binding capacity seems to have same clinical implications, since it is increased in the recurrent uric acid stone formers, besides that 42.10 % of this group had not an increase of the uric acid urinary concentration. This may suggest that the uric acid precipitation in urine could be affected by the uricine urinary excretion or the potential capacity to be bound by uric acid. The uric acid - uricine binding could be considered as a factor affecting the uric acid precipitation.

Acknowledgement

We are indebted to Dr. Sole - Balcells, head of this Division, for his continue support and patronage.

SITES OF URATE TRANSPORT IN THE RAT NEPHRON*

Greger, R., Lang, F., Deetjen, P., Knox, F.G.

Institut of Physiology, University of Innsbruck
Department of Physiology, Mayo Med.School,Rochester,Mn
A - 6010 Innsbruck/Austria, 3 Fritz-Pregl-Str.

The rat has been used frequently for studying renal urate excretion (1-14,16). This is clearly correlated with the fact that the overall renal handling of urate in this animal is comparable to that in man (5). Previous clearance studies have revealed that the net reabsorption of urate in the mammalian nephron is the result of bidirectional tubular transport (17). Recently several investigators have applied different micropuncture and microanalytical methods to the rat model to further elucidate the complicated pattern of bidirectional tubular transport (1,3,4,7-10,13,14, 16). As a result of these studies, it is generally accepted that both a secretory and a reabsorptive mechanism for urate exist in the proximal tubule. In the distal nephron very little, if any, urate is transported. However, some controversy still remains concerning the quantitative correlation of reabsorption and secretion in the proximal tubule and the quantitative importance of the post-proximal reabsorption, i.e. loop reabsorption (14).

To obtain answers to these questions, urate reabsorption was studied in different nephron segments, and, in addition, net transport of urate was measured with micropuncture and clearance techniques. Since there has been some discussion about the validity of the microanalytical techniques used previously (14) all studies reported here were done with 2-C^{14} uric acid. In recent metabolic studies it was shown that the rat kidney does not metabolize urate in vivo (11). Therefore, microperfusion data obtained with urate tracer are reliable. The same is true for microinjection data as long as the recovery is corrected for excretion by the contralateral kidney (11).

TABLE 1: PROTOCOLS

TYPE OF EXPERIMENT	N (ANIMAL)	STATE OF THE ANIMAL	N (OBSERVATION)	SOLUTIONS USED
COMBINED MICROPUNCTURE AND CLEARANCE	7	ANTIDIURECTIC A) CONTROL B) OXONAT (200 MG/KG) C) "HEPATECTOMY" (6)	PROX. TUBULES 65 DIST. TUBULES 9 URINES 22	2-C14 UA, 250 μCI/HR 2-C14 UA, 50 μCI,3-H IN,250μCI/HR 2-C14 UA, 50 μCI,2-H IN,250μCI/HR
PROXIMAL MICROPERFUSION 20 NL/MIN	8	ANTIDIURETIC	41	"EQUILIBRIUM SOL.(9)" + 0.3 MMOL/L 2-C14 UA + 50 MCI/L 3-H IN
LOOP PERFUSION 10 NL/MIN	11	ANTIDIURETIC	21	RINGER + 0.3 MMOL/L 2-C14 UA + 50 MCI/L 3-H IN
DISTAL MICROPERFUSION 10 NL/MIN	5	ANTIDIURETIC	19	RINGER + 0.3 MMOL/L 2-C14 UA + 50 MCI/L 3-H IN
DISTAL MICROINJECTION 3', 10 NL/MIN	6	ANTIDIURETIC	15	RINGER + 0.3 MMOL/L 2-C14 UA + 50 MCI/L 3-H IN

On the other hand, if urate tracer is infused systemically, it will be oxidized rapidly by hepatic uricase. Therefore, micropuncture and clearance data can be obtained with the use of labelled urate only if either the hepatic uricase is blocked (Oxonate, "hepatectomy" (6)) or if the purity of the tracer in the different fluids is measured. The latter can be done with a new thin layer chromatographic separation procedure (11) which discriminates urate from any of its catabolites.

The different experimental series are summarized in table 1. Free flow micropunctures and clearances were obtained in three groups of animals. In the first group, rats were hepatectomized as described previously (6). In the second group, rats were treated with oxonate (200 mg/kg, i.p.). Both methods of abolishing uricase increase plasma urate concentration five-fold but do not alter renal urate clearance significantly (6). In the third group, no effort was made to block endogenous uricase activity. Therefore, a larger amount of labelled urate had to be infused in that group and each sample was chromatographed. This group is labelled "con-

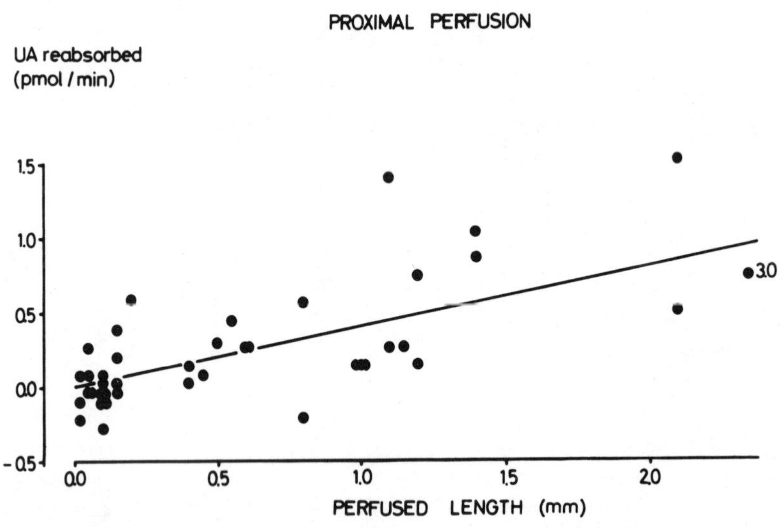

Fig.1: Microperfusions of proximal convoluted tubules: The absolute amount of urate reabsorbed is plotted on the ordinate versus the length of the perfused segment measured by Latex casts on the abscissa. The least square regression has the function: y=0.002 + 0.399x, r=0.672, p<.001.

trol". The techniques of microperfusion and microinjection have been described elsewhere (4,15).

The results of the proximal microperfusions are depicted in Fig. 1. A highly significant ($p<.001$) correlation exists between tubule length perfused and amount of urate reabsorbed. A mean of 0.4 pmol/min·mm urate is reabsorbed. This value is slightly below that reported previously from our laboratory (9) and almost identical to that reported recently by Roch-Ramel et al.(14). Since the size of the cellular urate pool is minor compared to the amount of urate perfused this outflux of urate tracer, in fact, represents transtubular transport.

Fig. 2 summarizes the loop perfusion data. The loop of Henle is defined arbitarily as the nephron segment between the last proximal and the first distal loop. It is apparent from this figure that the magnitude of urate reabsorption varies from loop to loop. A mean of 43% of the urate load delivered to the end of the proximal tubule is reabsorbed in the loop of Henle. This value is higher than that reported by other authors (7,8,14,16). The dif-

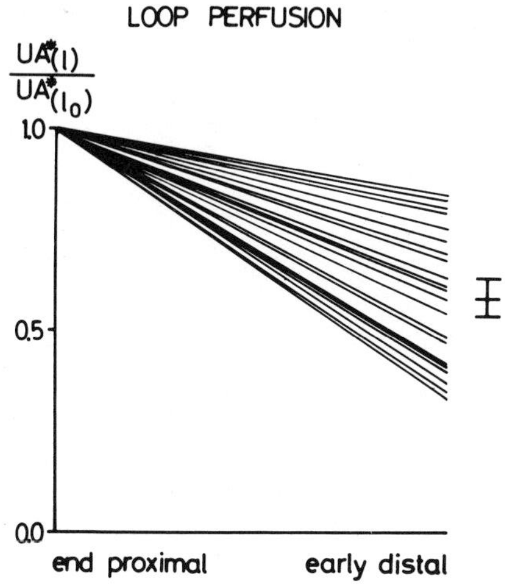

Fig.2: Microperfusions of loops of Henle: The urate load present at the distal collection site as a fraction of the load delivered to the loop ($UA_{(1)}/UA_{(1_o)}$) is shown for each single loop perfusion.

ference, however, is fully accounted for by the higher tubular flow rates used in those studies, since it has been shown that the loop reabsorption is inversely related to the tubular flow rate (4). Under conditions of moderate antidiuresis the flow rate at the end of the proximal tubule is about 10 nl/min. This is the flow rate we used to obtain the above data.

The distal microperfusions are summarized in Fig. 3. In contrast to the proximal perfusion data, no correlation between amount of urate reabsorbed and tubule length perfused could be found ($p>.2$). In fact the data scatter around zero with as many negative as positive data points. This finding is in close agreement with previous microperfusions (12) at pH 5 and 8. The results of the microinjections into distal tubules are shown in Fig. 4. A mean of 95% of the microinjected urate load is recovered in urine. This value is similar to those reported by other groups for the total recovery (7,8,16). The distal microperfusion and microinjection data make it unlikely that any major urate outflux exists in the distal nephron.

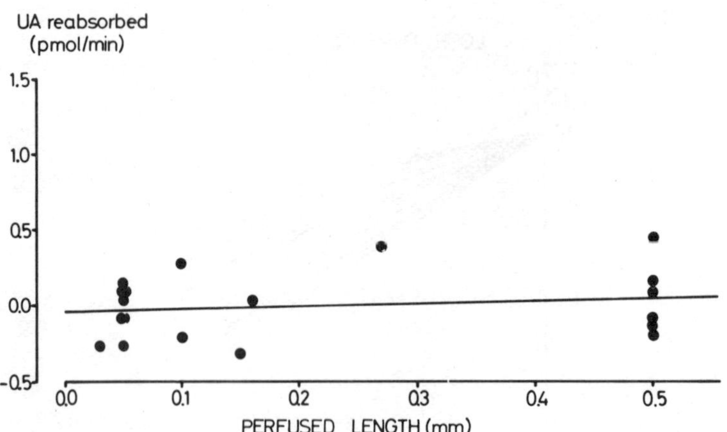

Fig. 3: Microperfusions of distal convoluted tubules: The method of plotting is the same as for Fig.1. The scale on the abscissa is expanded. The least square regression line has the function: $y= -0.047 + 0.259x$, $r= -0.205$, $p>.2$.

The second aim of this study was to compare the net fluxes of urate to the outward unidirectional fluxes and, by this means, to get some information about the quantitative importance of urate secretion to the tubular concentration profile for urate. The fraction of the filtered load of urate present at different sites in the proximal tubule is shown in Fig. 5. As is evident in this figure, there is little net flux of urate along the length of the proximal tubule in all three groups of animals. The least square regression line for all data is defined by: y=.902 -.001x, r=.006. The slope of this function is not significantly different from zero. In view of the data indicating unidirectional urate outflux (Fig.1) this high recovery of urate at the end of the proximal tubule has to be ascribed to urate secretion. The present value of about 90% of the filtered load of urate remaining at the end of the proximal tubule is slightly lower than the value (130% and 110%) obtained in previous micropuncture studies from our laboratory (3,4). It is, on the other hand, clearly higher than the 55% of the filtered load of urate found remaining at the end of the proximal tubule in the recent micropuncture study of Roch-Ramel et al. (14). The causes for this discrepancy are unresolved. It seems unlikely that dif-

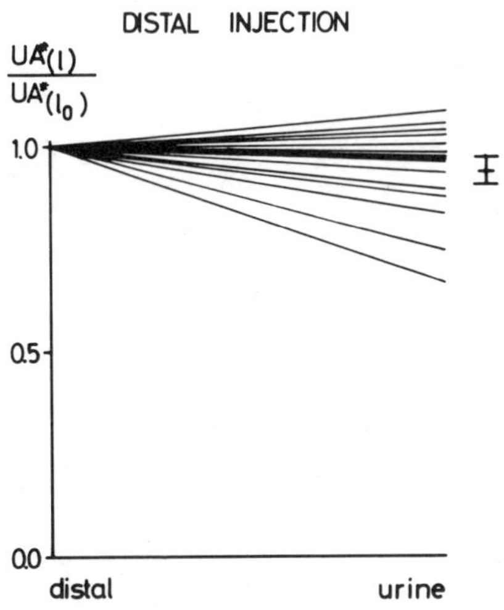

Fig. 4: Microinjections into distal convoluted tubules: The method of presentation is the same as for Fig.2. The mean of 0.95 is not significantly different from 1.0 ($p>.1<.05$).

ferences in plasma urate concentration are the cause since no differences in proximal tubular urate handling became apparent for the control and the hepatectomized group (Fig.5) despite a five-fold difference in plasma urate concentration (50 vs.250μmol/l).

Fig.6 summarizes the distal micropuncture and clearance values in the same animals. Only 51% of the filtered urate load is present in the early distal tubule. This value is almost identical to that obtained in previous studies using different analytical techniques (4,13,14). From the difference between the fraction of the filtered load recovered at the end of the proximal tubule and that recovered in the distal tubule, a reabsorption of 43 % of the delivered load in the loop can be calculated. This value is identical to the unidirectional reabsorption of urate found in loop perfusions at a flow rate of 10nl/min (Fig.2). About 48% of the filtered urate was excreted in the final urine in these animals. This is not significantly different from the percentage recovered in the distal tubule. From these data, therefore, there is no need to postulate any significant urate reabsorption in the distal nephron. Furthermore, in view of the absence of a unidirectional outflux of urate (Fig.3,4)

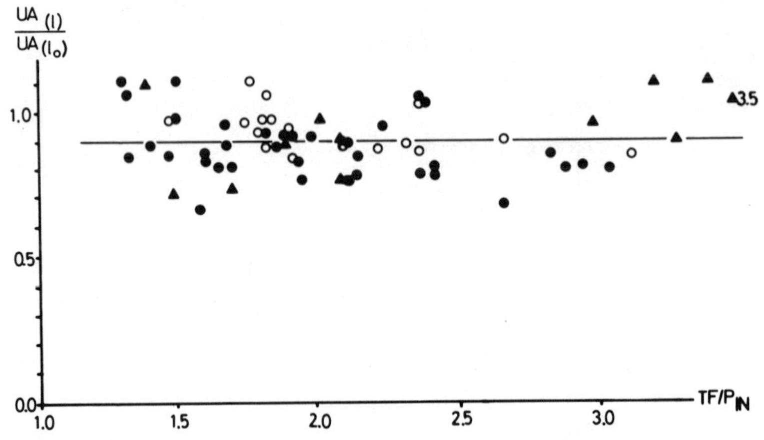

Fig. 5: Free flow micropunctures of proximal convoluted tubules in control animals (triangles), hepatectomized animals (closed circles), and oxonate treated animals (open circles): The urate load delivered to the puncture site as a fraction of the filtered load ($UA_{(l)}/UA_{(l_o)}$) is plotted on the ordinate versus the concentration ratio of inulin in tubular fluid and plasma (TF/P_{IN}) as a measure for tubular length on the abscissa. The least square regression has the function: $y = 0.900 - 0.001x$, $r = .006$, $p > .4$.

in this nephron segment, the free flow data exclude any significant secretion of urate at this site.

Fig.7 summarizes both the unidirectional reabsorption and the net transport data. In the proximal nephron a marked reabsorption of urate but almost no net transport was found. This has to be interpreted as a bidirectional transport in which the secretory and reabsorptive rates balance each other (9,10,14). In the loop segment a marked reabsorption of urate takes place. Here, the unidirectional outflux is identical to the netflux. At present, there is no direct clue as to what portion of the loop segment contributes to this reabsorption. However, the absence of urate reabsorption in the distal convoluted tubule renders it rather unlikely that the ascending portion of the loop is the main site of reabsorption. On the other hand, in view of the proximal urate reabsorption, some urate outflux in the straight portion of the descending limg of the loop seems likely. In the present studies as well as in the previous reports (7,12,14) there is no evidence for significant urate transport beyond the loop segment.

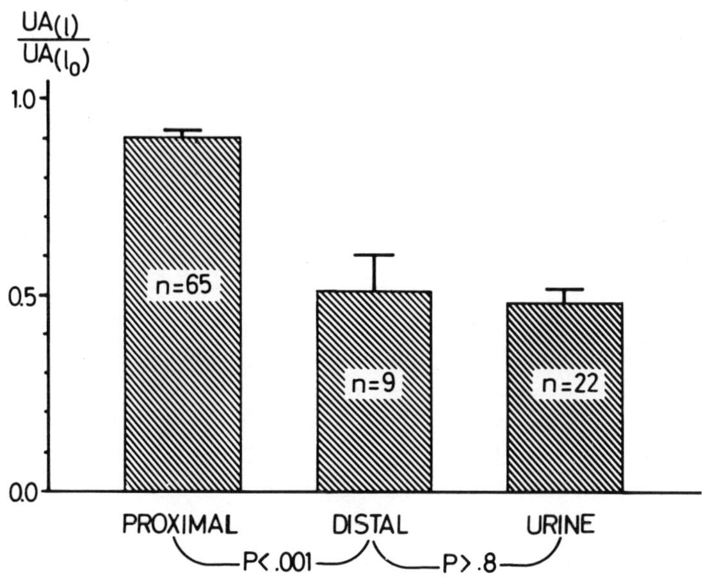

Fig. 6: Free flow micropunctures of distal convoluted tubules and clearance data. The legend for the ordinate is the same as in Fig. 5.

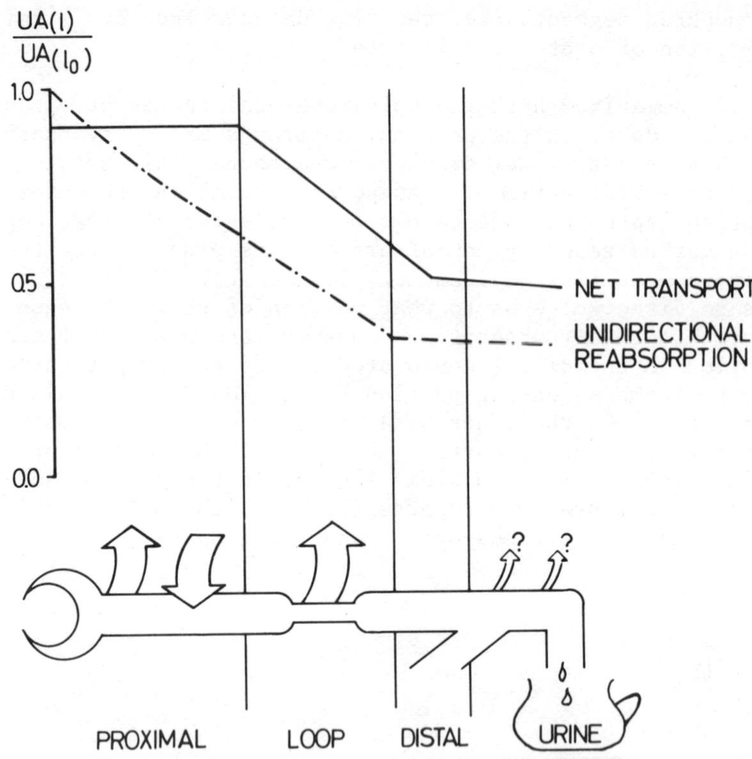

Fig. 7: Profiles of fractional urate delivery ($UA_{(l)}/UA_{(l_o)}$) along the nephron as measured from unidirectional urate outflux (dotted line) and from free flow micropunctures (solid line).

Footnotes:
* This study was supported by "Legerlotz Stiftung". Part of this study has been presented at the Intern. Symp. on Amino Acid and Uric Acid Transport, Innsbruck 1975. Georg Thieme Verlag, Stuttgart 1976.

References:

1. Abramson, R.G., Levitt, M.F.: Micropuncture study of uric acid transport in rat kidney. Am. J. Physiol. 228:1597, 1975.
2. Cook, M.A., Adkinson, J.T., Lassiter, W.E., Gottschalk, C.W.: Uric acid excretion by the rat kidney. Am.J.Physiol. 229:586,1975.
3. Greger, R., Lang, F., Deetjen, P.: Handling of uric acid by the rat kidney. I. Microanalysis of uric acid in proximal tubular fluid. Pflügers Arch. 324:279, 1971.

4. Greger, R., Lang, F., Deetjen, P.: Urate handling by the rat kidney. IV. Reabsorption in the loops of Henle. Pflügers Arch. 352:115, 1974.
5. Greger, R., Lang, F., Deetjen, P.: Renal excretion of purine metabolites: urate and allantoin by the mammalian kidney. Ed.: K. Thurau. Medical Technical Publishing House, London, in press.
6. Greger, R.: Purine excretion by the rat kidney. Intern. Symp. on amino acid transport and uric acid transport. Ed.: S.Silbernagl, F. Lang, R. Greger. Georg Thieme Verlag, Stuttgart, in press.
7. Kramp, R.A., Lassiter, W.E., Gottschalk, C.W.: Urate-2-^{14}C transport in the rat nephron. J. Clin. Invest. 50:35, 1971.
8. Kramp, R.A., Lenoir, R.: Distal permeability to urate and effects of benzofuran derivates in the rat kidney. Am. J. Physiol. 228:875, 1975.
9. Lang, F., Greger, R., Deetjen,P.: Handling of uric acid by the rat kidney. II. Microperfusion studies on bidirectional transport of uric acid in the proximal tubule. Pflügers Arch. 335:257, 1972.
10. Lang, F., Greger, R., Deetjen, P.: Handling of uric acid by the rat kidney. III. Microperfusion studies on steady state concentration of uric acid in the proximal tubule. Consideration of free flow conditions. Pflügers Arch. 338:295, 1973.
11. Lang, F., Greger, R., Deetjen, P.: In vivo studies on uricase acitivity in the rat. Pflügers Arch. 351:323, 1974.
12. Oelert, H., Baumann, K., Gekle, D.: Permeabilitätsmessungen einiger schwacher organischer Säuren aus dem distalen Konvolut der Rattenniere. Pflügers Arch. ges. Physiol. 307:178, 1969.
13. Roch-Ramel, F., de Rougemont, D., Peters, G., Weiner, I.M.: Micropuncture study of urate excretion by the kidney of rat and cebus monkey. Intern. Symp. on amino acid transport and uric acid transport. Ed.: S. Silbernagl, F. Lang, R. Greger. Georg Thieme Verlag, Stuttgart, in press.
14. Roch-Ramel, F., Diezi-Chomety, F., de Rougemont, D., Tellier, M., Widmer, J., Peters, G.: Renal excretion of uric acid in the rat: a micropuncture and microperfusion study. Am. J. Physiol. 230:768, 1976.
15. Sonnenberg, H., Deetjen, P.: Methode zur Durchströmung einzelner Nephronabschnitte. Pflügers Arch. ges. Physiol. 278:171,1965.
16. Weinman, E.J., Eknoyan, G., Suki, W.N.: The influence of the extracellular fluid volume on the tubular reabsorption of uric acid. J. Clin. Invest. 55:283, 1975.
17. Zins, G.R., Weiner, I.M.: Bidirectional urate transport limited to the proximal tubule in dogs. Am. J. Physiol. 215:411, 1968.

FACTORS AFFECTING URATE REABSORPTION IN THE RAT KIDNEY

Lang, F., Greger, R., Deetjen, P., Knox, F.G.

Institut of Physiology, University of Innsbruck
Dept. of Physiology, Mayo Med. School, Rochester,Mn,USA
A - 6010 Innsbruck/Austria, 3 Fritz-Pregl-Str.

Urate handling in the rat kidney has been shown to involve bidirectional transport in the proximal tubule, net reabsorption in the loops of Henle and negligible transport in the distal nephron. In this paper, results will be discussed describing the effects of luminal urate concentration, pH and flow rate on urate reabsorption.

Luminal Concentration

It is a well known fact that in man urate clearance increases with rising plasma concentration (1,23,33). This phaenomenon might be due to a secretory process, operating partially or completely beyond urate reabsorption, which increases with rising plasma concentration, or it might be due to saturation of the reabsorptive process. To study saturation kinetics of urate reabsorption in the rat, single proximal tubules were perfused in vivo with varying luminal urate concentrations. The perfusate was made up to prevent net water fluxes (80 mM mannitol, 110 mM Na^+) and was adjusted to the pH of 6.8 (13 mM HPO_4^{--}, 14 mM $H_2PO_4^-$) that is usually found in the proximal tubule of the rat. 2-^{14}C-urate (0.08, 0.3 and 1.0 mM, respectively) and ^{3}H-inulin were added to the perfusate, and the fractional urate recovery ($R = \{UA/UA_o\}/\{In/In_o\}$) was determined where UA and In are the tracer activities in the collected tubular fluid and UA_o and In_o the activities in the initial perfusate. The length of perfused segments was measured with the use of latex casts. The single data points are given in Fig. 1. Analysis of linear regressions allowed calculation of the reabsorption rate (I) from the slope (b) and initial concentration ($I = c_o b$). Assuming that the urate reabsorption rate is in linear proportion to the concentration of a hypothetical substrate-carrier complex and that the concentration of this substrate-carrier complex is in linear proportion to the product of the concentrations of free carrier and free substrate, we can apply Michaelis Menten kinetics for the description of transport. The transport rates can be plotted versus the respective concentrations with the different linear plots underlying Michaelis Menten kinetics. Fig. 2 a-c shows the result of

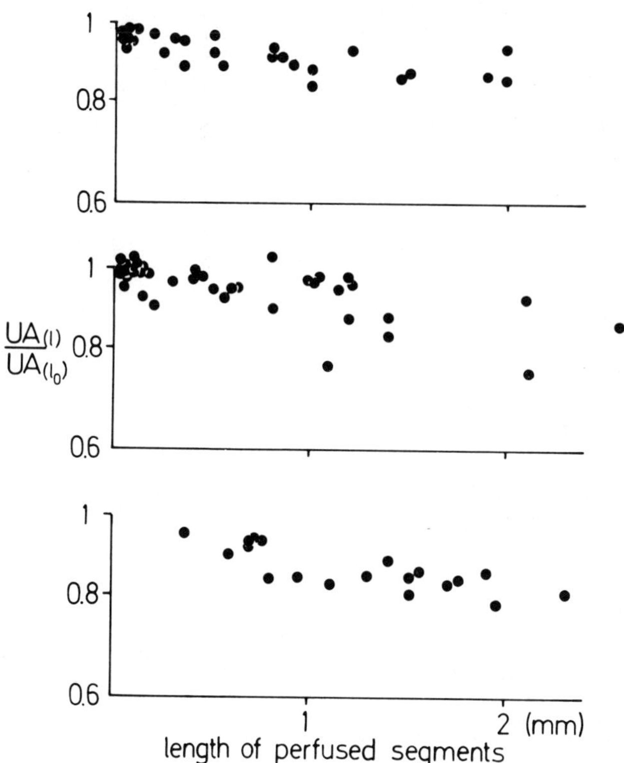

Fig. 1: Saturation of urate reabsorption. Single data points. Ordinate: Fractional urate recovery ($R = \{UA/UA_o\}/\{In/In_o\}$), as calculated from urate and inulin concentrations in collected tubule fluid (UA,In) and initial perfusate (UA_o/In_o). Abscissa: Length of perfused segments. Urate concentrations in initial perfusates were (from above) 1 mM, 0.3 mM and 0.08 mM.

such an analysis. The slopes and intercepts allow calculation of a maximal transport rate (I_{max} = 1.8 pmol/mm min or some 9 pmol/min prox nephron) and the concentration at which the transport rate is half maximal (K_M = 1 mM). The scatter of the data points is considerable and precludes an accurate estimate of the parameters. Furthermore, the limited solubility of urate renders it impossible to increase urate concentrations above the K_M apparent from these studies. Therefore, maximal transport rate must be obtained by extrapolation from values below K_M. In any case, the affinity of the

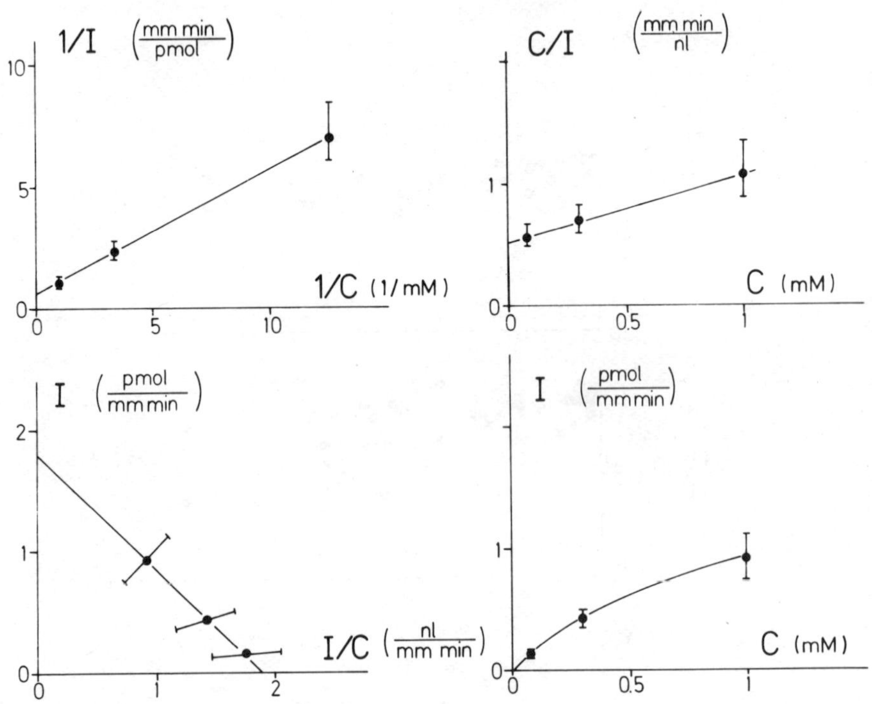

Fig. 2: Saturation of urate reabsorption. Correlation between reabsorptive transport rate (I) and luminal concentration (c). The transport rates (± SEM) are calculated from the regressions of the data given in Figure 1 and the urate concentration in the perfusate. The three linear transformations of the Michaelis Menten equation ($I = c\ I_{max}/\{c + K_M\}$) allow calculation of the maximal transport rate (I_{max}) and concentration at half saturation (K_M) from the intercepts and slopes. The curvilinear function in the right lower plot is based on an I_{max} of 1.8 pmol/mm min and a K_M of 1.0 mM derived from the preceeding plots.

transport system for urate appears to be very low and urate reabsorption seems to operate far from saturation in the rat at physiological plasma concentrations (<0.1 mM). Clearance studies in man suggest that the affinity of the reabsorption system for urate is as low as in the rat. Such a low affinity results in poor regulation of the plasma concentration by the kidney on the one hand, but on the other prevents peak urinary concentrations of urate when plasma concentration undergoes fluctuations.

Luminal pH

Acidosis is known to decrease, alkalosis to increase urate clearance (10,12,17,21). This could be due to the sensitivity of the urate transport to alterations in acid base balance or to the availability of more nonionized uric acid for reabsorption with an acid luminal pH. Fig. 3 displays the data points from microperfusion studies where the luminal pH was buffered to 5.8 (120 mM Morpholinethansulfonic acid (MES), 60 MES^-Na^+, 40 NaCl) instead of the physiological 6.8. Again, the fractional urate recovery ($\{UA/UA_o\}/\{In/In_o\}$) was determined from the tracer activities of ^{14}C urate and ^3H inulin in the collected tubular fluid (UA, In) and initial perfusate (UA_o, In_o), and the length of perfused segments was determined with the use of Latex casts. Since the concentration of nonionized uric acid was increased by a factor of five, an enhanced urate reabsorption should have become apparent if uric acid is reabsorbed by nonionic diffusion. Similarly, carrier mediated transport involving nonionized uric acid should have resulted in doubling of the transport rate if the affinity and maximal transport rate are

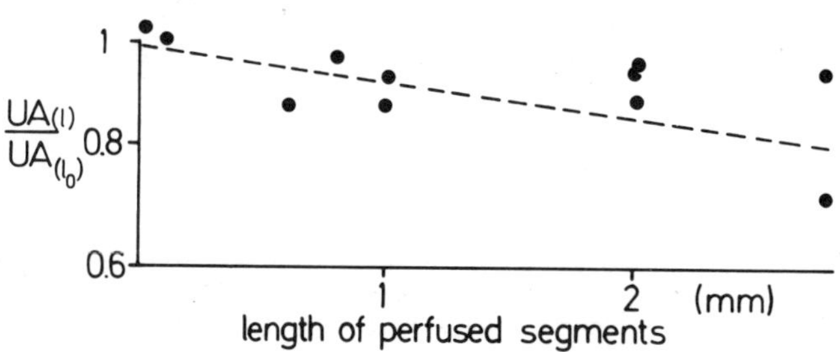

Fig. 3: Effect of luminal pH on urate reabsorption. Ordinate: Fractional urate recovery ($R = \{UA/UA_o\}/\{In/In_o\}$) calculated from urate and inulin concentrations in collected tubule fluid (UA,In) and initial perfusate (UA_o/In_o). Abscissa: length of perfused segments. The single data points at pH 5.8 are compared to the linear regression of the data at pH 6.8 (urate concentration 0.3 mM).

taken into account. However, analysis of the linear regression of the data points reveals that urate reabsorption actually decreased slightly but not significantly. Of course, inhibition of transport by the buffer base might have prevented an increase of urate transport with increasing concentrations of nonionized uric acid. However, microperfusion with acid solution with a phosphate buffer similarly failed to increase urate reabsorption (27). Therefore, it is unlikely that urate transport involves only the nonionized form. Interestingly, the opposite is true for urate transport in erythrocytes, which, therefore, cannot serve as a model for urate transport in the proximal tubule (8). The finding that urate clearance is enhanced during alkalosis and reduced during acidosis must be related to parameters other than the concentration of nonionized uric acid in the tubule fluid. Such mechanisms include the possible interaction of lactate and/or other metabolites with urate transport (6,19,20,22,24,26,32), a direct or indirect coupling of urate transport to tubular acidification, and the increased uptake of urate into erythrocytes in the renal medulla, hence facilitating urate reabsorption in the loop of Henle (8, 15).

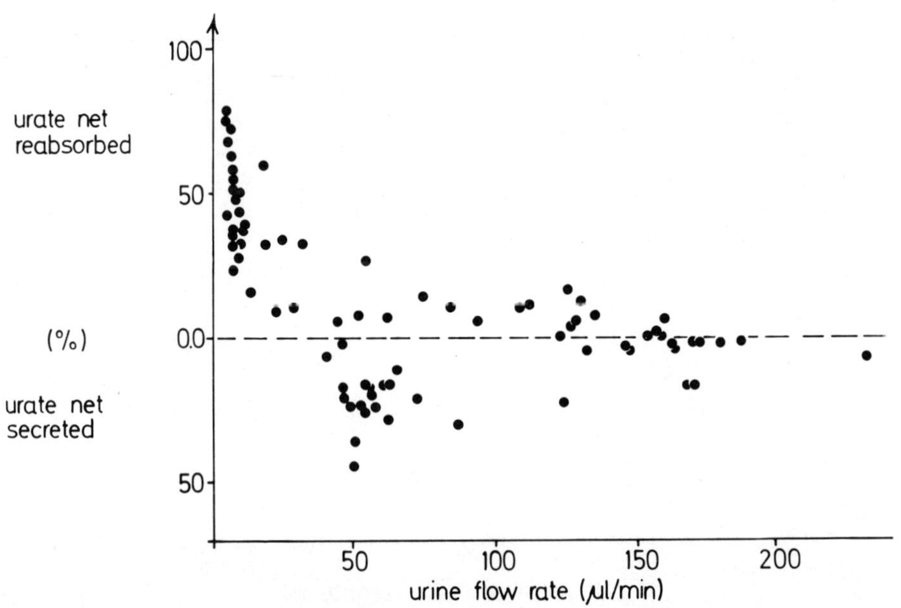

Fig. 4: urate reabsorption as a function of urinary flow rate. Ordinate: Net urate transport in percent of filtered load calculated from the difference of excreted amount of labelled urate and filtered load (glomerular filtration rate x plasma concentration of labelled urate). Abscissa: urinary flow rate.

Luminal Flow Rate

In man and other species, urate clearance has frequently been shown to increase with several maneuvers enhancing urinary flow rate (3, 4,18,28,29,31). As demonstrated in Fig. 5, the same is true for the rat if osmotic or saline diuresis is induced. At high flow rates even net secretion becomes apparent. Urate clearance was measured using 2-^{14}C urate. Break down of urate was prevented either by hepatectomy or oxonate and purity of the tracer was determined by chromatography (14). To test further the effects of luminal flow rate on urate reabsorption, loops of Henle were perfused at rates of 10 to 50 nl/min. As is apparent from the figure, increasing luminal flow rate results in a reciprocal change in fractional urate reabsorption. At 40 nl/min fractional urate reabsorption is only 1/3 of the value at physiological flow rates (10 nl/min(11)). Therefore, reabsorption of urate at least in the loop of Henle is highly sensitive to luminal flow rate. A similar conclusion can be derived from the comparison of microinjection studies at different flow rates (13, 31). The high sensitivity of urate reabsorption in

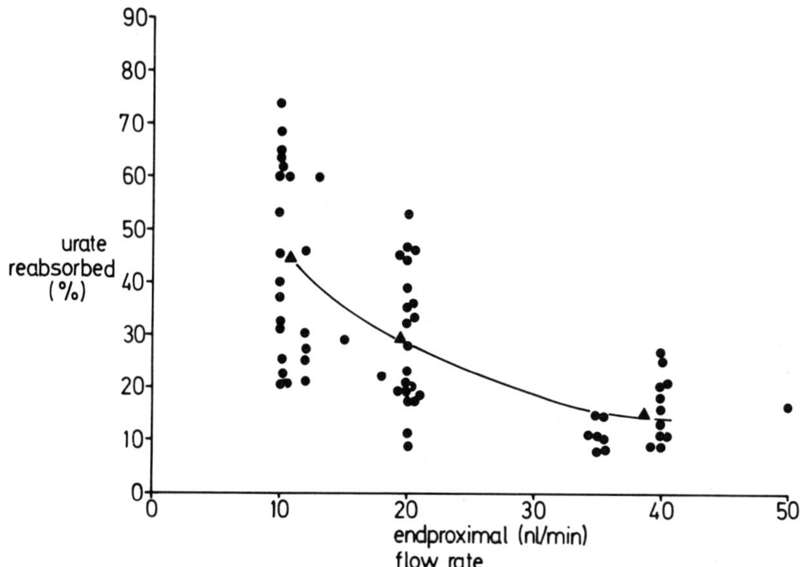

Fig. 5: Flow dependence of urate reabsorption in loops of Henle. Ordinate: urate reabsorption in percentage of delivery to late proximal tubules calculated from the concentrations of urate and inulin in the collected tubule fluid (UA, In) and initial perfusate (UA_o, In_o):100 ({UA_o/In_o}-{UA/In})/(UA_o/In_o). Abscissa: Perfusion rate.The triangles represent means of the recovery at a given flow rate.

the loop of Henle gives a ready explanation for the uricosuric effects of maneuvers which increase flow rate at the end of the proximal tubule. The effects of volume contraction could be related to enhanced fluid reabsorption in the proximal tubule resulting in decreases of luminal flow rate. This in turn would facilitate urate reabsorption. Since effective diuretics frequently result in volume contraction, their antiuricosuric effect is at least partially related to the occurence of volume contraction (16). This is further supported by the fact that antiuricosuria is prevented by replacement of urinary losses (30). A similar situation might arise in renal insufficiency where the functioning nephrons have grossly decreased fluid reabsorption in the proximal nephron and, therefore drastically increased flow rates at the end of the proximal tubule. Of course the possibility must be kept in mind that decreased sodium reabsorption in the proximal convoluted tubule might per se depress urate reabsorption in the same segment, e.g. if a co-transport operates for both species. However, proximal tubule urate reabsorption was not appreciably influenced by the presence or absence of mannitol sufficient to completely abolish net sodium reabsorption (15,25). Urate reabsorbed in the loops of Henle might undergo countercurrent trapping in the kidney medulla. As a matter of fact, accumulation of urate in the medulla has been

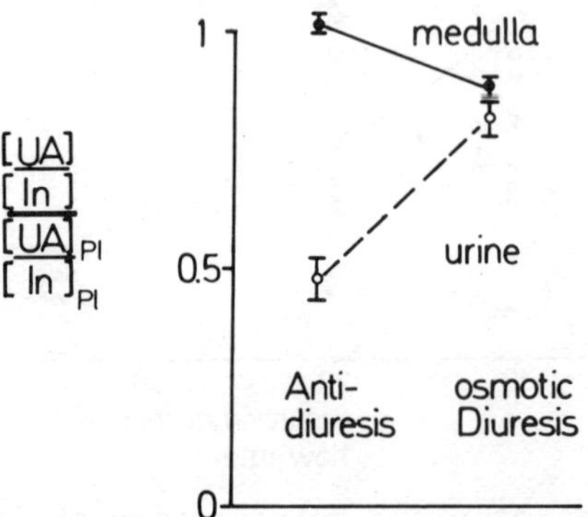

Fig.. 6: Accumulation of urate in kidney medulla. Ordinate: Urate concentration in papilla and urine, respectively, (S D) relative to that of inulin expressed as fraction of that in plasma. Open symbols and dotted line represent the values of urine, closed symbols and solid line the papilla.

demonstrated in various species,including man (2,5,7). Since urate reabsorption in the loops of Henle is reduced by increasing luminal flow rate, induction of osmotic diuresis should result in decreased urate reabsorption in the loop of Henle and hence decreased accumulation in the medullary tissue. To test this hypothesis, the concentration of ^{14}C labelled urate was determined in the renal papilla with or without induction of osmotic diuresis. However, the demonstration of a decline of urate concentration per se would be trivial since it might reflect simply dilution of urate in the collecting ducts. Therefore, ^{3}H inulin was determined simultaneously and the values given in Fig. 6 are expressed as concentration of urate over inulin in the renal papillary tissue as compared to that in plasma. Fig. 6 clearly demonstrates that the ratio of urate over inulin decreases under osmotic diuresis in the papillary tissue, despite the expected increase of that ratio in the urine. These data therefore support the view that accumulation in the renal medulla is caused by reabsorption in the loops of Henle which is drastically reduced during osmotic diuresis. It should be pointed out that the results presented here are based solely on observations in the rat and that urate excretion in man differs from that in the rat, at least in quantitative terms. Therefore, extrapolation to the human kidney may be hazardous. On the other hand, the bulk of indirect evidence indicates that urate excretion in man involves similar parameters, and, therefore, the present results might serve as a model which helps to understand similar phaenomena in man.

Conclusions:

1. Urate transport in the rat appears to be saturable.However, affinity of the transport system for urate is very low and transport far from saturated at physiological plasma concentrations.
2. Since increase of the nonionized fraction of uric acid by a factor of five failed to increase urate reabsorption, transport cannot be due to nonionic diffusion but rather involves ionized urate.
3. Increases in luminal flow rate markedly depress urate reabsorption in the loop of Henle, which results in wash out of medullary urate.

****

Part of the material has been presented in Pflügers Arch. (9) and at the international Symposium of amino acid transport and uric acid transport.
The work has been supported by Deutsche Forschungsgemeinschaft and Legerlotz-Stiftung.

References:
1. Berliner, R.W., Hilton,J.G., Yü, T.F., and Kennedy, T.J.: The renal mechanism for urate excretion in man. J.Clin.Invest. 29:396, 1950.
2. Cannon, P.J., Symchych,P.S., and Demartini,F.E.: The distribution of urate in human and primate kidney. Proc.Soc.Exptl.Biol. Med. 129:278, 1968.
3. Cannon, P.J., Svahn,D.S., and Demartini,F.E.: The influence of hypertonic saline infusions upon the fractional reabsorption of urate and other ions in normal and hypertensive man. Circulation 41:97, 1970.
4. Diamond, H.S. and Meisel,A.: Influence of volume expansion, serum sodium and fractional excretion of sodium on urate excretion. Pflügers Arch. 356:47, 1975.
5. Epstein, F.H. and Pigeon,G.: Experimental urate nephropathy. Studies of the distribution of urate in renal tissue. Nephron 1:144, 1964.
6. Fanelli, G.M., Bohn,D.L., and Stafford,S.: Functional characteristics of renal urate transport in the cebus monkey. Am.J.Physiol. 218:627, 1970.
7. Friedmann,M. and Byers, S.O.: Distribution of uric acid in rat tissues and its production in tissue homogenates. Am.J. Physiol. 172:29, 1953.
8. Greger, R., Lang,F., Puls,F., Deetjen,P.: Urate interaction with plasma proteins and erythrocytes. Pflügers Arch. 352, 121 (1974).
9. Greger, R., Lang,F., and Deetjen,P.: Urate handling by the rat kidney. IV. Reabsorption in the loops of Henle. Pflügers Arch. 352:115, 1974.
10. Gutman, A.B., Yü,T.F. and Sirota,J.H.: Contrasting effects of HCO_3^- and diamox with equivalent alkalization of urine on salicylate uricosuria in man.Fed.Proc.15:85, 1956.
11. Hierholzer, K., Mueller-Suur,R., Gutsche,H.U., Butz,M., and Lichtenstein,I.: Filtration in surface glomeruli as regulated by flow rate through the loop of Henle. Pflügers Arch. 352:315, 1974.
12. Isomaeki, H. and Kreus, K.E.: Serum uric acid in respiratory acidosis. Acta Med.Scand. 184:293, 1968.
13. Kramp, R.A., Lassiter,W.E., and Gottschalk,C.W.: Urate-2^{14}C transport in the rat nephron. J.Clin.Invest.50:35,1971
14. Lang, F., Greger, R ., and Deetjen,P.: In vivo studies on uricase activity in the rat. Pflügers Arch. 351:323,1974.
15. Lang, F.: Parameters and mechanisms of urate transport in the rat kidney. Intern.Symp.on amino acid and uric acid. Innsbruck,1975. Ed.:S.Silbernagl,F.Lang,R.Greger. Thieme Stuttgart, New York
16. Lang, F., Greger,R., Deetjen,P.: Effect of diuretics on uric acid metabolism and excretion. Intern.Symp.on diuretics in research and clinics. Buergenstock 1975. Ed.:W.Siegenthaler. Thieme, Stuttgart, New York.

17. Lecocq, F.R. and McPhaul,J.J.,Jr.: The effects of starvation, high fat diets and ketone infusions on uric acid balance. Metabolism 14:186, 1965.
18. Manuel, M.A. and Steele,T.H.: Pyrazinamide suppression of the uricosuric response to sodium chloride infusion. J.Lab. Clin.Med. 83, 417, 1974.
19. Michael, S.T.: The relation of uric acid excretion to blood lactic acid in man. Am.J.Physiol.141:71,1944.
20. Moller, J.V.: The effects of D- and L- lactate and osmotic diuretics on uric acid clearance in the rabbit. Acta Physiol. Scand. 54:30, 1962.
21. Musil, J. and Granzer, E.: Purine metabolism during experimental disturbances of the homeostasis (Abstr.). Experientia 30:6, 681, 1974.
22. Nolan, R.P. and Foulkes,E.C.: Studies on renal urate secretion in the dog. J.Pharmacol. Exptl. Therap. 179:429, 1971.
23. Nugent,C.A.: Renal urate excretion in gout studied by feeding ribonucleic acid. Arthr.& Rheum. 8:671, 1965.
24. Reem, G.H. and Vanamee,P.: Effect of sodium lactate on urate clearance in the dalmatian and in the mongrel. Am.J. Physiol. 207:113, 1964.
25. Roch-Ramel,F., Diezi-Chomety, F.,de Rougemont,D., Tellier,M., Widmer,J., Peters, G.: Renal excretion of uric acid in the rat: a micropuncture and microperfusion study. Am.J.Physiol. 230:768 (1976).
26. Skeith,M.D. and Healey, L.A.: Urate clearance in cebus monkeys, Am.J.Physiol. 214:582, 1968.
27. Sonnenberg, H., Oelert,H., and Baumann,K.: Proximal tubular reabsorption of some organic acids in the rat kidney in vivo. Pflügers Arch. 286:171, 1965.
28. Steele,T.H. and Oppenheimer,S.: Factors affecting urate excretion following diuretic administration in man. Am.J.Med. 47:564, 1969.
29. Steele,T.H.: Evidence for altered renal urate reabsorption during changes in volume of the extracellular fluid. J.Lab.Clin.Med. 74:288, 1969.
30. Steele, T.H.: Pyrazinamide suppressiblity of urate excretion in health and disease. International symposium on amino acid transport and uric acid. Ed.: Silbernagl, S.,Lang,F., Greger, R. Georg Thieme Verlag, Stuttgart, 1975.
31. Weinman, E.J., Eknoyan, G., and Suki, W.N.: The influence of the extracellular fluid volume on the tubular reabsorption of uric acid. J.Clin.Invest. 55:283, 1975.
32. Yü, T.F., Sirota, J.H., Berger, L., Halpern, M., and Gutman, A.B.: Effect of sodium lactate infusion on urate clearance in man. Proc. Soc. Exp. Biol. Med. 96:809, 1957.
33. Yü, T.F., Berger, L., and Gutman, A.B.: Renal function in gout. Effect of uric acid loading on renal excretion of uric acid. Am. J. Med. 33:829, 1962.

SOME CHARACTERISTICS OF URIC ACID UPTAKE BY SEPARATED RENAL TUBULES OF THE RABBIT

Ian Kippen and James R. Klinenberg

Departments of Pharmacology and Medicine, UCLA School of Medicine, and Department of Medicine, Cedars-Sinai Medical Center, Los Angeles, Calif.

Uric acid is generally believed to be both secreted and reabsorbed by the mammalian renal proximal tubule. The degree to which reabsorption and secretion take place is highly species dependent. Uric acid uptake by slices of rabbit renal cortex has been used by previous investigators as a model of secretory transport of this compound. In 1961, Platts and Mudge (1) demonstrated that uptake of uric acid into rabbit kidney cortical slices is influenced by a variety of factors including hypoxia and the presence of various metabolic substrates and drugs. Berndt and Beechwood (2) subsequently showed that this uptake was pH dependent, and was influenced by the medium concentrations of sodium and potassium. Both groups of investigators obtained uric acid slice to medium ratios of approximately 2.5 indicating concentrative uptake of uric acid.

However, a recent report by Jones and Despopoulos (3) suggested that uric acid uptake by rabbit kidney is non-concentrative and is unaffected by anoxia and drugs. We thus considered it of interest to further characterize the transport of uric acid by rabbit kidney cortex. We report here on the effects of various physico-chemical factors on the rate of uptake of uric acid by separated renal cortical tubules of the rabbit.

The renal tubules were prepared by a method similar to that described by Burg and Orloff (4). The cortex from a rabbit kidney was put through a tissue press and then digested with collagenase to remove interstitial connective tissue. The tubules were washed

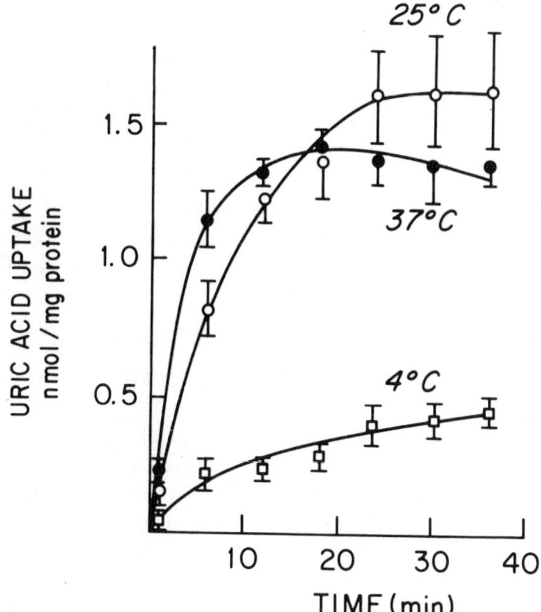

Figure 1. Time course of uric acid uptake by separated renal cortical tubules of the rabbit at three temperatures.

free of collagenase, resuspended in buffer and then incubated with uric acid-C^{14} under various conditions. At the end of the incubation period, which was usually 15 minutes, the tubules were rapidly filtered through a Nucleopore filter (8 μ pore size), washed, and counted in a liquid scintillation spectrometer. The length of time for filtering and washing each sample was under 10 seconds.

Shown in Figure 1 is the time course of uric acid uptake into the tubules at three different temperatures. The uptake was most rapid at 37° C and very slow at 4° C. A steady state was reached at about 20 minutes at 37° C and 30 minutes at 25° C. On the basis of these time course experiments, the uric acid uptake in 15 minutes at 25° C was considered to be a reasonable indicator of the rate of uric acid uptake in the following experiments.

We determined the effect of uric acid concentration on the rate of uric acid uptake. The rate of increase of uric acid uptake clearly decreased as the uric acid concentration was increased to 2.5 mM, which was the maximum concentration of uric acid attainable under the conditions of these experiments. Analysis of our data by nonlinear regression procedures showed a good fit of the data to

Table 1

	% Change from control	P
Citrate	+37	<.025
Pyruvate	none	----
Succinate	-11	N.S.
Acetate	-12	N.S.
α-Ketoglutarate	-48	<.01

the Michaelis-Menten model, with a Km for uptake of 3.2 mM.

Under the routine conditions described, alterations in medium pH had little effect on the rate of uric acid uptake. However, when the tubules were preincubated for 2 hours prior to addition of the uric acid, a clear maximum at pH 7.1 became apparent.

In tubules incubated under oxygen for 30 minutes prior to the addition of uric acid, the mean uric acid uptake was 0.80 pmol/mg protein. In tubules incubated under nitrogen for 30 minutes prior to the addition of uric acid, the uric acid uptake was 0.41 pmol/mg protein. The uptake under nitrogen was 50 percent lower than under oxygen, thus at least some of the uric acid uptake by the tubules is dependent on the presence of oxygen.

Shown in Table 1 are the effects of some metabolic substrates on uric acid uptake. Citrate was the only one of the substrates tested (all substrate concentrations were 1 mM) to increase the rate of uric acid uptake. The other substrates tested either had no significant effect or decreased the rate of uptake. In contrast to the data of Platts and Mudge, acetate did not stimulate uric acid uptake in these experiments.

In order to determine whether the uptake system is of the equilibrative or concentrative types, we determined steady-state tissue to medium ratios at equilibrium, as indicated by the time course experiments. The mean steady-state tissue to medium ratio for uric acid at 25° C was 0.86. This ratio is substantially lower than the 2.5 to 3.0 reported by earlier investigators. Murthy and Foulkes (5) reported finding substantially lower tissue to medium ratios for PAH in rabbit kidney tubules as compared with slices. They concluded that the lower ratio in tubules resulted from increased leaks across the luminal membrane in the tubules. It is thus likely that the differences in our steady-state data from that obtained in previous studies is related to the differences between the tubule and slice preparations. It is not possible to say at this point which of these preparations, and hence which of the estimates of uric acid steady-state tissue to medium ratio, more truely reflects the *in vivo* situation.

In summary, we conclude from our data that uric acid uptake by separated rabbit renal tubules occurs at least in part by some form of carrier mediated active transport. The uptake saturates as the uric acid concentration is increased. At least part of the transport is dependent on an oxygen requiring metabolic process. That the transport is inhibited by drugs will be shown in the following paper. In general, our data support the previous work of Platts and Mudge and Berndt and Beechwood, and disagree with that of Jones and Despopoulos.

REFERENCES

1. Platts, M.M. and Mudge, G.H.: Accumulation of uric acid by slices of kidney cortex. Am. J. Physiol. 200:387-391, 1961.

2. Berndt, W.O. and Beechwood, E.C.: Influence of inorganic electrolytes and ouabain on uric acid transport. Am. J. Physiol. 208:642-648, 1965.

3. Jones, V.D. and Despopoulos, A.: Is uric acid transported by the hippurate transport system? Pflugers Arch. 349:183-190, 1974.

4. Burg, M.B. and Orloff, J.: Oxygen consumption and active transport in separated renal tubules. Am. J. Physiol. 203:327-330, 1962.

5. Murthy, L. and Foulkes, E.C.: Movement of solutes across luminal cell membranes in kidney tubules of the rabbit. Nature 213:180-181, 1967.

EFFECTS OF DRUGS ON THE UPTAKE OF URIC ACID BY SEPARATED RENAL TUBULES OF THE RABBIT

Ian Kippen and James R. Klinenberg

Departments of Pharmacology and Medicine, UCLA School of Medicine, and Department of Medicine, Cedars-Sinai Medical Center, Los Angeles, Calif.

The effects of various compounds on the transport of uric acid by rabbit kidney have been investigated in several previous studies. Experiments utilizing clearance techniques have established that uric acid clearance is reduced by probenecid, salicylate, PAH, chlorothiazide and a number of other compounds (1,2). Stop-flow analysis has indicated that probenecid reduces uric acid clearance by blocking its secretion in the proximal tubule (3).

The effect of drugs on renal uric acid transport has also been studied *in vitro* using slices of rabbit renal cortex. Platts and Mudge (4) determined that a large number of compounds reduce the steady-state uric acid slice to medium ratio.

Information relating to the mechanism of action of drugs in inhibiting uric acid transport is not yet available. In order to further examine these drug effects on uric acid transport, we studied the effects of a number of compounds on the rate of uric acid uptake by separated renal tubules of the rabbit. We characterized the inhibitory effects of drugs with respect to I_{50} values (the drug concentration causing 50 percent inhibition of uric acid uptake at a medium uric acid concentration of 1 mM) and with respect to the slopes of the inhibition curves.

The tubules were prepared as described in the preceding paper. In each experiment, 100 μl of the tubule suspension containing approximately 10 mg of tubule protein were added to 1 ml of uptake buffer in a polypropylene test tube. The buffer con-

Table 1

	I_{50} (mM)	Slope
Sulfinpyrazone	0.02 ± .01	-0.73 ± .12
Benzbromarone	0.18 ± .01	-2.06 ± .09
Probenecid	0.26 ± .07	-0.60 ± .07
p-Diethylsulfamylbenzoic acid	0.28 ± .05	-0.77 ± .06
Phenylbutazone	0.38 ± .11	-0.93 ± .08
p-Dimethylsulfamylbenzoic acid	0.75 ± .11	-0.77 ± .06
Sodium salicylate	8.29 ± 2.1	-1.38 ± .09

tained 1 mM uric acid with a tracer amount of uric acid-C^{14}, and various concentrations of the drugs being tested. After 15 minutes of incubation, an aliquot of the suspension was filtered through a Nucleopore filter (8 μ pore size), washed, and the filter counted in a liquid scintillation spectrometer.

Shown in Table 1 are the effects of some uricosuric and related drugs on uric acid uptake by the tubules. The most potent inhibitor of uptake was sulfinpyrazone with an I_{50} of 0.02 mM. The next most potent inhibitor was benzbromarone with an I_{50} of 0.18 mM. Probenecid and p-diethylsulfamylbenzoic acid were slightly less potent. The least potent compound was sodium salicylate. Of this group of drugs, only salicylate and benzbromarone had inhibition curves with slopes greater than 1.0.

Shown in Table 2 are the effects on uric acid uptake of some miscellaneous compounds which we have studied. Calcium ipodate was the most potent inhibitor of uric acid uptake in this group of drugs. Iodipamide, iopanoic acid, ethacrynic acid and chlorothiazide were of about the same potency. The remaining compounds were substantially less potent. There was substantial variation in the slopes of the inhibition curves. Iodipamide, phloridzin, Penicillin G and PAH had shallow inhibition slopes. Iopanoic acid, ethacrynic acid and chlorothiazide had steep inhibition slopes. The inhibition curve for ouabain did not fit a straight line, as did the inhibition curves for the other compounds, but rather reached an asymptote at about 50 percent inhibition. This accounts for the unusually low value for its inhibition slope, and may suggest a different mechanism of action for ouabain.

We postulate from our data that drugs may inhibit uric acid uptake by separated renal tubules by more than one mechanism. Most of the compounds which would be expected *a priori* to be competitive inhibitors of uric acid uptake such as probenecid, PAH and Penicillin had shallow inhibition slopes. A shallow inhibition slope might thus indicate competitive inhibition of uric acid up-

Table 2

	I_{50}-(mM)	Slope
Calcium ipodate	0.08 ± .02	-1.08 ± .26
Iodipamide	0.37 ± .09	-0.68 ± .04
Iopanoic acid	0.38 ± .03	-1.81 ± .22
Ethacrynic acid	0.38 ± .08	-1.36 ± .26
Chlorothiazide	0.42 ± .10	-1.36 ± .04
Phloridzin	1.40 ± .24	-0.82 ± .04
Penicillin G	2.15 ± .72	-0.71 ± .07
Sulfamethoxypyridazine	3.51 ± .75	-1.08 ± .12
Ouabain	4.60 ± 2.90	-0.31 ± .03
PAH	9.97 ± 1.90	-0.59 ± .08

take. A steep inhibition slope may represent an allosteric effect or an effect on a metabolic process linked to the transport system. Further studies will be required to determine if this hypothesis is justified. Preliminary experiments in our laboratory indicate that differences do exist between drugs having a shallow inhibition slope and those having a steep inhibition slope, with respect to the kinetics of inhibition.

In these experiments, there was a strong correlation between a drugs potency in inhibiting uric acid uptake by the tubules (as expressed by the I_{50} value) and its uricosuric potency in man. We suggest that a compounds potency in inhibiting uric acid uptake by rabbit renal tubules could readily be used to estimate its potential uricosuric activity in man prior to human administration.

REFERENCES

1. Poulsen, H.: Inhibition of uric acid excretion in rabbits given probenecid or salicylic acid. Acta Pharmacol. Toxicol. 11:277-286, 1955.

2. Møller, J.V.: Clearance experiments on the effect of probenecid on urate excretion in the rabbit. Acta Pharmacol. Toxicol. 23:321-328, 1965.

3. Beechwood, E.C., Berndt, W.O. and Mudge, G.H.: Stop-flow analysis of tubular transport of uric acid in rabbits. Am. J. Physiol. 207:1265-1272, 1964.

4. Platts, M.M. and Mudge, G.H.: Accumulation of uric acid by slices of kidney cortex. Am. J. Physiol. 200:387-391, 1961.

TUBULAR HANDLING OF ALLANTOIN IN THE RAT KIDNEY

Kramp R.A., F. Diézi-Chométy, R. Lenoir and F. Roch-Ramel
Department of Medicine, University of Louvain School of Medicine, Louvain, Belgium and Institut de pharmacologie de l'Université de Lausanne, Lausanne, Switzerland

Allantoin is the end product of purine metabolism in lower mammals, which, in contrast, to man and apes possess uricase activity. In I948, Friedman and Byers (I) measured the urinary clearance of allantoin in the dog and the rat. They found that the clearance of allantoin and creatinine were similar and therefore proposed the use of allantoin as a glomerular marker. More recently, Yü and coll. (2) demonstrated reabsorption of allantoin in mongrel and Dalmatian dogs under non-diuretic conditions. The clearance of allantoin, however, approached that of inulin when the rate of urine flow was increased. Conversely, Greger, Lang and Deetjen (3) could not find evidence for tubular transports of allantoin in the rat kidney.

In an attempt to determine the characteristics of the renal excretion of allantoin in the rat, we have used different experimental procedures : the urinary precession method derived from Chinard's technique (4,5,6) , micropuncture and clearance methods.

METHODS

Anesthetized rats were prepared for micropuncture as previously described (7, 8). Two micropuncture techniques were employed in this study : the microinjection technique (7) and free-flow micropuncture (8).

Microinjection experiments. (Louvain) Rats were infused with a 5% solution of mannitol in isotonic saline at a rate yielding urine flows of 40 or 60 µl/min/kidney for tubular or capillary microinjections, respectively. Tubular microinjections of allantoin-^{14}C and inulin-^{3}H were carried out at three sites of the tubule : early or late proximal tubule and a distal convoluted tubule (5). Two rates of injection were chosen either to replace the glomerular ultrafiltrate with the injected solution (0,3 nl/sec.) or to dilute the injectate in the tubular fluid (0,I nl/sec.).

Radioactive solutions were also injected into the larger segments of the peritubular capillaries at a rate avoiding dilution of the injected solution (9). From the beginning of microinjections urine was collected sequentially from each kidney into scintillation vials at I5 and 30 sec. intervals for IO min. Percent recoveries of each isotope were calculated, as previously described (5, 9).

Free-flow micropuncture and clearance experiments . (Lausanne). Following an appropriate priming injection, rats were infused at a rate of 0.05 ml/min. with an isotonic saline solution containing allantoin-^{14}C and inulin, so that 0,6 µCi of allantoin-^{14}C was infused each minute. Free-flow micropuncture was performed on proximal and distal tubules. No difference in overall function was found between the punctured and the contralateral kidney.

Precession experiments. (Lausanne and Louvain). Small volumes of a solution composed of allantoin-^{14}C and inulin-^{3}H were deposited on the renal capsule of diuretic rats. Urine was collected serially from each kidney into scintillation vials and counted for ^{14}C and ^{3}H. The urinary appearance and recovery of the two isotopes were compared for the treated and the contralateral kidney.

Radioactive allantoin was prepared by mixing a solution of urate-2-^{14}C with an appropriate buffer and by adding uricase. Following an equilibration period, the activity of the enzyme was controlled by chemical and chromatographic methods. No degradation of the formed allantoin-^{14}C occurred during these incubations, or within the experimented animals.

RESULTS AND COMMENTS

In 14 precession experiments, significantly more allantoin-^{14}C than inulin-^{3}H was excreted in the first urine samples whereas the reverse occurred in later samples. This difference was more pronounced in the experimental than in the contralateral kidney and suggested the occurrence of an influx of allantoin into the tubular fluid of the rat kidney.

The mean recovery (± 1 S.D.) of allantoin-^{14}C injected into early proximal, late proximal and distal tubules was 85 ± 3% (n = 7), 90 ± 2% (n = 9) and 96 ± 1% (n = 10) respectively when the rate of injection was 0.3 nl/sec. Recoveries of allantoin-^{14}C did not significantly change when the rate of injection was reduced to 0.1 nl/sec. or less. These results indicate the presence of a small efflux of allantoin-^{14}C from tubular fluid. Although allantoin-^{14}C did not appear in the urine before inulin-^{3}H when the two isotopes were injected simultaneously into the peritubular capillaries, the percentage of allantoin-^{14}C recovered in the urine of the injected kidney was higher than in the urine of the contralateral kidney in 13 of 14 injections. Furthermore, the load of allantoin secreted increased proportionally to the load injected. These findings suggest an influx of allantoin-^{14}C into the lumen of the nephron.

In free-flow micropuncture, the F/P allantoin-^{14}C : F/P inulin ratio was slightly above units in 36/53 collections of proximal fluid and below units in 17/53 samples. The mean ratio (± 1 S.E.) averaged 1.12 ± 0.02 in 53 proximal samples, suggesting a secretory movement of allantoin through walls of proximal tubules. The F/P allantoin-^{14}C : F/P inulin ratio in distal convolutions was 1.0 ± 0.03 (n = 14) and did not appear to increase with increasing F/P inulin ratios. Allantoin, thus appeared to be reabsorbed at sites between the end of the proximal convoluted tubules. On the other hand, no transtubular net movement of allantoin could be demonstrated along the distal tubule. The overall fractional excretion of allantoin averaged 0.89 ± 0.04 in 10 rats demonstrating net reabsorption of allantoin under these experimental conditions. This reabsorptive process is probably located in the lowest portions of the nephron non accessible to micropuncture.

The reabsorptive movements of allantoin may be presumed to be diffusional, since in clearance experiments the fractional excretion of allantoin was positively correlated with urine flow. On the other hand, the fractional excretion of allantoin was not influenced by probenecid, para-amino-hippurate, creatinine, N-methyl nicotinamide, salicylate or allantoin. This finding may indicate a lack of sensitivity of active transport mechanisms to these agents, or, alternatively a simultaneous inhibition of secretory and of reabsorptive transports.

Our results differ from those reported by Greger, Lang and Deetjen (3). These differences are not clear but may be presumed to be methodological.

ACKNOWLEDGEMENTS

This study was supported by a Grant-in-Aid (Nr 3.3730. 74) from the Fonds National Suisse de la Recherche Scientifique and from the Fonds National de la Recherche Scientifique of Belgium. Dr. Kramp was Chargé de Recherche du F.N.S.R.

REFERENCES

I. Friedman M. and Byers J.O. - Clearance of allantoin in the rat and dog as a measure of glomerular filtrate rates. Amer. J. Physiol. I5I : I92, I948.

2. Yü T.S., Gutman A.B., Berger L. and Kaung C. - Low uricase activity in the Dalmatian dog simulated in mongrels given oxonic acid. Amer. J. Physiol. 220 : 973, I97I.

3. Greger R., Lang F. and Deetjen P. - Handling of allantoin by the rat kidney. Clearance and micropuncture data. Pflügers Arch. 357 : 20I, I975.

4. Chinard F.P. and Enns T. - Relative renal excretion patterns of sodium ion chloride,ion urea, water and glomerular substances. Amer. J. Physiol. I82:247,I955

5. Kramp R.A., Lassiter W.E. and Gottschalk C.W. - Urate-2-^{I4}C transport in the rat nephron. J. Clin. Invest. 50 : 35, I97I.

6. Roch-Ramel F. and Weiner I.M. - Inhibition of urate excretion by pyrazinoate , a micropuncture study. Amer. J. Physiol. 229 : I604, I975.

7. Kramp R.A. and Lenoir R. - Distal permeability to urate and effects of benzofuran derivatives in the rat kidney. Amer. J. Physiol. 228 : 875, I975.

8. Roch-Ramel F., F. Chométy and G. Peters - Urea concentrations in tubular fluid and in renal tissue of non-diuretic rats. Amer. J. Physiol. 2I5 : 429, I968.

9. Kramp R.A. and Lenoir R. - Characteristics of urate influx in the rat nephron. Amer. J. Physiol. 229 : I654, I975.

CONTROL OF NET CELLULAR UPTAKE OF ADENOSINE FROM THE LUMEN SIDE OF THE GUINEA PIG JEJUNAL EPITHELIUM

N.Kolassa, R.Stengg and K.Turnheim

Department of Pharmacology

University of Vienna, 1090 Vienna, Austria

The mechanism of intestinal purine absorption is not well defined. The assumption that nucleotides are not able to permeate cell membranes renders nucleosides and nucleobases as the only chemical forms for the absorption of the purine moiety. While some although contraversial reports exist on the intestinal permeation of purine bases (Schanker et al., 1963; Oh et al., 1967; Berlin and Hawkins, 1968; Khan et al., 1975), no data are available concerning the absorptive mechanism of purine nucleosides. Therefore we investigated this problem focusing the main interest on the permeation process of adenosine.

The study was carried out on isolated jejunal epithelia of guinea pigs. The epithelium was mounted between two fenestrated foils as a separating membrane between two incubation chambers (Figure 1). The preparation consists mainly of the epithelial layer and some adhering connective tissue (Lauterbach, 1975). (8-^{14}C)-labeled adenosine was administered together with ^{3}H-inulin as an extracellular marker in a modified Krebs-Henseleit solution (mM: 98.3 NaCl, 7.0 KCl, 3.0 $CaCl_2$, 1.0 $MgSO_4$, 14.0 glucose, 14.0 mannitol, 24.2 $NaHCO_3$, 0.9 sodium-phosphate buffer pH 7.4) at either the lumen or the blood side of the epithelium. To minimize the effect of extracellular adenosine degradation non-recirculatory perfusion was performed. By this procedure more than 95% of the adenosine administered were recovered unchanged in the perfusate. At the end of the experiments the radioactivity in a perchloric acid tissue extract was measured by liquid scintillation counting.

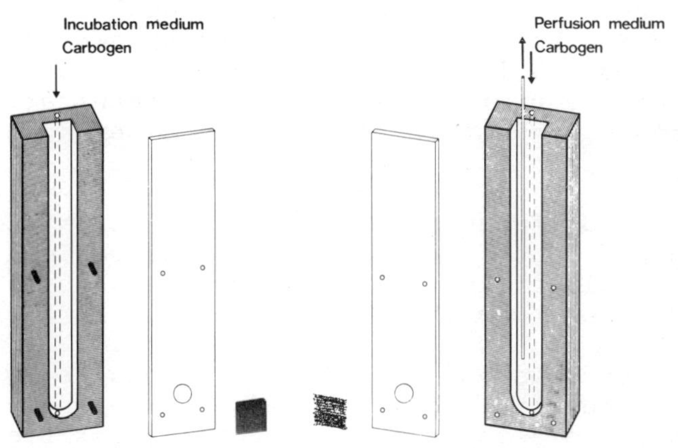

FIGURE 1: Incubation system for the isolated intestinal epithelium (Lauterbach, 1975). The jejunal epithelium of guinea pigs was mounted on thin nylon mesh-work between two fenestrated (diameter 5 mm) polyvinylchloride foils. The epithelium and the foils together served as a separating membrane between two Plexiglas^{R} chambers, which were firmly locked together with a clamp and submerged in a water bath of 37° C. Both chambers were filled with 0.2 ml incubation medium; stirring was achieved by continuous gassing with carbogen (O_2:CO_2 = 95:5 %, v/v). On the side of substrate administration the medium was infused into the chamber and sucked off above the epithelium (perfusion rate: 1 ml.min^{-1}).

Figure 2 shows the cellular ^{14}C-concentration in dependence on the time of incubation. The concentration of ^{14}C-adenosine in the perfusion medium was 10^{-3} M. During the first two minutes the rise of cellular ^{14}C-concentration was almost linear with time indicating constant cellular ^{14}C-uptake. Thereafter the net uptake from the lumen side of the epithelium decreased rapidly and the cellular ^{14}C-concentrations at 3 and 5 min were not different anymore. In contrast, with substrate administration at the blood side the cellular ^{14}C-concentration increased throughout the observation period.

Since net cellular uptake is the result of influx and efflux, the decrease in net uptake with time is most likely due to an increasing ^{14}C-efflux. This is expected to occur, when the endogenous purine pool is progressively labeled in the course of cellular uptake and metabolism of ^{14}C-adenosine. Therefore, considering the different time dependence of the cellular ^{14}C-concentration, a different metabolic pattern of ^{14}C-adenosine depending on the side of administration may be expected.

Another difference between adenosine uptake from the lumen and blood side of the epithelium is apparent in the initial uptake velocity, which was higher from the blood side than from the lumen side (Figure 2).

The dependence of the initial adenosine uptake velocity on the substrate concentration is shown in Figure 3. In the upper left the uptake at small substrate concentrations up to 10^{-5} M adenosine is depicted. This corresponds to the extreme left area of the lower coordinate system. With low substrate concentrations the adenosine uptake from the lumen side proceeded at a higher rate than that from the blood side of the epithelium. But with

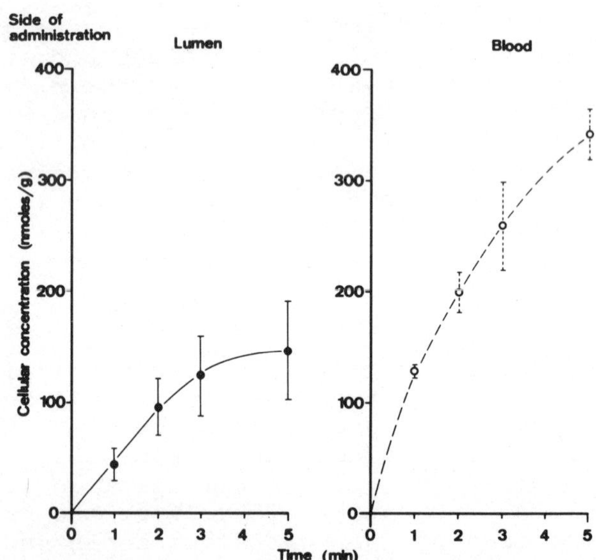

FIGURE 2: Cellular ^{14}C-concentration in dependence on the time of incubation after administration of 10^{-3} M ^{14}C-adenosine at either the lumen or the blood side of the epithelium. ^{14}C-activity in the tissue was corrected for the inulin space and expressed in nmoles.g^{-1} wet weight. Means $\pm$ SEM, n = 8.

increasing concentrations this ratio was reversed: a higher uptake rate was found from the blood side compared to that from the lumen side. From the uptake rates under the influence of p-nitrobenzylthioguanosine (generously provided by Dr.A.R.P.Paterson, University of Alberta, Canada), which has been shown to block the mediated transfer of adenosine (Paterson and Oliver, 1971; Olsson et al., 1972; Agarwal and Parks, 1975;), it may be inferred that both the apparent passive permeability and the apparent maximum velocity of adenosine uptake are lower at the luminal than at the basolateral membrane of the intestinal epithelium.

Following the concept that the metabolic fate of ^{14}C-adenosine after luminal and basolateral uptake may be different we studied the distribution of ^{14}C among the individual adenosine metabolites. After a 2 min perfusion with 5×10^{-6} M or 10^{-3} M ^{14}C-adenosine the epithelium was immersed in liquid nitrogen, the tissue radioactivity

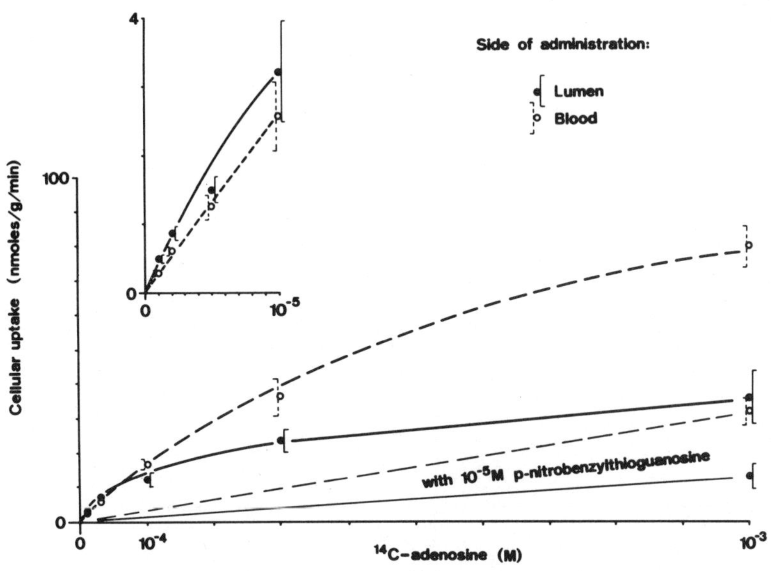

FIGURE 3: Concentration dependence of adenosine uptake velocity. After a 2 min perfusion with ^{14}C-adenosine at either the lumen or the blood side the ^{14}C-activity in the epithelium corrected for the inulin space was taken as an estimate of the initial uptake velocity. Control values (thick lines); values with 10^{-5} M p-nitrobenzylthioguanosine (thin lines). Means ± SEM, n = 5 - 10.

TABLE 1: Concentrations of total ^{14}C and ^{14}C-nucleotides in epithelium ($nmoles.g^{-1}$) after a 2 min perfusion with ^{14}C-adenosine at either the lumen or the blood side. Separation of ^{14}C-nucleotides in tissue extracts was performed by means of thin-layer chromatography. Means ± SEM, n=5-8

side of administration	total ^{14}C	^{14}C-AMP	^{14}C-ADP	^{14}C-ATP	^{14}C-IMP
^{14}C-adenosine concentration: 5×10^{-6}M					
blood	2.09 ± 0.37	0.17 ± 0.02	0.38 ± 0.06	0.79 ± 0.14	0.10 ± 0.01
lumen	2.17 ± 0.27	0.19 ± 0.04	0.26 ± 0.05	0.41 ± 0.08	0.77 ± 0.10
^{14}C-adenosine concentration: 10^{-3} M					
blood	118.8 ± 5.2	4.9 ± 1.2	7.5 ± 0.9	15.0 ± 1.3	5.5 ± 0.6
lumen	67.9 ± 21.6	5.4 ± 1.2	4.7 ± 1.7	7.7 ± 2.9	8.4 ± 2.0

extracted in ice-cold perchloric acid and separated by means of thin-layer chromatography (Kolassa et al.,1972).

In Table 1 the amount of total cellular ^{14}C-activity is compared with that of the labeled nucleotides. A significant difference with respect to the side of substrate administration was found in the distribution pattern of ^{14}C-label among adenine and hypoxanthine nucleotides: 80-90 % of the labeled nucleotides were identified as ^{14}C-adenine nucleotides, when ^{14}C-adenosine was taken up from the blood side, whereas this portion amounted to only about 50 % with luminal substrate administration. Of the remaining labeled nucleotides ^{14}C-IMP constituted the main fraction, while ^{14}C-guanine nucleotides were not detected in measurable amounts.

The formation of ^{14}C-IMP up to values of 0.5 nmoles. $g^{-1}.min^{-1}$ after the administration of 5×10^{-6} M ^{14}C-adenosine in dependence on total ^{14}C-nucleotide formation

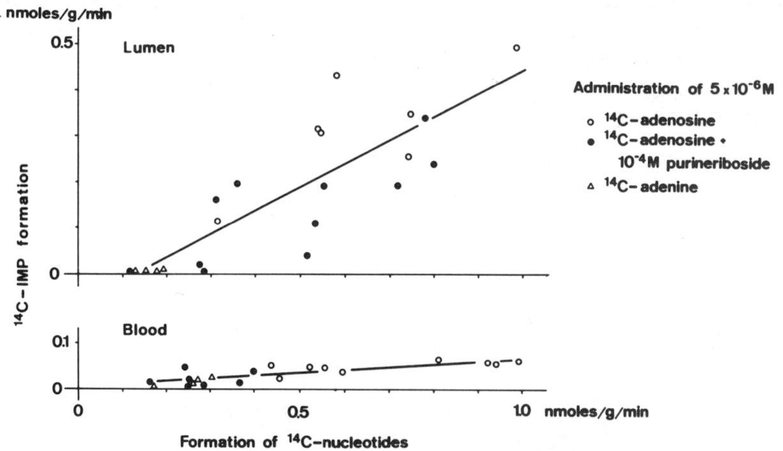

FIGURE 4: Formation of ^{14}C-IMP in dependence on total ^{14}C-nucleotide formation. After a 2 min perfusion with 5×10^{-6} M ^{14}C-adenosine without and with the addition of 10^{-4} M purineriboside and with 5×10^{-6} M ^{14}C-adenine, respectively, at either the lumen or the blood side the ^{14}C-activity in tissue extracts was separated by means of thin-layer chromatography. Each symbol represents the result of a single experiment. Linear regression lines were calculated with respect to nucleotide formation after luminal ($y = -0.07 + 0.51\ x$; $r = 0.84$) and basolateral adenosine uptake ($y = 0.003 + 0.06\ x$; $r = 0.81$).

is shown in Figure 4. In addition, results with the administration of 5 x 10^{-6} M ^{14}C-adenosine together with purineriboside, which inhibits adenosine deaminase (Cory and Suhadolnik, 1965; Baer et al., 1968), and with 5 x 10^{-6} M ^{14}C-adenine are included. A positive linear regression may be calculated between IMP and total nucleotide formation during luminal uptake, when the formation of total nucleotides exceeded 0.1 nmoles.g^{-1}.min^{-1}. At the blood side only a small part of the labeled nucleotides was identified as ^{14}C-IMP.

Beside the substrate, ^{14}C-adenosine, and ^{14}C-nucleotides also labeled inosine, hypoxanthine and xanthine were identified in the tissue extracts. Since the amount of ^{14}C-adenosine never exceeded the inulin space, it may be assumed that the substrate is extensively metabolized in the cell by phosphorylation yielding nucleotides and by deamination yielding oxypurine nucleosides and bases, which may leave the cells, thus decreasing net cellular ^{14}C-uptake with time (Figure 5). To account for a higher efflux of ^{14}C-activity at the lumen side in comparison to that at the blood side we would like to propose the following hypothesis: a relatively high activity of AMP-deaminase compared to a relatively slow AMP-phosphorylation leads to the marked formation of IMP in a luminal compartment of the cell. IMP may be catabolized to oxypurine

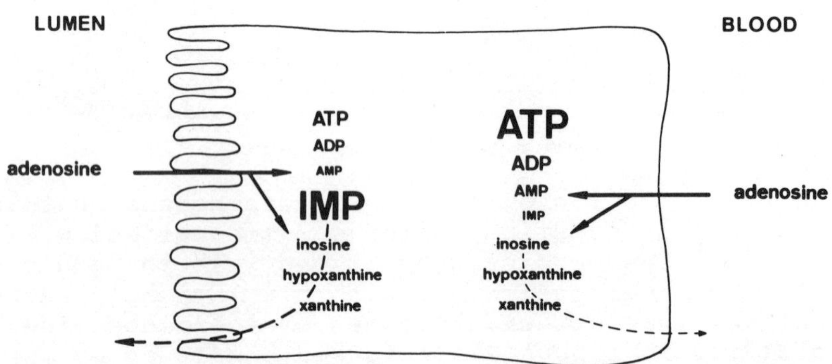

FIGURE 5: Schematic illustration of the incorporation of ^{14}C-label after administration of ^{14}C-adenosine at either the lumen or the blood side of the epithelial cell.

bases and nucleosides thus increasing ^{14}-efflux and limiting net cellular ^{14}C-uptake from ^{14}C-adenosine.

REFERENCES

Agarwal,R.P. and Parks,Jr.,R.E. (1975): A possible association between the nucleoside transport system of human erythrocytes and adenosine deaminase. Biochem. Pharmacol. 24, 547.

Baer,H.P., Drummond,I. and Gillis,J. (1968): Studies on the specificity and mechanism of action of adenosine deaminase. Arch.Biochem.Biophys. 123, 172.

Berlin,R.D. and Hawkins,R.A. (1968): Secretion of purines by the small intestine: general characteristics. Am.J. Physiol. 215, 932.

Berlin,R.D. and Hawkins,R.A. (1968): Secretion of purines by the small intestine: transport mechanism. Am.J. Physiol. 215, 942.

Cory,J.G. and Suhadolnik,R.J. (1965): Structural requirements of nucleosides for binding by adenosine deaminase. Biochemistry 4, 1729.

Khan,A.H., Wilson,S. and Crawhall,J.C. (1975): The influx of uric acid and other purines into everted jejunal sacs of the rat and hamster.Can.J.Physiol.Pharmacol. 53, 113.

Kolassa,N., Roos,H. and Pfleger,K. (1972): A separation of purine derivatives by thin-layer chromatography on silica gel plates suitable for metabolic studies. J.Chromatogr. 66, 175.

Lauterbach,F. (1975): Resorption und Sekretion von Arzneistoffen durch die Mukosaepithelien des Gastrointestinaltraktes. Arzneim.-Forsch. (Drug Res.) 25, 479.

Oh,J.B., Dossetor,J.H. and Beck,I.T. (1967): Kinetics of uric acid transport and its production in rat small intestine. Can.J.Physiol.Pharmacol. 45, 121.

Olsson,R.A., Snow,J.A., Gentry,M.K. and Frick,G.P. (1972): Adenosine uptake by canine heart. Circ.Res. 31, 767.

Paterson,A.R.P. and Oliver,J.M. (1971): Nucleoside transport. II. Inhibition by p-nitrobenzylthioguanosine and related compounds. Can.J.Biochem. 49, 271.

Schanker,L.S., Jeffrey,J.J. and Tocco,D.J. (1963): Interaction of purines with the pyrimidine transport process of the small intestine. Biochem.Pharmacol. 12, 1047

The investigation was supported by the "Fonds zur Förderung der wissenschaftlichen Forschung in Österreich.

UPTAKE OF HYPOXANTHINE IN HUMAN ERYTHROCYTES

M. M. Müller and G. Falkner

1st Department of Medical Chemistry

University of Vienna, Austria

INTRODUCTION

The concentration of oxypurines in the plasma is known to be rather low (10 - 40 μM) (6). Since the formation of purines is accomplished to a great extent by the liver (7, 8), a transport of purines by erythrocytes from liver to tissues with limited or no capacity of de novo purine synthesis has been postulated (1, 3). For this function erythrocytes should possess the ability to perform a fast uptake and release of these purines operating selectively under different conditions.

It was therefore interesting to measure directly the uptake of hypoxanthine by isolated erythrocytes, since so far a characterization of the hypoxanthine uptake into the cell has been attempted only in an indirect manner by equilibration studies (4). We have therefore tried to study the mechanism of hypoxanthine uptake, measuring the kinetics of the transport through the cell membrane by the use of a filtering-centrifugation technique, which allows to resolve the time-dependence of this transport into the erythrocytes.

MATERIALS AND METHODS

Erythrocytes were prepared in the following way: heparinized blood was centrifuged for 10 minutes at

2.000 rpm, plasma and the buffy coat were sucked off, and the erythrocytes were further washed with Krebs-Ringer solution (pH 7.4) for 3 minutes, and finally resuspended in a Krebs-Ringer solution containing 16.7 mM glucose up to a concentration of 4 % .

The filtering-centrifugation (2) was carried out with a Beckman Microfuge centrifuge using 0.4 ml polyethylene tubes. The tubes contained 20 µl 10 % $HClO_4$ at the bottom, above this was a layer of 70 µl silicone oil (AR 75, Wacker-Chemie, München, GFR). On top of this 300 µl erythrocyte suspension were added.

The incubation at 20°C in triplicates were started by addition of either 8-14C-hypoxanthine (52 mCi/mmole; Amersham, UK) or 3H-hypoxanthine (>1 Ci/mmole; Amersham, UK) and terminated by centrifuging the erythrocytes through the silicon oil, which occurred within 1 second. In every experiment incubations with THO (2.5 mCI/ml; Reaktorzentrum Seibersdorf, Austria) and 14C-sucrose (394 mCi/mmole; Amersham, UK) were carried out to determine the intracellular space: The THO in the pellet corresponds with the sum of inter- and intracellular space, whereas the 14C-sucrose in the pellet reflects the intercellular space only. The amount of hypoxanthine in this "sucrose space" had to be substracted from the hypoxanthine in the pellet to give the hypoxanthine taken up by erythrocytes. After separation of the erythrocytes from the incubation medium the radioactivity of the medium and the pellet was determined separately.

RESULTS AND DISCUSSION

The uptake of hypoxanthine at 37°C and 30°C was so fast that the kinetics of the transport through the cell membrane could not be resolved. Therefore all measurements have been performed at 20°C. Figure 1 shows the time dependence of hypoxanthine uptake at 3 different concentrations of hypoxanthine, indicating that initial velocities could be obtained at this temperature with the method used. The data demonstrate that the uptake process comes to an end within 2 minutes at physiological concentrations of hypoxanthine. After 2 minutes no metabolic conversion of hypoxanthine could be demonstrated, indicating that under the experimental conditions the uptake process into the cell has been measured only.

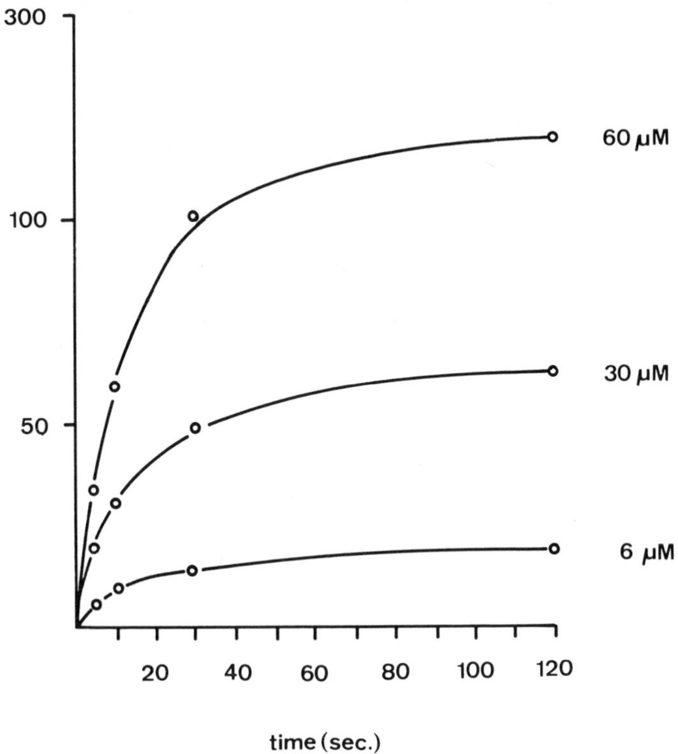

Figure 1. Time dependence of hypoxanthine uptake.

The translocation of hypoxanthine through the membrane of erythrocytes proceeds in both directions: Erythrocytes which were preloaded for 2 minutes with 3H-hypoxanthine immediately released partially this 3H-hypoxanthine back into the incubation medium, when further 14C-hypoxanthine was added (Figure 2). The data show that both forward and backward transport through the cell membrane depend on the relative concentrations of purine bases inside and outside the cell. Subsequently any hypoxanthine within erythrocytes can be removed by washing with hypoxanthine free medium (unpublished).

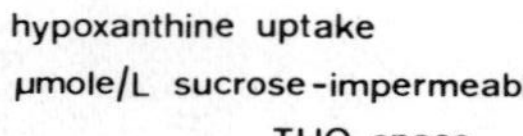

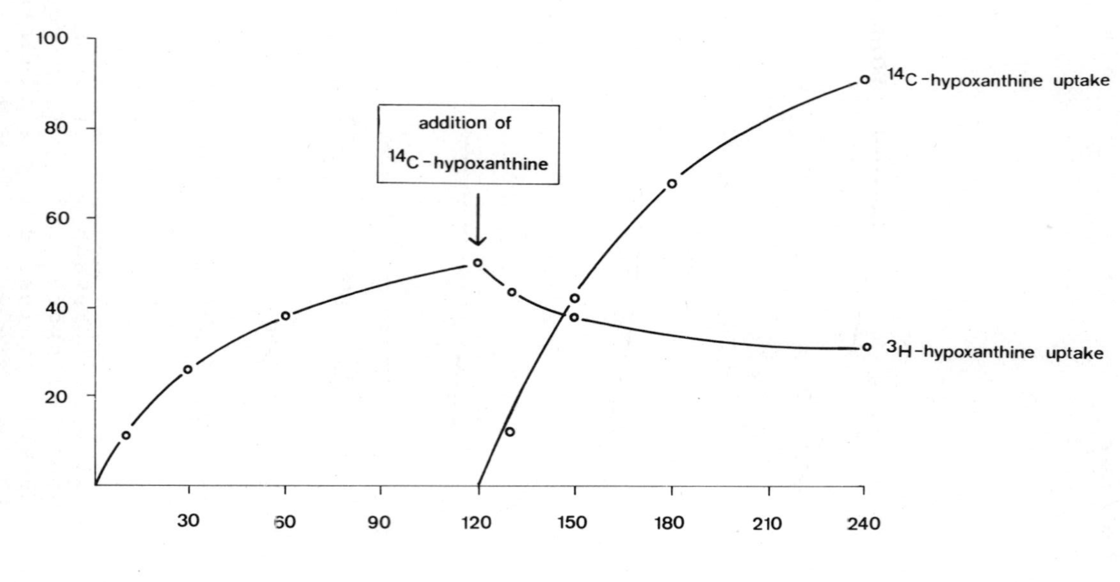

Figure 2. Hypoxanthine uptake and release from erythrocytes. Hypoxanthine concentrations in the incubation medium: 3H-hypoxanthine: 30 µM, 14C-hypoxanthine: 27 µM.

Compound added	Hypoxanthine uptake	Inhibition %
---	5.5	0
Adenine	1.3	76
Guanine	4.1	25
Uric acid	5.3	4
Xanthine	5.4	2

Table 1. Effect of purines on hypoxanthine uptake. Hypoxanthine concentration in the incubation medium: 11 μM. Concentrations of added compounds in the incubation medium: 13 μM. Incubation time: 10 seconds. Hypoxanthine uptake expressed as umoles/l sucrose impermeable THO-space.

After measurement of the uptake velocity it was interesting to see, whether this hypoxanthine uptake can be influenced by other purines. The results listed in table 1 demonstrate that both adenine and guanine inhibit hypoxanthine uptake. In order to study the mode of this inhibition the concentration dependence of the hypoxanthine uptake in the presence and absence of adenine was measured. The double recipical plot of these data (Figure 3) reveal saturation kinetics of hypoxanthine uptake (K_m = 140 uM). Both saturation kinetics and inhibition pattern can be regarded as evidence for the existence of a special transport system for purines in the cell membrane of erythrocytes that has been postulated by LASSEN on basis of his equilibration studies (4). The fact that adenine is inhibiting stronger than guanine the hypoxanthine uptake can be explained by different affinity of these purines to this transport protein. Kinetic studies of adenine and guanine uptake have subsequently shown that the K_m for adenine transport into the cell is lower than that of guanine transport (unpublished data).

This transport of hypoxanthine seems not to be energy dependent although an accumulation of hypoxanthine of 1.5 to 2-fold can be measured in accordance with findings of LASSEN (4). He has tried to explain the

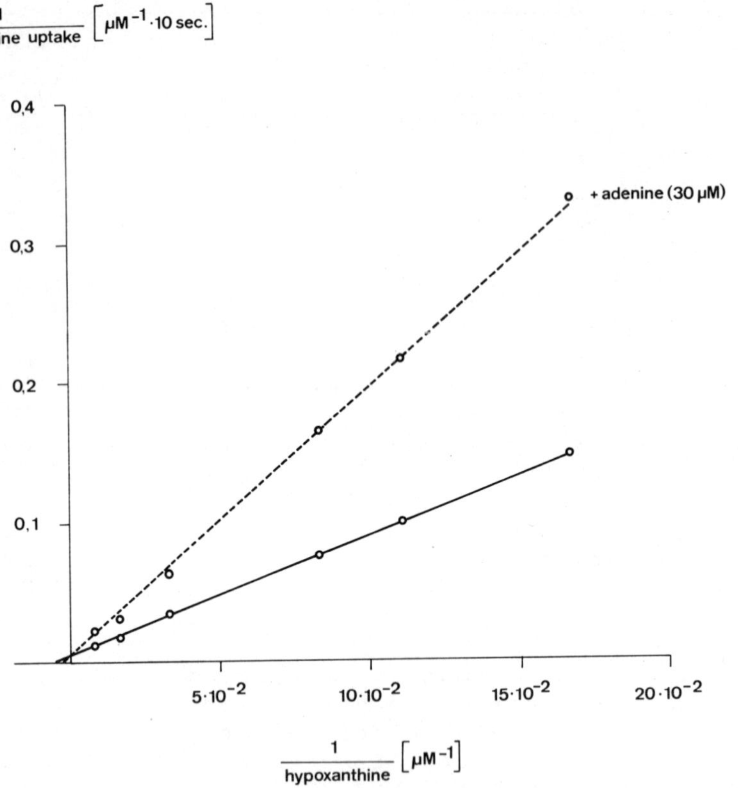

Figure 3. Lineweaver-Burk diagram of hypoxanthine uptake

pH	Endogenous hypoxanthine concentration (µM)
6.0	42
6.5	43
7.0	45
7.5	43
8.0	42

Table 2. Effect of pH on hypoxanthine accumulation. Hypoxanthine concentration in the incubation medium: 30 µM. Incubation time: 2 minutes.

the accumulation of hypoxanthine by the membrane potential. Since the pH in the incubation medium seems not to influence hypoxanthine accumulation in the erythrocytes significantly (Table 2), although the membrane potential is changed by this procedure (5), we would suggest that the accumulation of hypoxanthine in the erythrocytes is rather due to binding within the cell, which has found to be independent of Ca^{++}, Na^{+}, K^{+}, and Mg^{++}.

The data presented show the existence of a fast operating purine translocator in the cell membrane of erythrocytes. The high maximal velocity of this purine uptake system fullfills the requirement of a fast and permanent hypoxanthine supply for purine metabolism in erythrocytes. Transport function of erythrocytes for hypoxanthine itself on the other hand seems to be unlikely, because of the low accumulating capacity and the fact that influx and efflux is regulated by extracellular hypoxanthine concentration only. Studies are in progress to establish to what extent the "purine translocator" - interacting with oxypurines - is involved in uric acid accumulation by erythrocytes. These studies are supposed to bring more information about pathophysiological aspects of uric acid transport in gouty patients.

ACKNOWLEDGEMENT

This study was supported by a grant of "Fonds zur Förderung der wissenschaftlichen Forschung Österreichs" (project No. 3038).

REFERENCES

1. Henderson J. F., Le Page G. A. (1959) J. Biol. Chem. 234, 3219.

2. Klingenberg M., Pfaff E. (1967) In Methods of Enzymology, eds. Colowick S. P., Kaplan N. O., Vol. 10, p. 680, Academic Press.

3. Lajtha L. G., Vane J. R. (1958) Nature 182, 191.

4. Lassen U. V. (1967) Biochim. Biophys. Acta 135, 146.

5. Lassen U. V. (1972)
In "Oxygen Affinity of Hemoglobin and Red Cell Acid Base Status", eds. Astrup P., Rorth M., p. 291, Academic Press, New York.

6. Orsulak P. J., Haab W., Appleton M. D. (1968)
Analyt. Biochem. 23, 156.

7. Pritchard J. B., Chavez-Peon F., Berlin R. D. (1970)
Fed. Proc. Fed. Amer. Soc. Exp. Biol. 29, 799.

8. Sano L., Gamo T., Kakimoto Y., Taniguchi K., Takesada M., Nishinuma K. (1959)
Biochim. Biophys. Acta 32, 586.

INCORPORATION OF PURINE BASES BY INTACT RED BLOOD CELLS

C.H.M.M. de Bruyn and T.L. Oei

Dept.Hum.Genetics, University of Nijmegen,The Netherlands

INTRODUCTION

Mature human erythrocytes do not possess the ability to perform purine synthesis de novo (1-3). These cells meet their requirements for purine nucleotides by re-utilisation of preformed bases that are derived from the diet and those released from other tissues into the blood circulation (1,4).

In the presence of phosphoribosylpyrophosphate (PRPP) the purine bases hypoxanthine and guanine are metabolised in a one step reaction to their corresponding mononucleotides IMP and GMP, respectively. This reaction is catalysed by the enzyme hypoxanthine-guanine phosphoribosyl transferase (HG-PRT;EC 2.4.2.8). Adenine is converted to AMP by a related enzyme: adenine phosphoribosyl transferase (A-PRT; EC 2.4.2.7).

Hypoxanthine, guanine and adenine can rapidly penetrate the human erythrocyte membrane (5,6).

The dependence of the transport rate upon the concentration, as has been shown in the case of hypoxanthine, is suggestive of a two-component mechanism: the first component seems to be a saturable carrier system, the second non-saturable (6). Adenine uptake into human blood platelets appears to be a carrier-mediated process (7). The exact nature of purine transport in the membrane of the human erythrocyte, however, is still obscure.

In bacteria there is evidence for a role of purine phosphoribosyl transferase in the uptake of purine bases (8). A-PRT was demonstrated in cell free extracts and in membranes from E.coli. The membrane bound enzyme appeared to be identical with the soluble enzyme with respect to substrate affinities, pH optimum and inhibition by adenine nucleotides (8). Recently, evidence was obtained for the existence of purine phosphoribosyl transferase activities associated with human erythrocyte membranes apart from the activities in the soluble fraction (9).

In the present study the uptake of radioactive hypoxanthine, guanine and adenine in intact normal erythrocytes was studied in order to analyse the two component mechanism involved in purine transport. Also erythrocytes virtually complete deficient for HG-PRT were used to gain further insight into the role of purine phosphoribosyl transferases in red blood cell purine metabolism. Additionally, eventual dependence of purine uptake by red blood cells on PRPP was studied by incubating intact normal erythrocytes with radioactive purine bases in the presence of exogenous PRPP, followed by analysis of both cell content and incubation medium.

MATERIALS AND METHODS

1.Preparation of erythrocytes.Heparinised blood from normal and HG-PRT deficient individuals was obtained by venipuncture and immediately put into an ice bath for 10 minutes. After centrifugation (800 g; 10 min.), plasma and buffy coat were discarded. The packed erythrocytes were washed three times with 0.9% NaCl. To one volume of packed erythrocytes three volumes of 310 mOsm Na-phosphate buffer (pH 7.4) were added. In this buffer the cells were preïncubated for 30 minutes at 37°C and then recentrifuged.

2.Incubation experiments. In each experiment 1 volume of preincubated packed cells was added to 3 volumes of incubation medium giving an approximately 25% (v/v) cell suspension. 1 ml of incubation medium always contained 0.3 ml of 310 mOsm Na-phosphate buffer pH 7.4; 0.45 ml of 0.145 M NaCl, 0.05 ml of 0.115 M $MgCl_2$-6 H_2O and 0.2 ml of 0.05 M glucose. The cell suspension was put into a waterbath at 37°C and the incubation experiment was started by addition of one of the labeled purine bases, 8-^{14}C -guanine (spec.act.57 mCi/mmole); 8-^{14}C -hypoxanthine (spec.act.59 mCi/mmole); or 8-^{14}C -adenine (spec.act.59 mCi/mmole) in 10 μl of isotonic phosphate buffer, pH 7.4. All radiochemicals were obtained from Radiochemical Centre, Amersham. Samples (100 μl) were taken at 0,5,15,25 and 35 minutes.

3.Analysis of medium and cell content. The samples taken from the incubation mixture were rapidly chilled (0°C) and centrifuged. The supernatant (referred to as "medium") was carefully removed and kept for analysis. The cells were lysed with 75 μl ice cold 5% TCA, centrifuged at 1,000 g for 5 minutes and the supernatant referred to as 'cell content'. Medium and cell content were analysed

in two ways:
(a) 10-20 µl were spotted on Whatmann 3 MM paper strips (40x2 cm) and submitted to high-voltage electrophoresis (45 min.;60 V/cm) in acetic acid-pyridine-water-buffer (9:1:90; pH 3.6). Standard solutions with radioactive purine bases, nucleosides and nucleotides were co-electrophoresed. After localization and identification of radioactive spots with the aid of a radiochromatogram scanner (Actigraph III, Nuclear Chicago), ^{14}C-labeled spots were cut out and placed in 10 ml of scintillation fluid (toluene, containing 5,5 mg 2,5 diphenyloxazole (PPO) and 1 mg 1,4 bis 2 (-5 phenyloxazole-) benzene (POPOP) and 1 ml Triton X-100); radioactivity was quantified in a Packard Tri Carb liquid scintillation counter.
(b) 10-20 µl were spotted on Whatmann 3 MM paper and submitted to descending chromatrography in 5% Na_2HPO_4 (pH 8.45). Here again, standard solutions were co-chromatographed. Localisation, identification and counting were performed in the same way as mentioned above. Both methods gave identical results. Recovery of total radioactivity was 95% or more.

RESULTS

Time curve. A typical time curve for the uptake of hypoxanthine is shown in fig.1: a drop of radioactivity in the medium was accomponied by a rapid formation of IMP in the normal erythrocytes and a very slight initial rise of free intracellular hypoxanthine (fig.1, left).

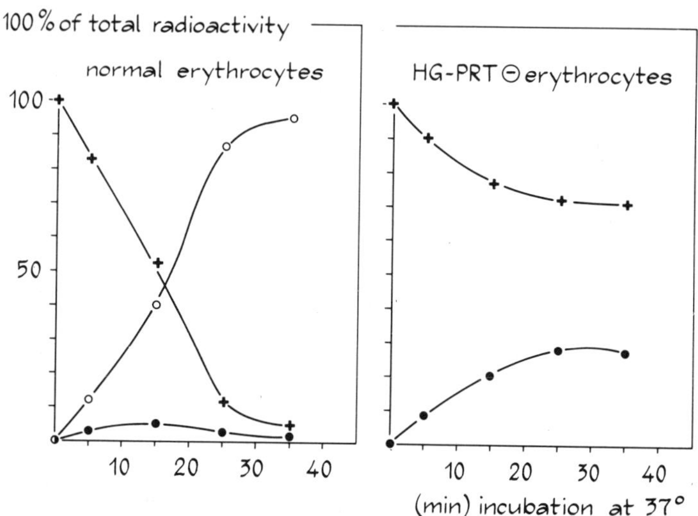

Figure 1.
Distribution of label in incubation medium and cell content as a result of uptake and metabolism of 8-^{14}C-hypoxanthine (10 µM) by intact human erythrocytes. + —— + extracellular hypoxanthine, ● —— ● intracellular hypoxanthine, o —— o intracellular IMP.

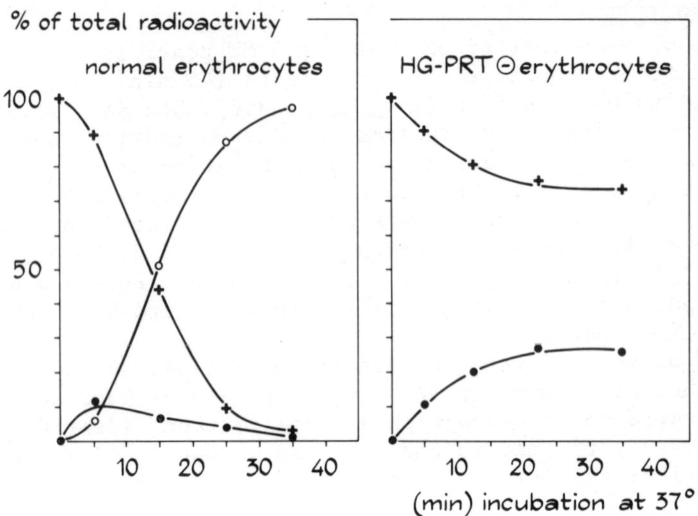

Figure 2.
Distribution of label in incubation medium and cell content as a result of uptake and metabolism of 8-^{14}C-guanine (10 μM) by intact human erythrocytes. + —— + extracellular guanine, • —— • intracellular guanine, o —— o intracellular guanine nucleotides.

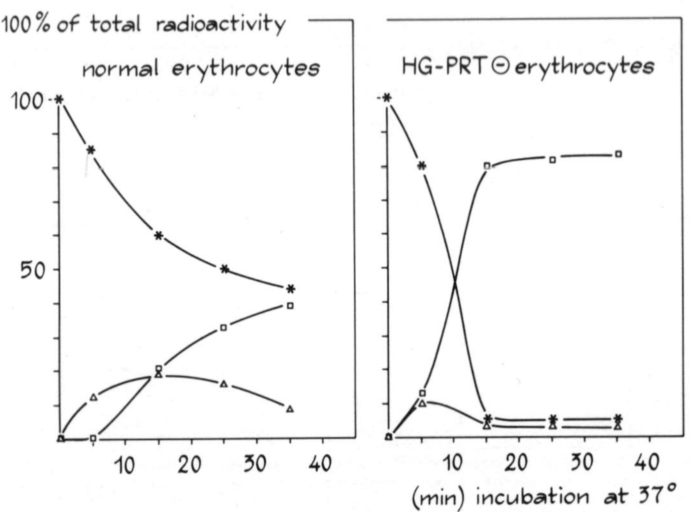

Figure 3.
Distribution of label in incubation medium and cell content as a result of uptake and metabolism of 8-^{14}C-adenine (10 μM) by intact human erythrocytes. ★ —— ★ extracellular adenine, △ —— △ intracellular adenine, □ —— □ intracellular adenine nucleotides.

35 Minutes after beginning of the reaction almost all radioactivity was recovered as intracellular IMP. In the incubations with HG-PRT deficient erythrocytes no compound other than hypoxanthine was identified; some 25% of the total radioactivity was recovered as intracellular hypoxanthine after about 20 minutes (fig.1, right). A similar picture was obtained with guanine (fig.2).

Uptake of adenine was much faster in the case of HG-PRT deficient cells than in normal erythrocytes. In the deficient cells after 15 minutes 80% of the total radioactivity was present in the form of adenine nucleotides, whereas in normal erythrocytes even after 35 minutes only some 40% of the ^{14}C-adenine had become converted to labeled nucleotides (fig.3). Much less radioactivity was recovered in the form of intracellular adenine in HG-PRT deficient cells than in normal erythrocytes (fig.3).

Concentration dependence. Fig.4 (left) shows the concentration dependence of the guanine uptake. At low concentrations, uptake linearly increases with the concentration. All label was recovered as intracellular 8-^{14}C -GMP. At higher concentrations there was also a linear correlation between concentration and uptake, but the slope of this part of the curve was considerable less steep.

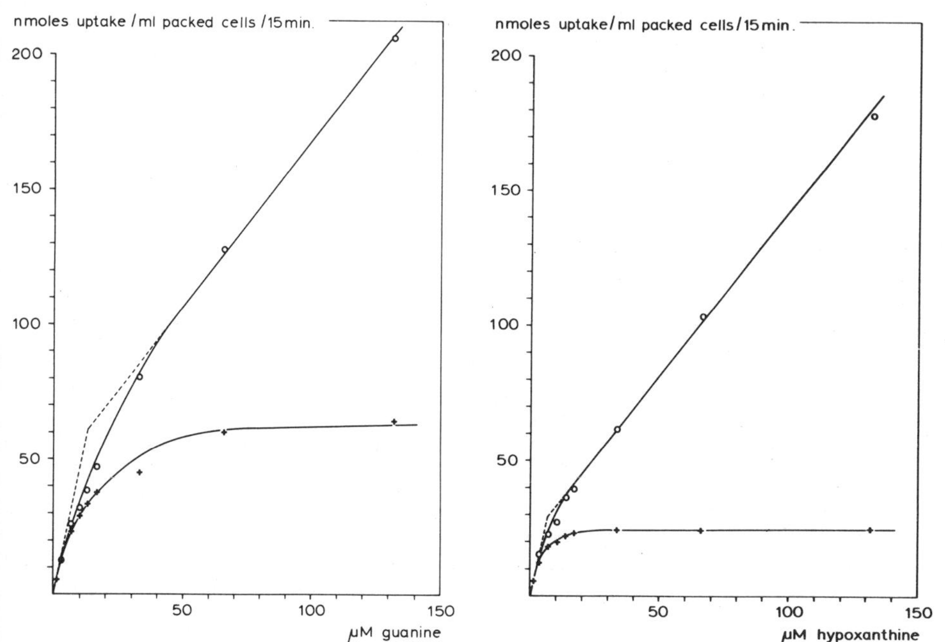

Figure 4.
Concentration dependence of guanine (left) and hypoxanthine (right) uptake by intact normal erythrocytes. o —— o total uptake, + —— + uptake in the form of nucleotide.

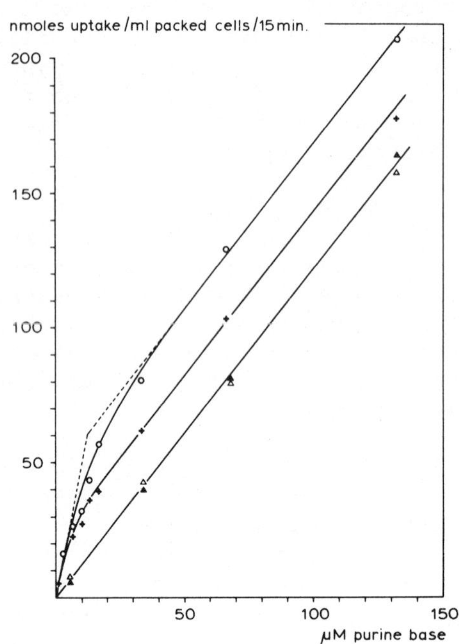

Figure 5.
Comparison between the concentration dependence of hypoxanthine and guanine uptake by intact normal and HG-PRT deficient erythrocytes. o —— o total guanine uptake by normal cells, + —— + total hypoxanthine uptake by normal cells, Δ——Δ total guanine uptake by HG-PRT deficient cells, ▲——▲ total hypoxanthine uptake by HG-PRT deficient cells.

At concentrations where the first part of the curve passed into the second part, the intracellular $8\text{-}^{14}C$-GMP formation reached a maximal level. Increasing the concentration of $8\text{-}^{14}C$-guanine above this level only resulted in increasing amounts of free guanine in the cell. In studies with erythrocytes obtained from individuals with a deficiency for HG-PRT, only labeled guanine was recovered from cells and medium.

$8\text{-}^{14}C$-Hypoxanthine transport showed the same features as guanine transport: at low concentrations all label was recovered as intracellular $8\text{-}^{14}C$-IMP. No IMP formation was observed in HG-PRT deficient erythrocytes (fig.4, right). As compared to GMP formation, IMP formation from hypoxanthine in control cells reached a maximum at lower concentrations (fig.4, right). Under the present conditions the maximum amount of IMP formed was about 12 nmoles/ml packed cells/15 minutes whereas for GMP the corresponding value was about 50 nmoles/ml packed cells/ 15 minutes.

In the same way as with guanine, free 8-^{14}C-hypoxanthine was observed in the cells as soon as 8-^{14}C-IMP formation leveled off.

In HG-PRT deficient red blood cells the relation between guanine or hypoxanthine concentration and purine uptake was represented by a straight line that seemed to be parallel to the second part of the curve obtained with control cells (fig.5). Incubation of normal red blood cells with varying amounts of 8-^{14}C-hypoxanthine and 8-^{14}C-guanine at about 4°C revealed a similar linear relationship between concentration and uptake. No nucleotide formation inside the cells was noticed. With the methods used in these experiments this straight line could not be distinghuised from that obtained with HG-PRT deficient erythrocytes (fig.5).

Adenine uptake resembled that of guanine and hypoxanthine: at low concentration all label was recovered in the form of adenine nucleotides. In comparison with the GMP- and IMP forming systems saturation of the AMP forming system was observed at lower levels in normal red blood cells, but not in HG-PRT deficient erythrocytes.

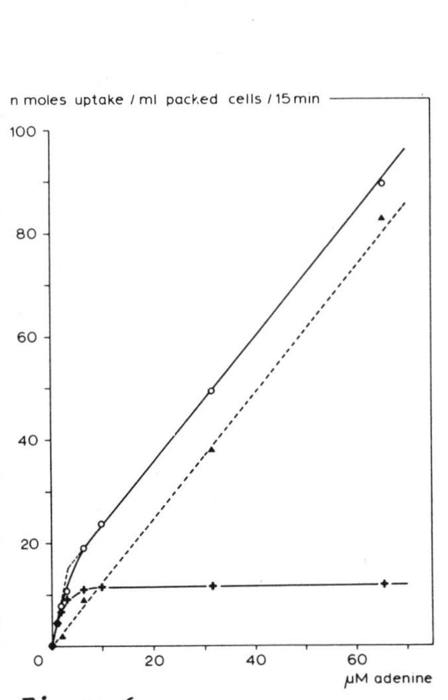

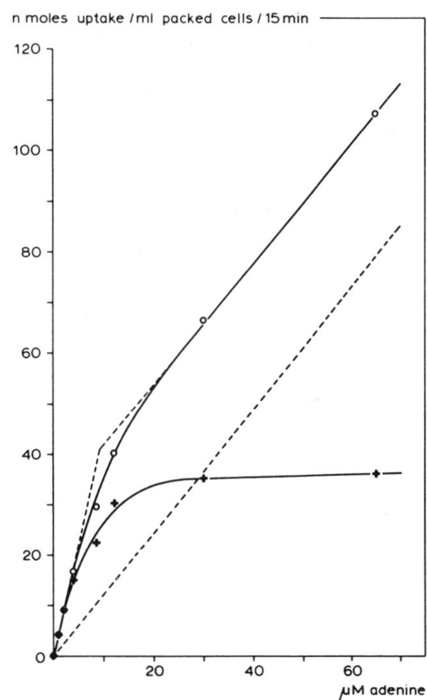

Figure 6.
Concentration dependence of adenine uptake by intact normal (left) and HG-PRT deficient (right) erythrocytes. o —— o total uptake, + —— + uptake in the form of nucleotide, ▲ ----▲ total uptake at 4°C by normal cells.

Fig.6 also shows a two component transport system for adenine. Here again, saturation of the AMP forming system is attended by the appearance of increasing amounts of free 8-^{14}C-adenine in the cells. HG-PRT deficient erythrocytes show a higher capacity with respect to 8-^{14}C-AMP formation: under the conditions used the maximal amount of 8-^{14}C-AMP formed in HG-PRT deficient red blood cells was about 50 nmoles/ml packed cells/15 minutes, in normal red blood cells 12 nmoles/ml packed cells/15 minutes.

Effect of exogenous PRPP. Fig.7 shows the effect of varying the external PRPP concentration on extra- and intracellular nucleotide formation. There is an absolute dependency of extracellular nucleotide formation on the presence of exogenous PRPP. Extracellular GMP and IMP formation is much enhanced by increasing the exogenous PRPP from 0 to 1 mM. This is much less the case with extracellular AMP formation. Maximal extracellular GMP and IMP formation was about 20-22 nmoles per ml packed cells; maximal extracellular AMP formation was about

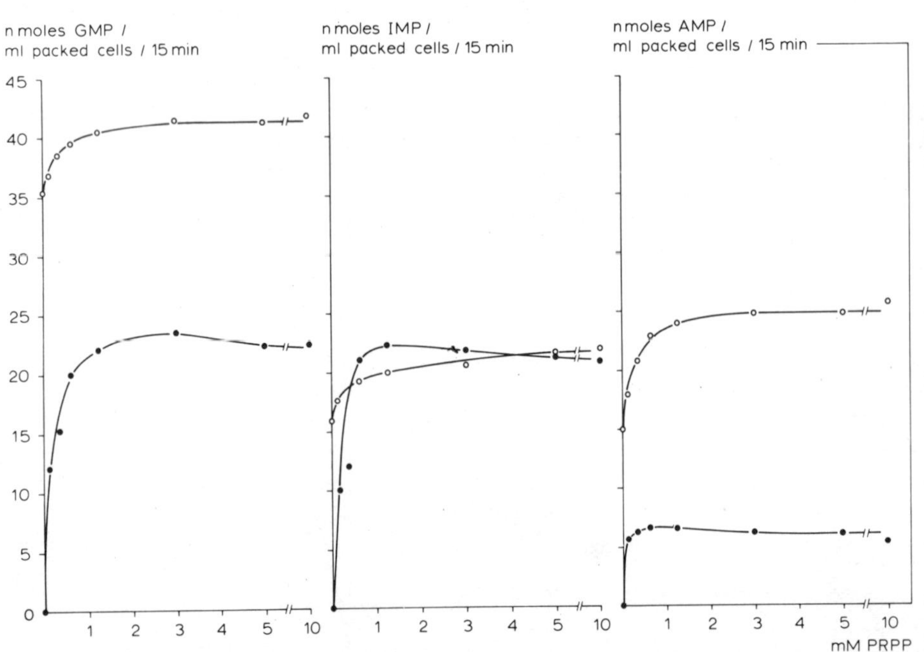

Figure 7.
Effect of varying concentrations of exogenous PRPP on intracellular and extracellular nucleotide formation from extracellular purine bases by intact normal erythrocytes. o —— o intracellular nucleotides, • —— • extracellular mononucleotides.

6 nmoles per ml packed cells (fig.7). Intracellular nucleotide formation is only slightly stimulated by increasing extracellular PRPP levels from 0 to 1 mM. The effect was more pronounced on intracellular AMP formation, than on intracellular GMP and IMP formation.

DISCUSSION

The purine bases guanine and hypoxanthine are taken up rapidly from the medium and converted in a first step to their corresponding 5'-mononucleotides GMP and IMP, respectively, in normal erythrocytes. These reactions are dependent on the activity of HG-PRT, because no radioactive nucleotides were recovered from HG-PRT deficient red blood cells (figs.1 and 2).

Adenine uptake by HG-PRT deficient red blood cells is more rapid than by normal cells (fig.3). In hemolysates from HG-PRT deficient Lesch-Nyhan patients increased A-PRT activities have been measured (10). Also increased PRPP levels in HG-PRT deficient erythrocytes have been reported (11). In experiments with intact rabbit erythrocytes, uptake of purine bases and incorporation into ribonucleotides was markedly stimulated by increasing the organic phosphate content of the incubation medium and this enhancement of purine uptake was attended by parallel increase in the intracellular pool of PRPP (12). Our results suggest, that in intact human erythrocytes A-PRT is involved in the uptake of adenine and that its activity is influenced by endogenous PRPP levels.

In order to study the distribution of the free purine bases inside and outside the red blood cells, after separation of the incubation mixtures into medium and packed cells by centrifugation, the cells were immediately lysed with 5% TCA after carefull removal of the medium. The erythrocytes were not washed after centrifugation, because this procedure removes the free bases from the cell content. After several washings the remaining label in the cell was shown to be exclusively in the form of the corresponding nucleotides. In normal erythrocytes, thoroughly washed after incubation, the relation between concentration and uptake is identical to the curves of the saturable system in figs. 4 and 6. In HG-PRT deficient red blood cells, where no saturable GMP and IMP forming system was observed, all radioactivity disappeared from the cell content after several washings.

The results showed in figs. 4-6 reveal a two component system for the transport of the hypoxanthine and guanine and a separate one for adenine. The overall type of concentration dependence of both two component systems resembles a "diffusion-type system" superimposed on a saturable system. The results also show that the saturable system is in fact a nucleotide forming system (figs.4 and 6). Nucleotide formation from free purine bases in human erythrocytes is known to be HG-PRT or A-PRT dependent. The finding that the concentration dependence of hypoxanthine and guanine transport in HG-PRT deficient erythrocytes does not show a saturable component, is suggestive that HG-PRT is part of the saturable transport system. In addition, incubation at 4°C resulted in a similar concentration dependence as in the case of HG-PRT deficient red blood cells no saturable, nucleotide forming system, but only a straight line, indicating a diffusion-like transport.

It has been shown that part of the red blood cell HG-PRT and A-PRT activity is associated with the erythrocyte membrane (13). The extracellular nucleotide formation following addition of PRPP to the purine base-containing incubation medium demonstrates that in the intact erythrocyte, the membrane associated purine phosphoribosyl transferases are accessible to PRPP from the outside.These membrane-associated purine phosphoribosyl transferases might be involved in the saturable component of the transport system. Our data indicate that at low concentrations (below 10 μM), purine base uptake depends mainly on the saturable, nucleotide forming system, whereas at higher concentrations diffusion prevails. In human erythrocytes physiological concentrations of adenine have thus far escaped detection; hypoxanthine is present in concentrations ranging from 3 μM to 20 μM (14). It seems, therefore, that in vivo the saturable, nucleotide forming system plays a major role with respect to transport of purines. Comparable data have come from studies on rabbit polymorphonuclear leucocytes: when added at low concentrations almost all the adenine that entered the cells was converted to adenosine nucleotides, the limiting step to adenine incorporation occurring at the membrane (15).

In HG-PRT$^-$ red blood cells, elevated concentrations of PRPP have been found together with increased A-PRT activities (11). The results in figs. 3 and 6 show that 8-^{14}C-adenine uptake by HG-PRT deficient cells is more extensive than in normal cells. Apparently this is due to a higher capacity of the saturable, AMP forming component of the transport system. These findings provide further evidence for the role of A-PRT in the uptake of adenine and for the crucial role of PRPP: the saturating levels of the nucleotide forming components of the purine transport systems are probably highly dependent on endogenous PRPP concentrations in the erythrocytes.

The coupling of a first step in the metabolism of an exogenous compound with its transport provides a mechanism in which a single input of energy can drive both processes. Such a mechanism is involved in the uptake of sugars by bacterial cells, since a phospho-enolpyruvate-dependent phosphotransferase system catalyses the vectorial phosphorylation of glucose and other sugars (16,17). Studies on purine utilisation in bacteria suggest that purine phosphoribosyl transferases play a comparable role in purine uptake.

It has been shown that purine uptake in Bacillus subtilis is proportional to phosphoribosyl transferase activity for that purine (18) and studies on E.coli have shown that A-PRT is directly involved in the transport of adenine across the membrane (8).

The present results suggest a similar role of purine phosphoribosyl transferases in eukaryotic organisms. The finding that the saturable parts of purine transport are nucleotide forming systems, the fact that these systems are influenced by endogenous and exogenous PRPP levels and the observation that the saturable part of the transport system for hypoxanthine and guanine is lacking in HG-PRT deficient erythrocytes, make a strong case for the involvement of the phosphoribosyl transferase system in the translocation of purines across the human red blood cell membrane.

ACKNOWLEDGEMENTS

This study was supported by a grant from FUNGO, Foundation for Medical Scientific Research in the Netherlands, and the Medical Prevention Fund. The authors thank Mr.Cor van Bennekom and Mr.Antoon Janssen for skilfull technical assistance and Dr,R.Geerdink (Unit of Clinical Genetics, University of Utrecht) and Dr.M.Verstegen (De Winckelsteegh, Nijmegen) for supplying blood samples from HG-PRT deficient patients.

REFERENCES

1. Lowy B.A., Ramot B. and London I.M. (1960). J.Biol.Chem. 235, 2920-2924.

2. Lowy B.A., Williams M.K. and London I.M. (1962). J.Biol.Chem. 237, 1622-1626.

3. Fontenelle L.J. and Henderson J.F. (1969). Biochim. Biophys.Acta 177, 175-177.

4. Henderson J.F. and Le Page G.A. (1959). J.Biol.Chem. 234, 3219-3224.

5. Whittam R. (1960). J.Physiol.London 154, 614-623.

6. Lassen U.V. (1967). Biochim.Biophys.Acta 135, 146-154.

7. Sixma J.J., Holmsen H. and Trieschnigg A.C.M. (1973), Biochim. Biophys. Acta 298, 460-468.

8. Hochstadt-Ozer J. and Stadtman E.R. (1971). J.Biol.Chem. 246, 5304-5320.

9. de Bruyn C.H.M.M. and Oei T.L. (1974). In: Purine Metabolism in Man (Sperling O., de Vries A. and Wijngaarden J.B. Eds.) pp.223-227. Plenum Press, New York.

10. Seegmiller J.E., Rosenbloom F.M. and Kelley W.N. (1967). Science 155, 1682-1684.

11. Greene M.L., Boyle J.A. and Seegmiller J.E. (1970). Science 167, 887-889.

12. Hershko A., Razin A., Shoshani T. and Mager J. (1967). Biochim.Biophys. Acta 149, 59-73.

13. de Bruyn C.H.M.M. and Oei T.L. (1976). These Proceedings.

14. Henderson J.F. and Le Page (1959). Cancer Res. 19, 67-71.

15. Hawkins R.A. and Berlin R.D. (1969). Biochim. Biophys. Acta 173, 324-337.

16. Kundig W., Kundig F., Anderson B. and Roseman S. (1966). J.Biol.Chem. 241, 3243-3246.

17. Kaback H.R. (1968). J.Biol.Chem. 243, 3711-3724.

18. Berlin R.D. and Stadtman R.D. (1966). J.Biol.Chem 241, 2679-2686.

URIC ACID TRANSPORT CHARACTERISTICS IN HUMAN ERYTHROCYTES

B. LUCAS - HERON and C. FONTENAILLE

U.E.R. MEDICINE NANTES LABO. NEPHROLOGY

1, RUE GASTON WEIL 44035 NANTES FRANCE

We have studied the distribution at equilibrium and the Uric Acid (U.A) transport characteristics by the red blood cell membrane. We took in account the last data dealing with the general feature of various mineral anions transport through the human erythrocyte membrane.

MATERIAL AND METHOD

We used in these experiments heparinized blood from normal healthy adults. The sample were drawn immediatly before the experiment and centrifuged (3 000 g). Plasma and leucocytes were poured off. The pelet was washed three times in saline (pH 7.4) and the erythrocytes were kept in appropriate solution (see below) in order to obtain a final hematocrit of about 50 % . The cells were then stirred (roller tube) intermittently at 37 ° C for two hours to complete the equilibrium repartition of U.A between erythrocytes and supernatant. The U.A levels in the supernatant are those observed after equilibrium. After shaking, half a milliliter of suspension is centrifuged for 15 seconds at 5 000 g and the supernatant radioactivity is quantified. Once the time zero sample is done, serial samples are studied each five minutes for one hour, then each 20 minutes for two hours. The supernatant radioactivity decay is thus measured as a time function.

Medium

Several media are used in order to study various parameters. Osmolarity was at 300 mOsm and, exepted special mentions the U.A concentration was 0.3 mM/L. The U.A was solubilized using lithium carbonate which has been checked for lacking of any specific effect. The same concentration of lithium carbonate was used all along experiments dealing with the effect of U.A concentration upon his transport.

Control

A sample of R.B.C. drawn for the same donnor was used as a control for each experiment. The meduim was here a Ringer solution (pH 7.4).

pH

The solution is the same than in the control excepted for the pH which is ajusted at various levels ranging from 6.5 to 7.8. The pH was checked all over the experiments and only minor variations of 0.05 units were allowed to occur.

Ionic composition

In some experiments Nacl was partially replaced by non diffusible anion salt (Na citrate 68 mM/L) or saccharose (72g/L).

Ions transport inhibitors

The Na salicylate phenylbutazone, dinitrofluorobenzene and Na benzoate are known for the inhibitory effect upon various anions transport. These agents were thus added to the incubation medium in several experiments at the following concentrations : DNFB : 4mM/L, phenylbutazone : 10 mM/L, benzoate : 20 mM/L and Na salicylate : 30 mM/L. An equivalent amount of NaCl was substracted when the two last coumpounds were used.

U.A concentrations

U.A has been added to the incubation medium in order to obtain a final concentration equilibrium ranging from 0.3 to 8.4 mM/L in the supernatant.

Metabolism

Glycolysis inhibition is achieved either by suppression of glucose from the solutions and incubation of the cells at 37 ° C for one hour or by adding Na fluoride (10 mM/L).

Glycolysis metabolites

Without glucoe the effect of P.E.P. (30mM/L), pyruvate (36mM/L), lactate (38mM/L) , A.T.P. (6 mM/L) and N.A.D. (2mM/L) were successively tested.

Glycolysis inhibition and anions transport inhibitors

The RBC samples incubated with the following medium :1) Ringer without glucose ; 2) Ringer with salicylate (30 mM/L) ; 3) Ringer with salicylate without any glucose were compared with control sample.

NaF and U.A concentrations

The RBC samples were added to supernatant containing various amount of U.A (0.3 to 8.4 mM/L) and the transport kinetics was studied with or without NaF (10 mM/L).

Calculation

- partition coefficient U_H / U_S

Knowing the supernatant radioactivity a time zero (C_o),at the equilibrium (C_∞), and assuming that the water content of RBC equal 70 % of the total volume,(1) we have :

$$\frac{U_H}{U_S} = \frac{C_o - C_\infty}{C_\infty} \cdot \frac{1 - Ht}{Ht} \cdot 0{,}70$$

- permeability coefficient P and flux m

Preliminary work showed that the U.A transport into RBC obeyed apparently to a first order kinetics for concentration ranging from 0.3 to 3mM/L : at equilibrium, the decay of labeled U.A followed an exponential time function : $\frac{C - C_\infty}{C_o - C_\infty} = e^{-kt}$.

We then applied to this two compartments system the classical formula of compartimental analysis and used the concept of flux (m) and permeability (P) in agreement with Passow recommendation (2). We assume that the RBC surface average 140 microns^2 and that their number reach $1.1.10^{10} / cm^3$ of pellet (14). We thus used the permeability coefficient P (cm/sec.) to express data from experiments carried out with an U.A concentration of 0.3 mM/L and the flux m (mM/sec/cm^2) for all the others experiments.

RESULTS AND DISCUSSION

Partition coefficient

pH effect: When experiments were performed with Cl^- at a pH ranging from 6.5 to 7.5, partition coefficient of U.A decreases when the pH increases (fig. 1). The relation obeys to the following formula: $\log \frac{U_H}{U_S} = -0.344 \text{ pH} + 2.30$

Effect of citrate and saccharose: When the Cl^- of the incubation medium is partly replaced by compounds which do not enter the RBC (Citrate or saccharose), the pH influence is exhibited in the same direction but the values of U_H/U_S are significantly increased obeying to :

$$\log \frac{U_H}{U_S} = 0.376 \text{ pH} + 2.82$$

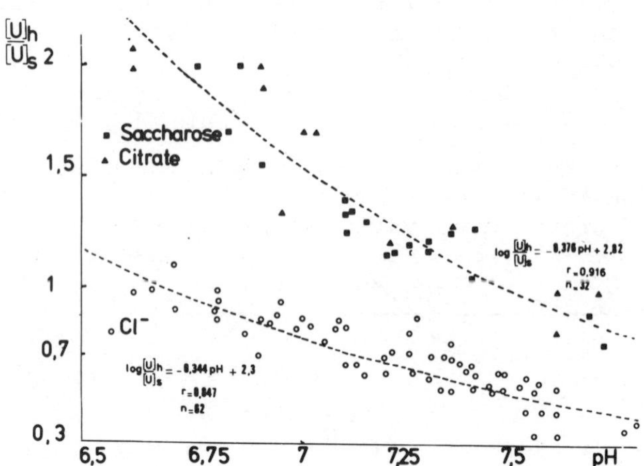

fig 1 : U.A partition coefficient between RBC and supernatant at equilibrium for various incubation media pH : (o) ringer solution (▲) medium supplemented with citrate, (■) or saccharose

The similarity between these data and those of others authors who studied the equilibrium distribution of the ions Cl^-, H^+, OH^- and SO_4^{--} (3,4,5,6) allowed us to think that the partition equilibrium of U.A between RBC and supernatant is in agreement with GIBBS-DONNAN law : U_H/U_S increases when the pH decreases and reaches one at pH 6.8 which is the isoelectric point of Hb, main non diffusible anion of the erythrocyte.
The potential difference through the both sides of the RBC membrane usually negative can be reversed by increasing the concentration of non diffusible compounds (citrate or saccharose) outside the RBC. These compounds do modifie the distribution of three diffusible anions by counteracting the effect of the intracellular on diffusible anion, the hemoglobin.
These results suggest a U.A repartition which is in agreement with DONNAN equilibrium and an identity between the physico-chemical state of U.A on both sides of the RBC membrane.
However, we cannot drawn any indication from these data on the mecanism of transport through the membrane.

KINETICS

Effect of pH In ringer solution the permeability P decreases when the pH increases (fig. 2)

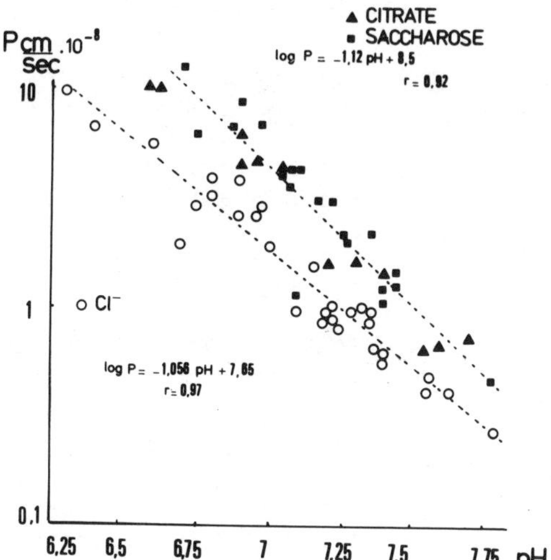

figure 2 : U.A permeability coefficient P (cm/sec) through human erythrocyte membrane as a function of the pH of the incubation medium. the symbols are the same as in fig. 1.

This relation obeys to : $\log P .10^{8} = - 1.056 \text{ pH} + 7.65$

Effect of citrate or saccharose When the Cl^- is partly replaced by citrate or saccharose acts in the same direction but P is significantly increased. This relation obeys to :

$$\log P.10^{8} = - 1.12 \text{ pH} + 8.5$$

The U.A whose pK is 5.8 (7) is a weak acid which is mainly ionized at the pH used in the experiments. Our data could , thus, be explained by the PASSOW 's theory (8) according which the High permeability of the RBC membrane results from fixed positive charges located inside the membrane and to which would stick some anions during the transfer. From that point of view the H^+ ions concentration increase triggers an increase of the fixed intramembranous positive charges and thus enhances anions transfer. Results obtained with incaubation in presence of citrate or saccharose would suggest a competing effect at the intramembranous level between two anions of different mobility : chloride and urate - Cl^- crossing RBC membrane very quickly (9). Similar observations have discribed for others anions as PO_4^{--} (10)

and SO_4^{--} (11).

Effect of mineral anions transport inhibitors
DNFB, salicylate, phenylbutazone and benzoate exibit an inhibitor effect upon U.A influx into RBC. Permeability coefficient is then decreased of 30 % for the concentration used.

Results :
- without DNFB : Pm = 5.6 ; n =4 ;Sm = 1.32) 0.01 $< p <$ 0.02
- with DNFB : Pm = 2.7 ; n =4 ;Sm = 1.24)

- without benzoate : Pm = 3.54 ; n = 3 ; Sm = 0.38) 0.001 $< p <$ 0.01
- with benzoate : Pm = 2.64 ; n = 3 ; Sm = 0.33)

- without salicylate : Pm= 3.23; n = 6; Sm = 0.48) 0.01 $< p <$ 0.02
- with salicylate : Pm= 2.1 ; n= 6 ; Sm = 0.43)

The similarity of the data obtained from passive mineral anion flux through erythrocyte membrane and from U.A influx allowed us to suggest that U.A crosses RBC membrane as an ion; obeying to a passive diffusion at least partially. Non ionic diffusion process which has been previously suggested (7) could neither account for the inhibiting influence of anions transport inhibitors nor the enhancing effect of the citrate or saccharose.

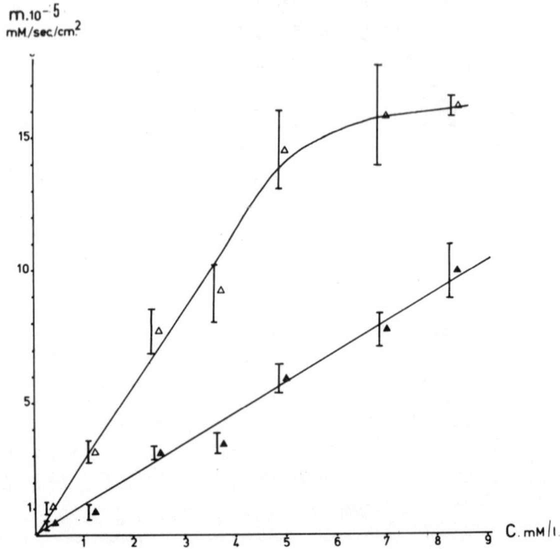

Figure 3 : the U.A flux m ($mM/sec/cm^2$) through the human RBC membrane as a function of U.A concentration of the supernatant without NaF (upper part) and with NaF (lower part)

We can confirme a saturation phenomenal of U.A transport already mentioned by LASSEN (7) , for increasing concentration of U.A (fig.3). These finding evokes the existence of membrane specific sites for U.A.

Metabolism influences

Without glycolysis (using medium lacking of glucose or supplemented with Na F) the permeability of U.A is significantly reduced.

Results :

- control : Pm = 3.29 ; n = 8 ; Sm = 0.18) $p < 0.001$
- without glucoe : Pm = 1.98 ; n = 8 ; Sm = 0.14)

- control : Pm = 2.49 ; n = 8 ; Sm = 0.20) $p < 0.001$
- with NaF : Pm = 0.97 ; n = 7 ; Sm = 0.074)

The $^{35}SO_4^{--}$ influx is not modified by absence of glycolysis in the same conditions .

The NaF acts upon glycolysis by enolase blockode (12) ; we thus thought that the U.A influx could be linked to the presence of compound coming from phospho 3 glycerate as P.E.P., pyruvate or lactate. Unfortunatly, none of these three compounds are able to avoid the inhibition effect triggered by the incubation without glucose. A secondary effect of NaF is to decrease both cellular ATP and NAD. Without glucose, the last compounds do not restitute an adequate U.A influx.

These results show a close relationship between the U.A influx and RBC metabolism but the exact mechanism remains to be clarified. However, the data confirm the existence of an active process as opposed to the SO_4^{--} behaviour, for instance (despite a close similarity) .

Glycolysis inhibitors and anions transport inhibitors

We obtain the following data :

- control : Pm = 2.76 ; n = 6 ; Sm = 0.33
- 1) : Pm = 1.66 ; n = 6 ; Sm = 0.17)
- 2) : Pm = 1.78 ; n = 6 ; Sm = 0.35) $0.001 < p < 0.01$
- 3) : Pm = 0.85 ; n = 6 ; Sm = 0.16)

These results give evidence of a synergistic effect of the two kinds of inhibition.

NaF and U.A concentrations

Results are shown in figure 3. A linear relationship exists between m and c. The only passive mecanism seems to be maintained in these experiments .

CONCLUSION

Taken all together the results show that the RBC transport of U.A seems to be partly of passive nature , closely similar to others mineral or organic anions, and partly of active nature related to compound of the membrane playing a role in cellular metabolism. It remains to be precised the role of the membrane itself and to characterize the membrane site which could be responsable of a part of the transport of U.A through erythrocyte membrane.

REFERENCES

1° B. VESTERGAARD - BOGIND and T. HESSELBO, biochim. biophys. Acta, 1960, 44, 117.

2. PASSOW H., Laboratory technics in membrane biophysics, H. Passow and R. Stampfli Edit., 1969.

3. E.J. HARRIS and M. MAIZELS - J. Physiol. (1952) -118 , 40, 53.

4. Edwards J. Fitzsimons and Julius Sendroy, J.R. Journal of biological Chemistry 1961 . 236 (5), 1595-1601

5. P.A. Brondberg, J. Theodore, E.D. Robin and W.N. Jensen J. Lab. and Clin. Med. 1965 - 66 (3) - 464, 475

6. J. Funder and J.O. Wieth Acta physiol. Scand. 1966 . 68 , 234, 235

7. U.V. Lassen . Biochim. Biophys. Acta 1961, 53, 557, 569

8. H. Passow Proc. XXIII Intern. congr. physiol. sci. Tokyo, 1965, 555,566

9. R.B. Gunn, M. Dalmark, D.C. Tosteson and J.O. Wieth J. gen. physiol. 1973. 61, 185,206

10 R. Whittam, Transport and diffusion in red blood cells, Edward Arnold , edit. 1964

11 Passow H. In progress in biophysics and molecular biology, Butler J.A.V., Noble D. edit.; Pergamon press London, 1969, 19 (2), 425-465

12 O. Warburg and W. Christian. Biochem. Z. 1941 310, 384

13 S.A. Feig, S.B. Shohet and D. Nathan. J.Clin. Invest. 1971 50, 1731, 1737

14 Wintrobe, M. M., Clinical hematology, 5° ed. Philadelphie (1961), p. 95

A RELATIONSHIP BETWEEN FREE URATE, PROTEIN-BOUND URATE, HYPERURICEMIA AND GOUT IN CAUCASIANS AND MAORIS

J. R. Klinenberg, D. S. Campion, R. W. Olsen,
D. Caughey, and R. Bluestone

Departments of Medicine, Cedars-Sinai Medical Center and UCLA School of Medicine; Medical & Research Service, VA Wadsworth Hosp. Center, Los Angeles, Ca., U.S.A.

Since clinical gout is characteristically associated with intra-articular urate crystals in hyperuricemic patients, much of the investigation into the pathogenesis of gout has involved studying the causes and effects of urate deposition (1). We have previously demonstrated that approximately 20% of urate in serum is bound to albumin (2). If the serum urate concentration per se is of paramount importance in the pathogenesis of gouty arthritis, then one would expect that in gouty subjects the free urate fraction should be elevated. It thus became of interest to determine the actual free concentration of urate in patients with gout and to compare this to the concentration of free urate in normal subjects and in patients with asymptomatic hyperuricemia.

It is well known that Maoris (New Zealand natives of Polynesian origin) have elevated serum urate concentrations and an increased incidence of gout as compared to Caucasians living in the same environment (3). However, a study of Maoris living on the remote island of Pukapuka revealed a high incidence of hyperuricemia but a low incidence of gout. These observations suggest that the Maoris as a race may have a genetically determined cause of hyperuricemia, and that other environmental factors may play a role in the pathogenesis of gout.

In order to investigate some of these possibilities, we have measured the free and protein-bound urate in gouty patients, subjects with asymptomatic hyperuricemia and normal control subjects - in both Caucasian and Maori populations.

Patients were defined as hyperuricemic if the serum urate was greater than 7 mg/100 ml. The diagnosis of gout was established clinically. Serum urate was measured by the uricase method (4), and the binding of urate to serum was determined by the ultrafiltration method (2). All studies were performed under physiologic conditions of pH and ionic strength and at 22.5°C.

Ten ml of serum was ultrafiltered at 10 lb/sq.in. in an Amicon ultrafiltration chamber, using a 10,000 m.w. cutoff membrane. Three 0.5 ml ultrafiltrate samples were collected and the urate determined. The difference between the original serum urate and that in the ultrafiltrate was considered to be the bound urate. The urate concentration of the ultrafiltrate was the free urate.

RESULTS

The results of these studies are shown in Table 1. Caucasian gouty and asymptomatic hyperuricemic patients had similar values for free urate (7.1 and 7.0 mg/100 ml, respectively), but both groups differed significantly from the control subjects ($p < 0.02$). There was no significant difference in urate binding observed in the 3 Caucasian groups (the percent bound varying from 17.7-23.7%).

The total urate concentration in Maoris with gout (11.6 mg/100 ml) was significantly higher than the urate of asymptomatic hyperuricemic Maoris (9.5 mg/100 ml) ($p < 0.05$), and both groups had much higher urate concentrations than did normal control Maoris ($p < 0.001$). Maori gouty and asymptomatic hyperuricemic subjects both had similar free urate concentrations and these differed significantly from the Maori control group ($p < 0.01$). The percent urate bound in Maori gout patients (51%) was significantly higher than the percent bound in Maori control subjects (35%).

Although the total serum urate of control Maoris was almost identical to that of Caucasian controls (6.2 mg/100 ml vs 6.1), the total serum urate of the gouty Maoris was significantly higher

Table 1

Caucasian	Control (12)	AHU (9)	Gout (10)
Total urate mg%	6.1 ± .2	9.0 ± 1	8.5 ± .3
Free urate mg%	4.7 ± .2	7.0 ± .8	7.1 ± .4
% bound	23.7 ± 2.6	21.6 ± 2.9	17.7 ± 2.2
Maori	(16)	(11)	(9)
Total urate mg%	6.2 ± .3	9.5 ± .3	11.6 ± .9
Free urate mg%	4.0 ± .2	5.3 ± .4	5.4 ± .4
% bound	35.2 ± 2.7	43 ± 5.1	51.3 ± 5

than that found in Caucasian gouty patients (11.6 mg/100 ml vs 8.5 - $p < 0.01$). However, the free urate in Maori gouty patients was significantly lower than the free urate of the Caucasian gouty patients (5.4 mg/100 ml vs 7.1 - $p < 0.01$), since all the Maori groups had much higher percentages of urate bound than the comparable Caucasian groups (35-51% vs 18-24% - $p < 0.005$).

DISCUSSION

Both the Maoris and Caucasians had elevated free urate concentrations in gouty and asymptomatic hyperuricemic subjects as compared to controls, but the concentrations of free urate were quite similar in the gouty and asymptomatic groups - both Maori and Caucasian. This strongly suggests that the free urate concentration alone is _not_ the major determinant for development of clinical manifestations of gout. Furthermore, the binding of urate to proteins does not appear to play a major role in protecting a hyperuricemic subject from gout.

The finding of significantly increased binding of urate in the Maori could be the major explanation for the hyperuricemia seen in this race. Preliminary studies show no evidence for urate over-production or under-excretion in these patients. Thus genetically determined increases in urate-binding protein, with consequent increase in _bound_ and _total_ urate but _not free_ urate, may be another mechanism for the development of hyperuricemia.

The source of the increased urate-binding to serum proteins in Maoris is currently under investigation. In the Caucasians, most urate which is bound binds to serum albumin. Studies of Maori serum reveal no quantitative or qualitative abnormalities of serum albumin. Rather, preliminary studies of serum components - separated by column chromatography - suggest that increased binding of urate in Maori serum is to a globulin. Further characterization of this binding protein is currently underway.

In summary, the studies of urate binding to serum proteins show no difference between gouty and asymptomatic hyperuricemic subjects, thus both have similar free urate concentrations. This suggests that the clinical manifestations of gout are _not_ directly related to the concentration of free urate.

The Maoris have markedly increased binding of urate to serum as compared to Caucasians. Fractionation of Maori serum demonstrated a urate binding globulin that does not seem to be present in Caucasians. These studies suggest that a genetically determined increase in urate-binding protein might be another mechanism for the development of hyperuricemia, in this instance the increased urate being bound rather than free.

REFERENCES

1. Klinenberg, J.R., Bluestone, R., Schlosstein, L., Waisman, J., and Whitehouse, M.W.: Urate deposition disease. How is it regulated and how can it be modified? Ann. Int. Med. 78: 99-111, 1973.
2. Campion, D.S., Bluestone, R., and Klinenberg, J.R.: Uric Acid: Characterization of its interaction with human serum albumin. J. Clin. Invest. 52: 2383-2386, 1973.
3. Prior, I.A.M., Rose, B.S., and Davidson, F.: Metabolic abnormalities in New Zealand Maoris. Brit. Med. J. 1: 1065, 1964.
4. Liddle, L., Seegmiller, J.E., and Laster, L.: The enzymatic spectrophotometric method for determination of uric acid. J. Lab. Clin. Med. 54: 903, 1959.

INFLUENCE OF ALLOPURINOL ON THE GENETIC MATERIALS OF EHRLICH ASCITES TUMOR CELLS

H. Becher and B. Puschendorf
Department of Medicine
University of Freiburg, Germany
and Institut of Clinical Biochemistry
University of Innsbruck, Austria

From the kinetic parameters of allopurinol with red cell Hypoxanthine-guanine-phosphoribosyltransferase, it is known that allopurinol is a very poor substrate for this enzyme (1) and the quantitative data by ELION (2) demonstrate the unliklihood that allopurinol ribonucleotides are responsible for the feedback inhibition of purine de novo synthesis. However a minimal concentration of nucleotide derivatives of allopurinol could potentially influence the chromatin metabolism of cells.

The present study was undertaken to examine the influence of allopurinol on the genetic material in Ehrlich-ascites tumor cells.

Methods

Chromatin from Ehrlich-ascites tumor cells was prepared by a modified procedure of MARUSHIGE and BONNER (3, 4). DNA-dependent RNA-polymerase from E. coli was prepared and tested according to the method of BURGESS (5). The isolation of DNA from tumor cells was carried out as described by MARMUR (6) with minor modifications and RNA synthesized in vitro is a modification of the procedure used by MELLI and BISHOP (8). The hybridization of RNA to denatured DNA is essentially that of GILLESPIE and SPEIGELMANN (9). For competitive hybridization studies the method by PAUL and GILMOUR (1O) was used.
Recently, IWAMOTO and MARTIN (11) demonstrated that allopuri-

nol inhibits the growth of hepatoma cells in culture as well as their de novo purine synthesis. To test this cytotoxic evidence, we studie the influence of allopurinol on the proliferation of Ehrlich-ascites tumor cells. No inhibition of cell proliferation or cell transplantability was detected after in vivo or in vitro preincubation of the tumor cells with allopurinol 120 mg/kg or 1 mM under in vitro conditions.
The influence of allopurinol on the incorporation of laballed nucleosides into the nucleic acids of tumor cells is shown in Table 1.

	experiment-nr.	allopurinol mg/kg	incorporation into nucleic acids cpm/mg DNA (x 10^3)		
			thymidine	uridine	guanosine
tumor cells	I	—	110	120	131
	II	60	115	114	141
	I	—	435	28	136
	II	4 x 60	657	26	159
mouse liver	I	—	26	1,6	13,3
	II	60	25	1,4	16,9
	I	—	8,2	4.8	0,92
	II	4 x 60	6,8	5.4	0,96

Table 1 Influence of allopurinol on the incorporation of thymidine, uridine and guanosine into the nucleic acids of tumor cells and mouse liver

After a 4 hr preincubation period, under in vivo conditions, with th indicated doses of allopurinol, 10 - 40 mC of laballed nucleoside were injected into the mice, and after 1 h the tumor cells were harvested and the incorporation of radioactivity into the acid insoluble material was determined.
In comparison with the untreated controls no inhibition of the incorporation of the nucleosides was observed, neither in the ascites tumor cells, nor in the mouse liver. Therefore we can assume tha allopurinol does not significantly change the intracellular nucleotide pools, as demonstrated by the studies of NELSON et al. (12). The absence of di- and triphosphates of the ribonucleosides of allo purinol and oxipurinol in the acid-soluble pools (13) and their lack of incorporation into nucleic acids (14) does not rule out qualitative

alterations in the nucleic acid synthesis. The chromatin in eucaryotic cells is composed mainly of DNA, RNA and protein. The DNA in chromatin is largely restricted so that only a small fraction is able to serve as a template for the DNA-dependent RNA-synthesis (transcription). However it is not completely known which components of chromatin determine the tissue-specific restriction of DNA.
Both chromosomal RNA and acidic nonhistone proteins were implicated as possible genetic repressors and regulators in cells of higher animals.
It could be possible that allopurinol or their derivatives are involved in these regulatory mechanisms and can change the genetic expression.
Therefore, at first we studied the influence of allopurinol on the DNA-directed RNA synthesis of in vitro or in vivo pretreated chromatin from tumor cells and rat liver in the DNA-dependent RNA-Polymerase system by RNA-Polymerase from E. coli. No inhibition of template activity of calf thymus DNA or chromatin from rat liver and tumor cells could be demonstrated. (Table 2)

allopurinol	incorporation of [^{3}H]-UMP (p moles) calf-thymus DNA	chromatin from rat liver	chromatin from tumor cells
——	3872	86	457
0,3 mM	3712	80	462
0,6 mM	4042	110	476
1,5 mM	3770	96	469
2,1 mM	3869	78	535

Table 2 Influence of allopurinol on the template activity of calf thymus DNA and tumor cells chromatin. 10 µg DNA was employed in each assay.

In vivo pretreatment of tumor cells with allopurinol (60 - 240 mg/kg i.p.) has also no effect on the RNA-synthesis from tumor cells chromatin and therefore we can suppose that allopurinol does not alter the template-activity from in vitro or in vivo pretreated chromatin. (Table 3)

	incorporation of [³H]-UMP (p moles)			
experiment-nr.	I	II	III	IV
control	319	365	310	228
60 mg/kg	273 (86)	394 (108)	336 (108)	177 (78)
120 mg/kg	269 (84)	307 (84)	—	233 (102)
240 mg/kg	282 (89)	366 (100)	340 (110)	234 (102)

Table 3 Template activity of chromatin following pretreatment of tumor cells with allopurinol in the RNA polymerase system. In each test 10 μg DNA was employed. The values in parenthesis are those of the untreated control.

In the next experiments these chromatin-fractions were used for the hybridization experiments. (^{3}H) RNA transcribed from tumor cells chromatin following in vivo treatment with allopurinol in the indicated doses was hybridized with denatured DNA isolated from ascites tumor cells (Figure 1).

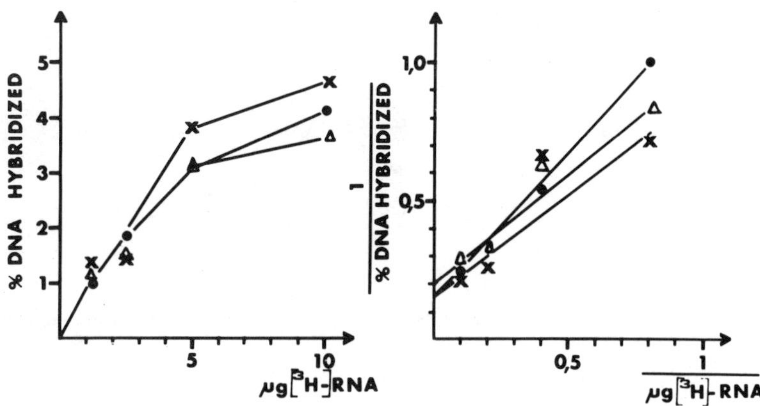

Figure 1 Hybridization of (^{3}H)-RNA templated from tumor cells chromatin following in vivo treatment with allopurinol versus 2 μg tumor cells DNA per filter.
x——x control; •——• allopurinol 60 mg/kg; Δ——Δ allopurinol 240 mg/kg

The maximal rates of hybridization are 6,7 % for the control, 6,7 % for allopurinol 60 mg/kg and 5,0 % for allopurinol 240 mg/kg. These results show that the RNA-fractions transcribed from controls and allopurinol treated chromatin are quantitatively similar, but it would be possible that qualitatively different species of RNA are transcribed. This question was investigated by means of competitive hybridization studies using (^{3}H-) RNA transcribed from untreated (A) or in vivo with allopurinol pretreated (B) chromatin (Figure 2), with increasing amounts of competing RNA from untreated tumor cells or from allopurinol pretreated tumor cells.

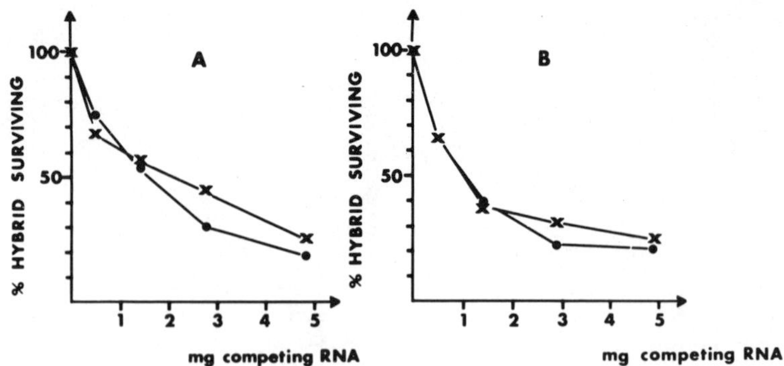

Figure 2 Competition experiments using (^{3}H)-RNA templated from untreated (A) or in vivo with allopurinol pretreated (B) chromatin and competing RNA isolated from untreated and with allopurinol pretreated tumor cells. In all cases 2 µg (^{3}H) RNA was hybridized with 2 µg DNA per filter isolated from tumor cells in the presence of increasing amounts of the competing non labelled RNA.

A) (^{3}H) RNA from untreated chromatin
x——x competing RNA from untreated tumor cells
•——• competing RNA from allopurinol treated cells 240 mg/kg

B) (^{3}H) RNA from allopurinol pretreated chromatin
x——x competing RNA from untreated tumor cells
•——• competing RNA from allopurinol treated cells 240 mg/kg

One can see that by the competition between radioactive and non radioactive RNA which was isolated from allopurinol treated and untreated cells, no difference was observed. We can therefore conclude that allopurinol does not cause different RNA-patterns in the cell and a mutagenous effect of allopurinol is not to be expected from these experiments.

However we are left with the only possibility that allopurinol causes alterations in the metabolism of the cell, for example by the blocking of a single specific messenger-RNA , which cannot be detected by the methods used. Above all, because of the absence of long-term investigations of allopurinol and the recently describe observations by WANG et al. (15) that allopurinol has an enhancing effect on bladder carcinogenicity by carcinogenic nitrothiophenes it should be a strong indication for allopurinol therapy.

REFERENCES

1. Krenitzky, T. A., R. Papaioannou and G. B. Elion J. Biol. Chem., 244, 1263 (1969)
2. Elion, G.B. and D. J. Nelson: Purine metabolism in man. Plenum Press, New York - London 1974, p. 639
3. Marushige, K., J. Bonner J. molec. Biol. 15, 160 (1966)
4. Puschendorf, B., H. Wolf, H. Grunicke, Biochem. Pharmacol. 20, 3039 (1971)
5. Burgess, R. R., J. biol. Chem. 244, 6166 (1969)
6. Marmur, J J., J. molec. Biol. 3, 208 (1961)
7. Girard, M.: In L. Grossman, K. Moldave (Hrsg.): Methods of Enzymology. Academic Press, New York - London 1967. S. 586
8. Melli, M., J. O. Bishop, J. molec. Biol. 40, 117 (1969).
9. Gillespie, S., D. Gillespie, Biochem. J. 125, 481 (1971)
10. Paul, G., S. Gilmour, J. molec. Biol. 34, 305 (1968)
11. Iwamoto, K., D. W. Martin jr., Biochem. Pharmacol. 23, 3199 (1974)
12. Nelson, D. J., C. J. Bugge, H C. Cransny, G.B. Ellion, Biochem. Pharmacol., 22, 2003 (1973)
13. Elion, G B., A. Kovensky, G. H. Hitchings, E. Metz and R. W. Rundless, Biochem. Pharmacol., 15, 863 (1966)
14. Hitchings, G. H. In: Biochemical Aspects of Antimetabo - lites and of Drug Hydroxylation. FEBS Symposium, 16, 11 (1969)
15. Wang, C. J , S. Hayashida, A.M. Pamukcu and G. T. Bryan Canc. Res. 36, 1551 (1976).

6-AZAURIDINE, AN INHIBITOR OF THE PURINE SALVAGE PATHWAY

G. Partsch and R. Eberl

Ludwig Boltzmann-Institute for Rheumatology and Balneology, Kurbadstrasse 1o, A-11o7 Vienna, Austria

Since the time when the knowledge about nucleic acid metabolism has been extended and the connection to malignant processes was recognized efforts were made to find new inhibitors interfering with different stages of nucleic acid metabolism. One of the pathways involved in these processes is the purine salvage pathway. Henderson (1) and Lau (2) examined different purine and pyrimidine derivatives and analogues for their ability to inhibit the enzymes of the purine-phosphoribosyltransferase. Henderson (1) found 13 compounds out of 116 chemicals which showed more than 75 % inhibition of the adenine-phosphoribosyltransferase. Most of them were adenine analogues. Lau (2) on the other hand, investigated different purine and pyrimidine derivatives to inhibit the hypoxanthine- guanine-phosphoribosyltransferase activity of human erythrocytes. There were only some purine and purine nucleoside analogues active inhibitors of these enzymes. Similar experiments were reported by Lau (3) of Ehrlich ascites tumor cells.

Azauracil and its riboside azauridine were of some interest in the chemotherapy of cancer (4). Since azauracil reflected some unexpected neurotoxic effects (5) azauridine did not show similar effects. Azauridine will be incorporated within the cell and phosphorylated to the azauridine-monophosphate. This compound is known to interfere with the de novo synthesis of pyrimidine because of blocking the decarboxylation of the orotidine-monophosphate (OMP) (6). Therefore, OMP accumulates and uridine will be utilized to a higher degree for nucleic acid synthesis. We tested azauridine for its behaviour to inhibit the purine-PRT in HeLa cell homogenates.

MATERIALS AND METHODS

Unpurified preparations of HeLa cells were obtained by homogenisation of cells in TRIS-HCl buffer (o,1 M, pH 7,4 with 5 mM $MgCl_2$). HeLa cells were cultivated in Roux flasks with MEM Medium complemented by 1o % foetal calf serum. The estimation of the purine-PRT activities was performed by measurement of transposed radioactive purine-bases to their nucleotides. The labelled compounds were purchased by NEN Chemicals. The specific activities were 1o,1 mCi/nMol for ^{14}C-8-adenine, 51,3 mCi/mMol for ^{14}C-8-guanine and 48,6 mCi/mMol for ^{14}C-8-hypoxanthine. The method followed the paper written by Partsch (7). Protein was measured according to the method of Lowry (8). The values obtained by liquid scintillation counting were calculated by regression analysis.

To get information of the inhibition mechanism of the azauridine (AU) on HeLa cell adenine-PRT we started an additional experiment. For each concentration of AU ($1o^{-5}$ M, $1o^{-2}$ M and $1o^{-1}$ M) and different PRPP-concentrations (5o µM, 1oo µM, 2oo µM, and 1ooo µM) a kinetic of the adenine-PRT was performed, the nMol nucleotide/mg protein/h were calculated by regression analysis and a double reciprocal plot was made.

When azauridine was tested as a DNA inhibitor in vivo, HeLa cells were cultivated in falcon plastic disks with MEM medium complemented by 1o % foetal calf serum. 1 µCi /ml ^{14}C-adenine or ^{3}H-adenosine (23,2 Ci/mMol) were added together with $2,5 \cdot 1o^{-3}$ Mol azauridine to each culture and incubated for 3 hours. After that time the DNA was extracted and determined according to the method of Schmidt-Tannhauser (9) and Schneider (1o).

RESULTS AND DISCUSSION

Figures 1 - 3 represent the results obtained with different concentrations of azauridine. Table 1 shows the calculation of purine-PRT activities and the percentage of inhibition by the compound. Controls exhibit a four times higher A-PRT activity than G- and H-PRT.

The addition of different concentrations of azauridine to the test system reflected in different reductions of the enzyme activities. At the same concentrations the adenine-PRT was much more stable than the G- and H-PRT. There was only a low reduction with $1o^{-5}$ M azauridine at A-PRT (6 % of the control) whereas the H-PRT was diminished by 25 %. With 1 mM azauridine the A-PRT activity was distinctly reduced (22 %) whereas the G- and H-PRT were inactivated to around 5o %. The effect of azauridine was extremely

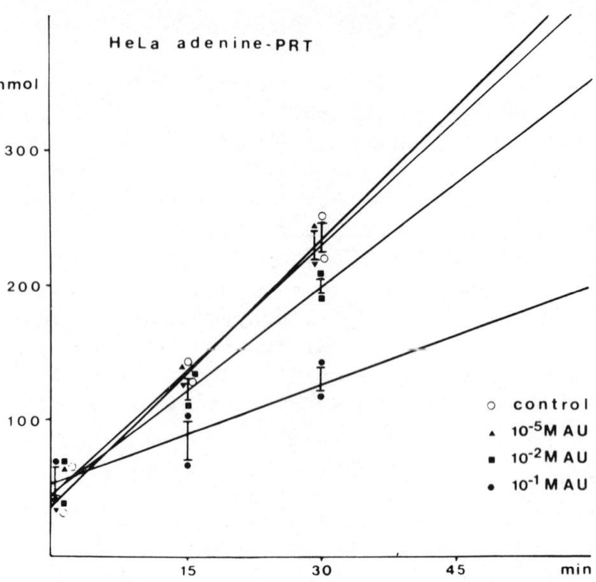

Figure 1. The influence of 6-azauridine on the adenine-phosphoribosyltransferase from HeLa cells.

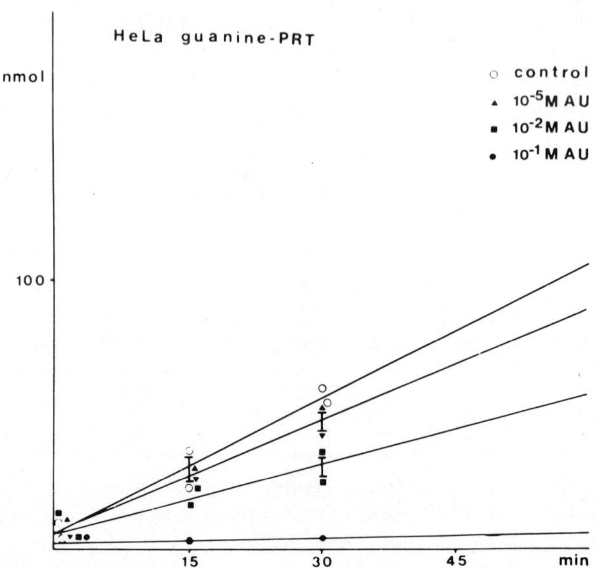

Figure 2. The influence of 6-azauridine on the guanine-phosphoribosyltransferase from HeLa cells.

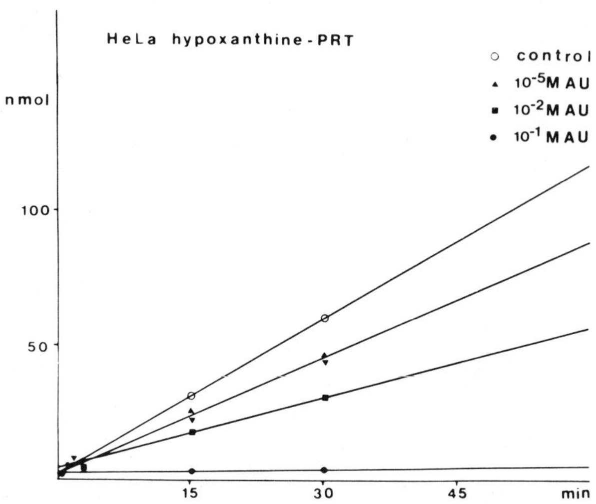

Figure 3. The influence of 6-azauridine on the hypoxanthine-phosphoribosyltransferase from HeLa cells.

	A-PRT		G-PRT		H-PRT	
0	394,76	loo %	loo,4o	loo %	115,36	loo %
$1o^{-5}$ M	372,oo	94,2o %	82,88	82,54 %	87,o4	75,45 %
$1o^{-3}$ M	3o8,oo	78,o2 %	51,8o	51,59 %	52,8o	45,77 %
$1o^{-1}$ M	145,88	36,95 %	3,4o	3,38 %	2,52	2,18 %

Table 1. Inhibition of purine-phosphoribosyltransferase of HeLa cells by azauridine (nMol/mg protein/h).

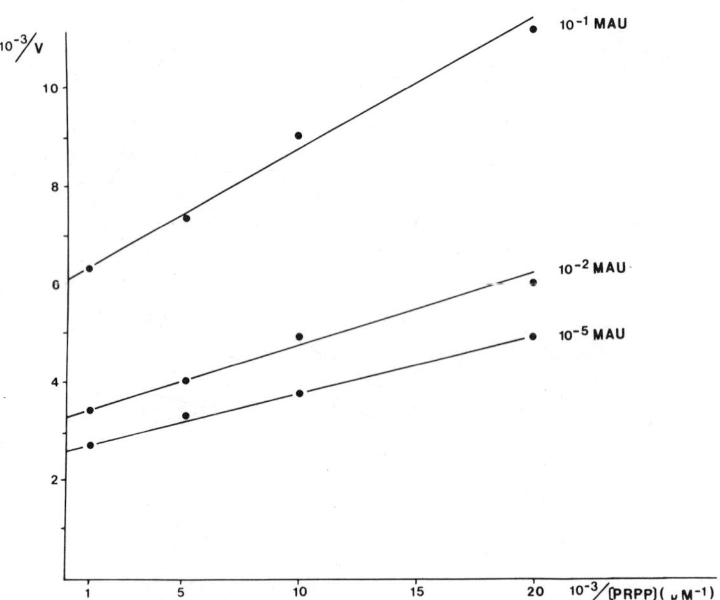

Figure 4. Double reciprocal plot of adenine-phosphoribosyltransferase reaction velocity (v) versus PRPP concentrations in the presence of 6-azauridine. Reaction velocity expressed in nMol of AMP formed/mg protein/h.

	DPM/ μG DNA	%
ADENINE		
CONTROL	644 ± 31	1oo
AZAURIDINE	4o4 ± 6	62,7
ADENOSINE		
CONTROL	5.961 ± 196	1oo
AZAURIDINE	6.o6o ± 13o	1o2

Table 2. Influence of azauridine ($2,5\cdot 1o^{-3}$ M) on the incorporation of adenine and adenosine into the DNA of HeLa cells.

high with 10^{-1} M (G-PRT and H-PRT were nearly completely inhibited). The A-PRT was reduced at this concentration to only 63 %.

The double reciprocal plot (11) of adenine-PRT reaction velocity versus PRPP concentrations in the presence of different AU-concentrations (Figure 4) indicated none-competitive or "mixed" inhibition with respect to PRPP. It is not known from these experiments if AU acts as the nucleoside or as the monophosphate.

Using the results reported before we tested the influence of ($2{,}5 \cdot 10^{-3}$ M) azauridine on the DNA synthesis in HeLa cells too. The results are summarized in Table 2.

Azauridine in the presence of labelled adenine decreased the activity in the DNA to about 37 %. There was no reduction of the labelling of the DNA when adenosine was used as a precursor of the DNA synthesis. The diminished incorporation of adenine into the DNA of HeLa cells in the presence of the A-PRT inhibitor seems to indicate that the purine salvage pathway is involved within the DNA synthesis.

Summarizing we can say that azauridine known to inhibit the de novo pyrimidine synthesis also acts on purine salvage pathway mechanisms of HeLa cells.

REFERENCES

1. Henderson, J.F., Gadd, R.E.A.: Cancer Chem.Rep. 1, 363 (1968).
2. Lau, K.F., Henderson, J.F.: Cancer Chem.Rep. 3, 87 (1972).
3. Lau, K.F., Henderson, J.F.: Cancer Chem.Rep. 3, 95 (1972).
4. Handschumacher, R.E., Calabresi, P., Welch, A.D., Bono, V., Fallon, H., Frei, E.: Cancer Chem.Rep. 21, 1 (1962).
5. Shnider, B.I., Frei, E., Ruohy, J.H., Gorman, J., Freireich, E.J., Brindley, C.O., jr., Clements, J.: Cancer Res. 2o, 28 (196o).
6. Pasternak, C.A., Handschumacher, R.E.: J.biol.Chem. 234, 2992 (1959).
7. Partsch, G., Sandtner, I., Eberl, R.: Wien.klin.Wschr. 88, 66 (1976).
8. Lowry, O.H., Rosenbrough, N.J., Farr, A.L., Randall, R.J.: J.biol.Chem. 193, 265 (1951).
9. Schmidt, G., Thannhauser, S.J.: J.biol.Chem. 161, 83 (1945).
1o. Schneider, W.C.: J. biol.Chem. 165, 747 (1946).
11. Lineweaver, H., Burk, D.J.: J.Amer.Chem.Soc. 56, 658 (1934).

SUPPRESSION OF EXPERIMENTAL URATE NEPHROPATHY BY SALICYLATE

J. R. Klinenberg, R. Bluestone, and J. Waisman

Departments of Medicine, Cedars-Sinai Medical Center and UCLA School of Medicine; Medical & Research Service, VA Wadsworth Hosp. Center, Los Angeles, Ca., U.S.A.

An experimental model for sustained hyperuricemia, uricosuria and associated nephropathy was produced by feeding rats a potent uricase inhibitor - sodium oxonate, together with dietary supplements of uric acid (1,2). In both human gouty nephropathy and in this experimental hyperuricemic nephropathy in rats, deposition of crystalline materials within the renal tubules and parenchyma have been observed (3). The deposition of urate crystals in the kidney could account for many of the inflammatory features of human gouty nephropathy, but the absence of a satisfactory experimental model has precluded detailed study of this possible relationship.

We have found that salicylate suppressed the early renal medullary neutrophilic response to urate deposition. In the present studies we have examined the role of the acute inflammatory response to urate crystals in the genesis of chronic hyperuricemic nephropathy.

Five groups of young rats were fed one of 5 daily diets for 26 weeks. Group I (5) received 20 g of chow containing 2% oxonic acid (OA) and 3% uric acid (UA); Group II (10) received the same diet with added salicylate 300 mg/Kg; in Group III (7) the salicylate was not added until the 4th week; Group IV (12) were fed chow with salicylate and Group V (6) were fed chow alone. Monthly assays were made of serum urate (SU), serum salicylate (SS) and urinary urate (UU). All rats were killed after 26 weeks. Deposited urate was eluted from one kidney (KE) by suspending the kidney in 10 ml of 0.5% lithium carbonate containing 2% sodium acid, homogenizing and dialyzing the homogenate and then measuring the uric acid in the dialysate. The results were expressed as milligrams of eluted urate per gram wet weight of kidney.

Table 1

GRP	SU*	UU*	SS*	EU**	UD	TE	Int.In+
I	5.8(.4)	133(11)	0	1.0(.2)	1.8(.2)	1.0(.6)	2.0(0)
II	5.3(.3)	121(6)	19.6(.4)	1.2(.3)	1.5(.2)	0.7(.2)	1.0(.1)
III	4.8(.3)	108(6)	20.3(.8)	3.4(1.1)	1.5(.2)	0.3(.2)	1.2(.1)
IV	0.9(0)	11(1)	16.7(.4)	0.4(0)	0	0	0
V	1.0(.2)	22(3)	0	0.7(0)	0	0	0

The above represent mean results (+ SEM) for each group.
*mean of 6 monthly determinations, mg%; **mg urate/G wet weight kidney; + $p < 0.005$ for Groups II and III compared to Group I.

The other kidney was processed for microscopic study following perfusion with 1.5% glutaraldehyde buffered with phosphate (pH 7.0). The histologic changes in the kidneys were graded according to the severity of the lesions as: absent 0, minimal 1, moderate 2, severe 3. Structures evaluated included glomeruli, tubules, interstitium and vessels. Mean grades for these various microscopic lesions were obtained for each group. For the purposes of this study data on urate deposition (UD), tubular exudate (TE) and interstitial inflammation (Int.In) are reported.

The results of these studies are shown in Table 1. Only rats in Groups I, II and III, fed oxonate and urate, developed hyperuricemia, uricosuria and increased eluted urate. The addition of salicylate to the regimen (Groups II and III) did not significantly alter these parameters, although adequate serum salicylate concentrations were obtained ($\sim$20 mg/100 ml). Similarly, only the kidneys from rats fed oxonate and urate (Groups I, II and III) showed renal urate deposition, chronic interstitial nephritis with scarring, tophi formation and urinary stone formation. Control rats (Group V) and rats fed salicylate alone (Group IV) showed none of these changes. Salicylate given for all 26 weeks (Group II), or beginning the 4th week (Group III), reduced the score for both tubular exudate and interstitial inflammation, the latter change being highly statistically significant when compared to Group I - oxonate-urate without salicylate ($p < 0.005$).

Thus it can be concluded that salicylate can alter the nature of the inflammatory response to urate crystal deposition in the rat kidney. These studies suggest that hyperuricemic nephropathy may be the result of differing types of cellular response and even raises the possibility of an immunologic reaction to urate crystals mediated by mononuclear cells. Based on these studies, one might also speculate that interference with the inflammatory response to urate crystals by use of anti-inflammatory drugs may be an important modality in the prevention of chronic hyperuricemic interstitial nephritis.

REFERENCES

1. Johnson, W.J., Stavric, B., Chartrand, A.: Uricase inhibition in the rat by S-triazines. An animal model for hyperuricemia and hyperuricosuria. Proc. Soc. Exp. Biol. Med. 131: 8, 1969.
2. Waisman, J., Bluestone, R., Klinenberg, J.: A prelminary report of nephropathy in hyperuricemic rats. Lab. Invest. 30: 716-722, 1974.
3. Bluestone, R., Brady, S., Waisman, J., and Klinenberg, J.R.: Experimental hyperuricemic nephropathy: A model for human urate deposition disease. Arth. Rheum. 18: 823-834, 1975.

PHYSIOLOGICAL CHARACTERISTICS OF VARIOUS EXPERIMENTAL MODELS FOR THE STUDY OF DISORDERS IN PURINE METABOLISM

J. Musil

HOECHST AG

6230 Frankfurt/M. 80

In recent years, decisive progress has been made in research on purine metabolism. Although at present highly effective drugs for the therapy of gout are available, the search for the therapeutic optimum is to be continued. Each form of gout (gouth with typical disorders in renal function, hyperproduction-induced gout and the hyperalimentation-induced gout, Lesh-Nyhan syndrome) should always be treated adequately. As in other fields of medicine, sufficient material from experimental investigations should be available for the detection of pathophysiological connexions. In the field of gout research, however, the animal experimental models are very inadequate. The main difficulty lies in the fact that the biochemical profile of purine metabolism in non-primate vertebrates shows distinct differences, i.e. these animals show relatively low uric acid values and excrete renally mainly allantoin instead of uric acid.

The fundamental biochemical difference between non-primates and primates is the fact that contrary to primates, where the last enzymatic process is the production of uric acid, in non-primates there is an additional enzymatic step, uricase, in which uric acid is changed into allantoin. This biochemical difference in laboratory animals which are cheaper to keep than primates can be overcome by administration of a uricase inhibitor. Several suggestions regarding mode of administration of a uricase inhibitor have been published. JOHNSON and others (1) administered orally to rats the uricase inhibitor potassium oxonate in feed which resulted in a significant increase in uric acid values in the serum and in a statistically significant increase in uric acid excretion with urine. STAVRIC and others (2) employed the same method for pharmacological examination of the influence of chlorprothixene on purine metabolism. BONARDI and VIDI (3) published a report on examinations in which they combined the administration of the uricase inhibitor (potassium oxonate) with water loading of experimental rats. This method, in which the animals are in a state of increased diuresis, has the advantage that unspecific effects of diuretically active substances do not simulate a uricosuric activity in the test. HATFIELD and others (4) proved additionally in their experiments that the administration of potassium oxonate not only inhibits uricase activity but that also an almost complete block of pyrimidine synthesis occurs which manifests itself in an increased excretion of orotic acid.

In previous publications (5, 6) we have reported on examinations of various experimental models for their influence on metabolism. One model which is pathophysiologically similar to gout in humans necessitates - more than in any other field of research - a fundamental simplification. Aim of our investigations was the achievement of a minimum of unneccessary metabolic disturbances and a maximum of pharmacological relevance in the models employed.

The experiments were performed in white male Wistar rats of our own breeding and lasted for three days each. The course of experiments is illustrated in slide 1.

Figure 1

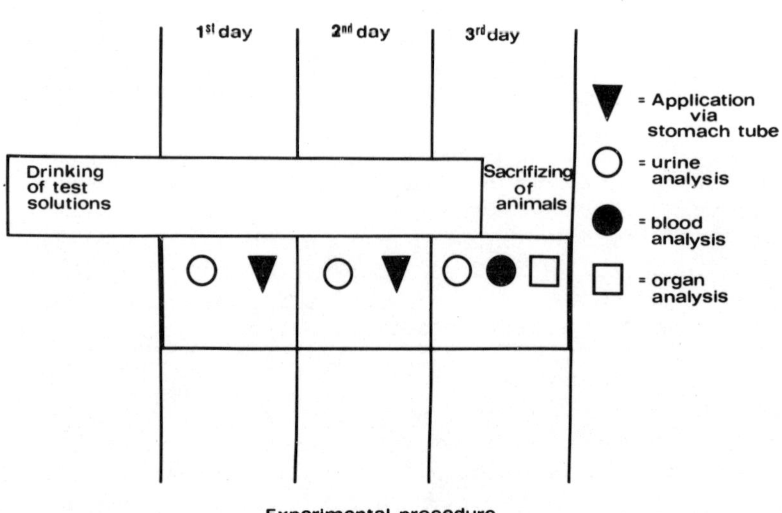

Experimental procedure

On the day prior to the first day of experiment, the experimental animals were placed separately into metabolism cages, and the respective drinking solutions were offered ad libitum to the individual groups. Liquid and food intake were measured throughout the experiment. On the first day of experiment, about 18 hours after the animals had been placed into the metabolism cages, the first urine sample was collected and analysed for parameters described in the following. Subsequently, the animals received the test solutions orally by means of a stomach tube. The same procedures were performed on the second day of experiment. On the third day of experiment the animals were killed after collection of urine samples. The animals were exsanguinated, and the blood samples and urine samples were analysed biochemically.

In the first test series the effects of different oxonate treatments were evaluated and compared with the unspecific effects of the shift in the acid-base balance. Metabolic acidosis was induced by administration of 2 % ammonium chloride, alkalosis was induced by administration of 3 % sodium bicarbonate. The group which had received water or isotonic saline solution in the above method served as control group. One group received the oxonate as a 1 % solution ad libitum as drinking solution and by means of a stomach tube according to the above method; the other group received, in addition to the solution, feed which contained 5 % oxonate. The results are illustrated in slide 2.

Figure 2

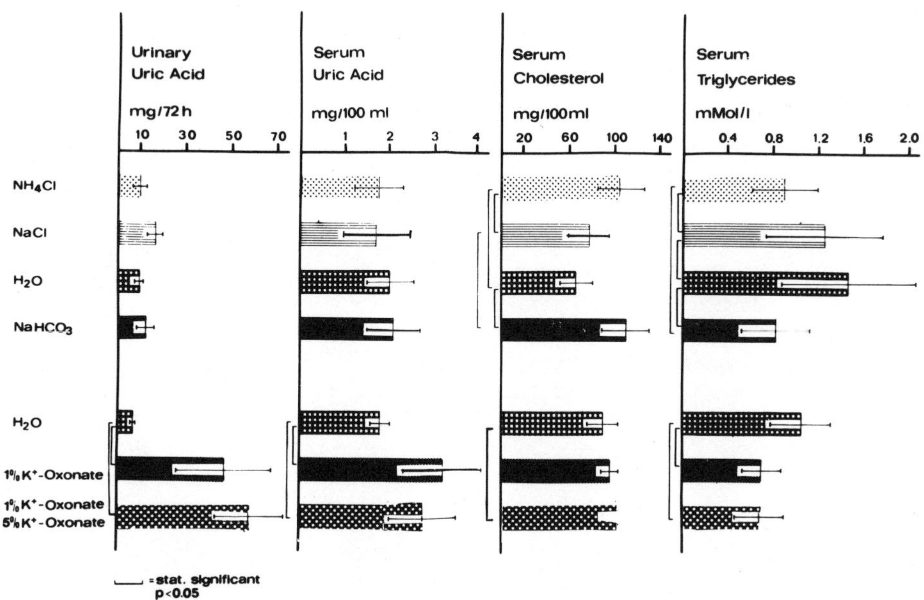

The slide shows, that acidosis and alkalosis distinctly influence the metabolism which is reflected in the significant hypertriglyceridemia; purine metabolism was not remarkably influenced. Only after the administration of a uricase inhibitor could a statistically significant hyperuricosuria and hyperuricemia be observed. Analogously to the results of the acid-base test the additional effects on the lipid metabolism were regarded as the consequence of a relative acidosis. This fact was confirmed in an equivalent 3-day test in which a control group was compared with water-loaded groups which had received oxonate (slide 3).

Figure 3

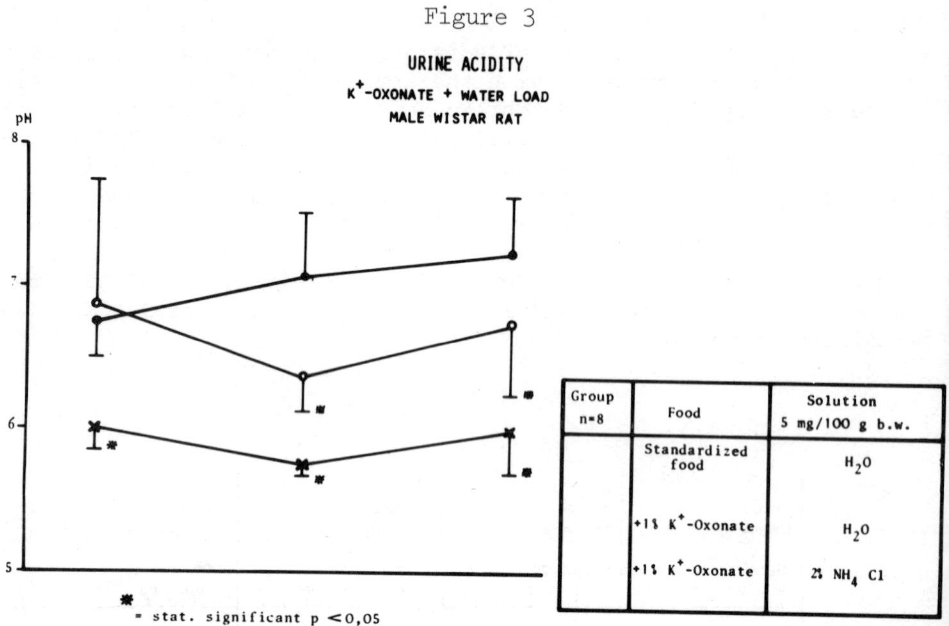

Group n=8	Food	Solution 5 mg/100 g b.w.
	Standardized food	H_2O
	+1% K^+-Oxonate	H_2O
	+1% K^+-Oxonate	2% NH_4 Cl

The table clearly shows that oxonate administration induced a relative acidosis; this fast must be taken into account particularly in pharmacological examinations.

In the development of a test model the question must be considered to what extent the negative influence on purine metabolism after increased carbohydrate administration observed in human experiments can be proved by means of the above method. For clarification of this problem, four groups were treated with fructose and oxonate in a 3-day experiment. Mode of administration and results are illustrated in slide 4.

Figure 4

Fructose load

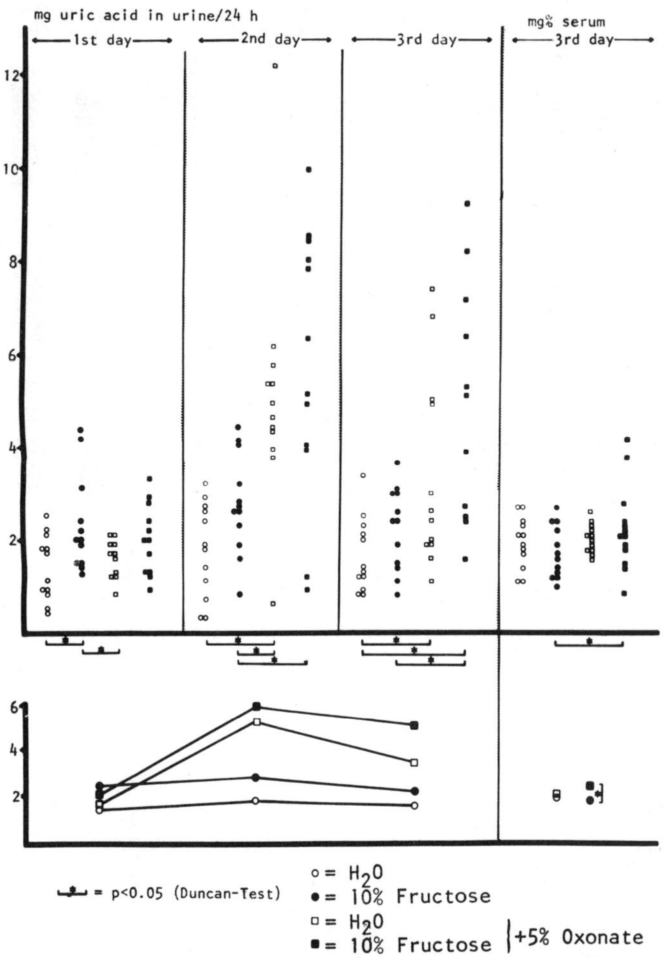

This illustration presents the clear physiological proof of an addition of the effects of fructose and oxonate on purine metabolism and of a possible potentiation in certain parameters. It is highly probable that the latter is true in the case of fructose-induced hyperuricosuria. A comparison of fructose treatment to osmotic load indicates that apart from the direct metabolic effect, which is the result of the increase in phosphoribosyl-pyrophosphate for uric acid de novo synthesis, the unspecific effect of osmotic load induced by fructose administration may increase renal excretion. This test is illustrated in slide 5.

Figure 5

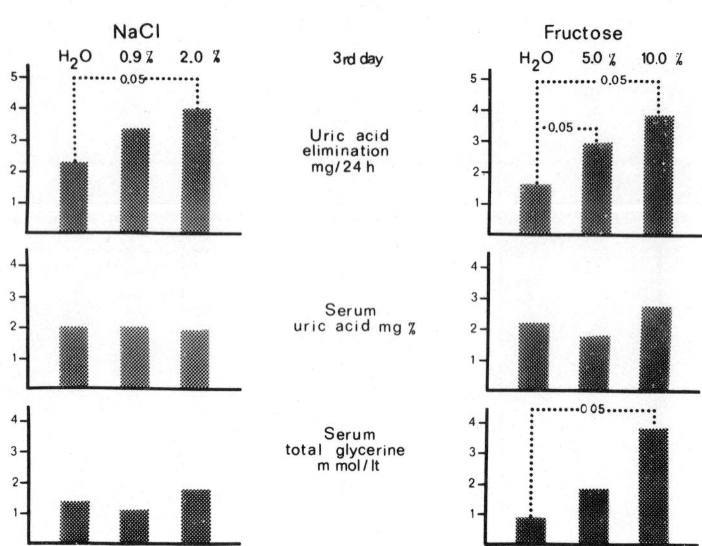

In this test, control groups received water, the other experimental groups reseived 0,9 % and 2,0 % NaCl and 5 % and 10 % fructose. As in the test method with oxonate, it is obvious that the influence of isolated carbohydrate administration on the metabolism, which is reflected in the uric acid values in the serum, is only slight, whereas the effect on uric acid excretion is almost identical with that after NaCl loading.

In al tests, the determination of basic parameters of purine metabolism and diuresis was complemented by measurements of indicating ions and of creatinine in urine and serum, in some tests it was completed by examination of uric acid in organs. This wide range of data available for the characterisation of reactivity of experimental animals permitted selection of the optimal combination of treatments, thus providing a model situation in the minimization of undesired effects which corresponds to the most important physiological parameters of the metabolism profile in humans. A comparison of possible modes of treatments is illustrated in slide 6.

Figure 6

EFFECTS OF TWO TYPES OF OXONATE TREATMENT

GROUP	TREATMENT			URIC ACID				
	Solutions to drink ad lib.	Solutions for p.o.applying	Food ad libitum	Elimination in Urine mg/24h/100g			Serum mg %	Clearence U / B 3rd day
				1st day	2nd day	3rd day		
	H_2O	H_2O	Standardized food	0.67 ±0.18	0.87 ±0.19	0.81 ±0.26	1.34 ±0.45	0.69 ±0.34
	0,5 % K^+-Oxonate	0,5 % K^+-Oxonate	Standardized food	2.89 ±1.46	3.04* ±1.53	3.32* ±1.21	3.50 ±1.15	1.09 ±0.58
	H_2O	0,5 % K^+-Oxonate	Standardized food + 5 % Fructose + 3 % Uric Acid + 2 % K^+-Oxonate + 0,1% Sweetener	0.81 ±0.61	2.68* ±1.73	3.45* ±1.26	6.70 ±0.89	0.51 ±0.16

*= stat. significant $p < 0,05$

This illustration shows that of all three ways of treatment the simultaneous administration of fructose, oxonate and uric acid proved to have the strongest effect on uric acid metabolism.

An easy test model for non-primates in which uric acid values in the serum comparable to those in primates were obtained was subsequently tested in a specific pharmacological experiment. The control group received the above described feed mixed with fructose, oxonate and uric acid. The test groups received allopurinol and probenecid. The results of this test are illustrated in slide 7.

Figure 7

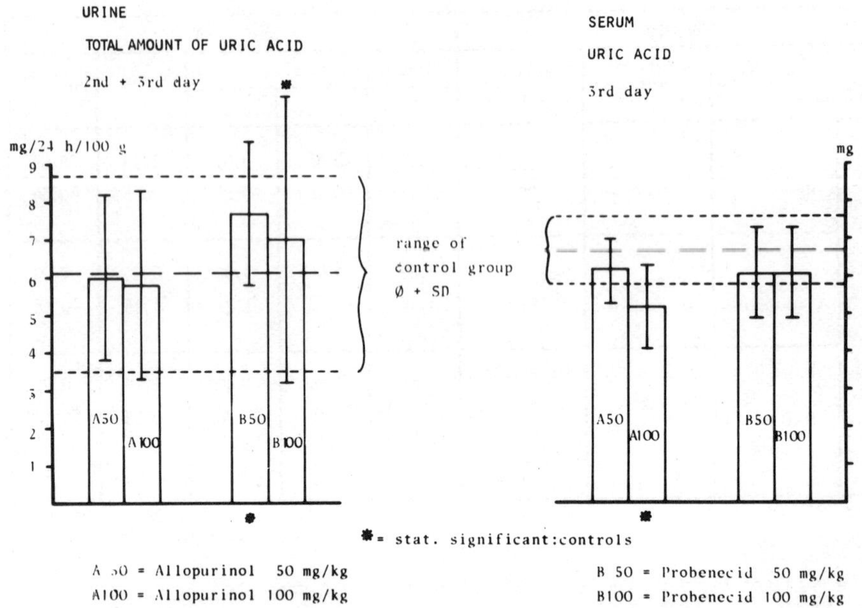

The curves and serum values of uric acid show clearly that in this 3-day experiment the administration of allopurinol has no marked influence on the amount of uric acid excreted, while the uric acid values in the serum are statistically significantly decreased. The administration of probenecid, however, resulted in a distinct increase in uric acid excretion, whereas the decrease of uric acid values in the serum was not statistically significant with values in the range of 15 %. The total pharmacological spectrum of administration of the substances tested in this experiment is illustrated in slide 8. This activity profile is obtained when the percent deviations of all parameters measured in experimental groups are compared to those of the control group.

Figure 8

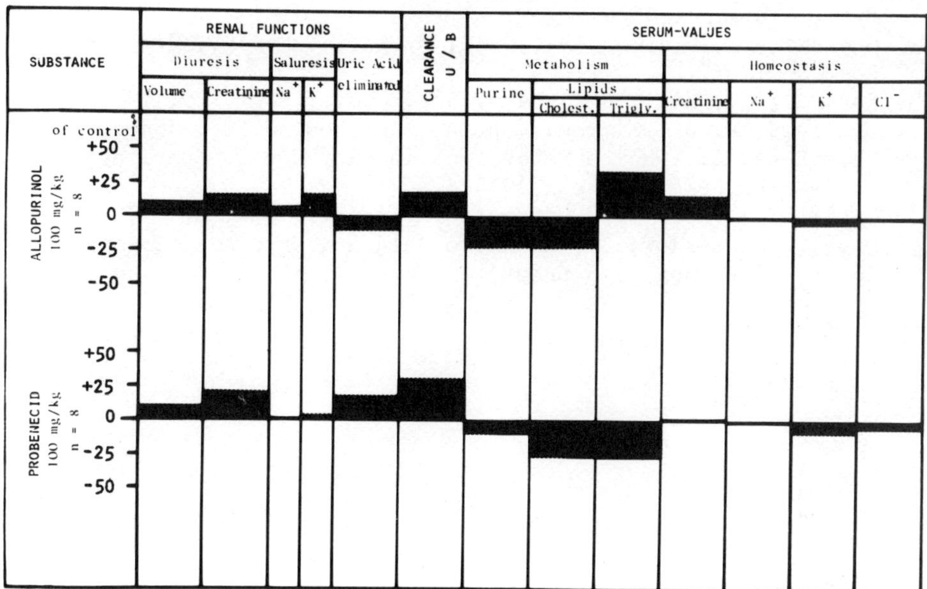

The curve of the activity profile shows clearly that the influence of probenecid on renal function is marked, whereas that of allopurinol is within the range of variability of the methods employed. Marked changes were observed in serum values. In previous publications we have frequently mentioned that parallel to the influence on purine metabolism, marked effects on lipid metabolism can be observed in animals experiments. These results indicate that the use of a uricase inhibitor is a clear progress in experimental work. It is evident that the administration of the uricase inhibitor alone does not provide the desired model situation but that the use of various physiological factors (increased diuresis, osmotic load, etc) offers on additional improvement of test facilities, thus opening the way to the clarification of pathophysiological connexions not only in purine metabolism disorders, but also in metabolism disorders in general.

Literature

1) Johnson, W. J., Stavric, B., Chartrand, A.: Proc. Soc. Exp. Biol. Med. 131, 1969
2) Stavric, B., Clayman, S., Gadd, R. e. A., Hébert, D.: Pharm. Res. Comm. 7, 2 : 117, 1975
3) Bonardi, G., Vidi, A., Pharm. Res. Com. 5, 2 : 125, 1973
4) Hatfield, P. J., Simmonds, H. A., Cameron, J. S., Jones, A. S., Cadenhead, A.: Purine Metabolism in Man. ED. O. Sperling, A. DeVries, Plenum Press 1974
5) Musil, J.: Proc. Int. Congress Physiology 1974, p. 1175
6) Musil, J., Sandow, J.: in Amino Acid and Uric Acid Transport, Ed. S. Silbernagl (in print)

IN VITRO INVESTIGATIONS ON THE INFLUENCE OF ANTIRHEUMATIC DRUGS ON PURINE PHOSPHORIBOSYLTRANSFERASES AND THEIR POSSIBLE CLINICAL CONSEQUENCE

G. Partsch, I. Sandtner, G. Tausch and R. Eberl

Ludwig Boltzmann-Institute for Rheumatology and Balneology, Kurbadstrasse 1o, A-11o7 Vienna, Austria

According to the chronic course of rheumatic diseases drug therapy is necessary for a long time. Because of the great variability of diseases which are summarized under the term "rheumatic diseases" different drugs have been developed. Most of them have analgesic and anti-inflammatory effects. Today these pharmaceuticals will be divided into nonsteroidal anti-inflammatory compounds, corticosteroids, drugs for chrysotherapy, anti-malarial drugs, antigout drugs and for special cases of septic arthritis antibiotics. The term antirheumatic is used in a general sense and indicates that the drug has a good effect in the treatment of human rheumatic disease, but its precise action is not completely clear. On the other hand, there is evidence of a lot of side effects being caused by the application of antirheumatic drugs (Mathies, 1). Most of the side effects have a clinical background, whereas biochemical effects are not usually seen. We therefore tested some antirheumatic drugs for their influence on the purine-phosphoribosyltransferases.

MATERIALS AND METHODS

The enzyme activities were estimated in erythrocyte lysate of pooled blood and leucocyte homogenates of patients without any joint diseases. Additional studies were performed with selected drugs to find out if there is any effect on these enzymes in whole cells.

Blood of the probands was collected in heparinized tubes and after gently mixing the erythrocytes or lymphocytes were prepared by a ficoll-urografin method (2). After washing the preparations with physiological saline the erythrocytes were lysed by dilution with water. The leucocytes were homogenized (3) together with TRIS-HCl buffer (o,1 M, pH 7,4 + 5 mM $MgCl_2$) and then frozen at - 2o° C. The homogenates were thawed and centrifuged for 11 min. at 12.ooo U/min., as well as the erythrocyte lysate. These preparations were used in the tests.

In those experiments where the effect of antirheumatic drugs was investigated on whole cells heparinized blood was incubated at 37° C for 2 hours with oxyphenbutazone (12o µg/ml) or flufenamic acid (4o µg/ml). Afterwards lymphocytes and erythrocytes were separated by the gradient technique and homogenized as described before.

The method used for the measurement of the enzyme activities was reported earlier (4). The values obtained from the liquid scintillation counter were calculated in nMol transposed purine base per mg protein per time and a regression analysis was performed. For each volunteer seven points were used for the calculation.

RESULTS AND DISCUSSION

As pointed out in Table 1 2 of the 5 pharmaceuticals tested on the purine-PRT showed a decrease of the enzyme activity. There was a moderate influence (Figure 1) on the A-PRT by oxyphenbutazone (reduction of 41 %) and flufenamic acid (reduction of 34 %) whereas the effect of both drugs on G-PRT and H-PRT was low. Oxyphenbutazone reduced the G-PRT activity to about 11 % and the H-PRT to 14 %, whereas the flufenamic acid diminished the G-PRT activity to 15 % and the H-PRT to 1o %.

Based on these results we wanted to know if there is also an effect of the purine-PRTs in human leucocyte homogenates. We therefore tested both drugs which were found to affect in some way the purine-PRTs in the erythrocytes on the homogenates of separated lymphocytes.

	A-PRT	G-PRT	H-PRT
AZAPROPAZONE	-	-	-
OXYPHENBUTAZONE	++	+	+
FLUFENAMIC ACID	++	+	+
PREDNISOLONE	-	-	-
INDOMETACINE	-	-	-

Table 1. The effect of tested antirheumatic drugs (5o µg %) on purine-PRTs of human erythrocytes.

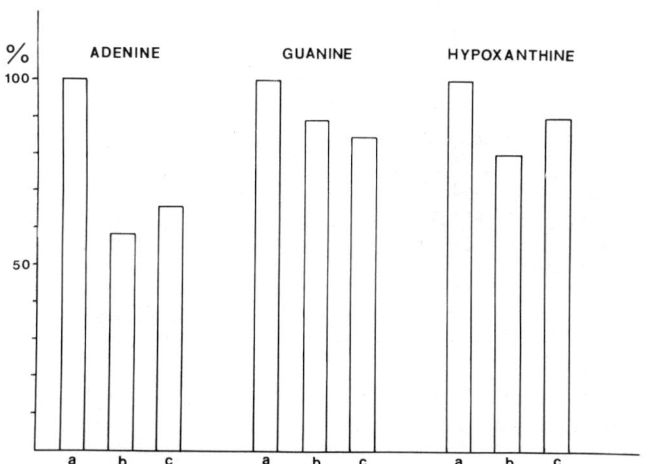

Figure 1. Effect of oxyphenbutazone (b) and flufenamic acid (c) on the purine-PRT from human erythrocytes in % of the control (a).

In Table 2 the purine-PRT activities of the untreated lymphocytes are shown. It can be seen that the A-PRT activity in lymphocytes has a dominant role. The A-PRT activity is 3,46 times higher than the G-PRT and 4,8 times higher than the H-PRT.

The effect of oxyphenbutazone and flufenamic acid on the enzymes of the human lymphocytes are summarized in Figure 2. It is apparent that as well as in erythrocytes the most effective decrease of the enzyme activity occurs in the A-PRT but the effect on G-PRT and H-PRT is much greater than in erythrocytes. Of both these drugs flufenamic acid exhibits the greater reduction of the enzyme activity.

These experiments done with the cell free extracts of erythrocytes and lymphocytes do not allow any conclusion of in vivo effects. First of all the used concentrations of the pharmaceuticals cannot be transposed in clinical terms and the experiments were done with disrupted cells.

To approximate clinical conditions we incubated heparinized blood with the pharmaceuticals. As one can see in Figures 3 and 4 both pharmaceuticals reduce the enzyme activities also under these test conditions. There is no great difference between oxyphenbutazone and flufenamic acid. The A-PRT similar to earlier studies exhibits a much greater decrease in enzyme activity than G-PRT and H-PRT do. The reduction is approx. 3o % for A-PRT in lymphocytes respectively 8o % in erythrocytes. Contrary to lymphocytes (around

	A-PRT	G-PRT	H-PRT
H.H. ♀ (49 A)	222,o	7o,4	55,2
I.S. ♀ (26 A)	187,o	66,8	49,6
S.G. ♂ (4o A)	226,o	7o,o	55,o
K.L. ♀ (59 A)	326,o	72,8	56,2
R.H. ♂ (31 A)	281,o	71,6	57,5
E.T. ♀ (38 A)	275,o	73,4	43,6
C.K. ♀ (37 A)	24o,o	79,6	47,3
W.M. ♀ (2o A)	216,o	64,8	44,2
K.S. ♂ (61 A)	236,o	68,3	49,6
	245,o ± 41,7	7o,8 ± 4,3	5o,9 ± 5,2

Table 2. The purine-PRT activities from human lymphocytes (nMol nucleotide/mg protein/h).

lo % inhibition) both compounds are more effective on G-PRT and H-PRT of erythrocytes (nearly 2o % inhibition).

In view of the steady increase of hyperuricemia and gout in the last years all factors which are able to attribute disturbances of the purine metabolism should be observed carefully. As we could show, some pharmaceuticals are able to produce a reduction of the purine-PRT activities in erythrocytes as well as in lymphocytes. Because of the interference of oxyphenbutazone and flufenamic acid

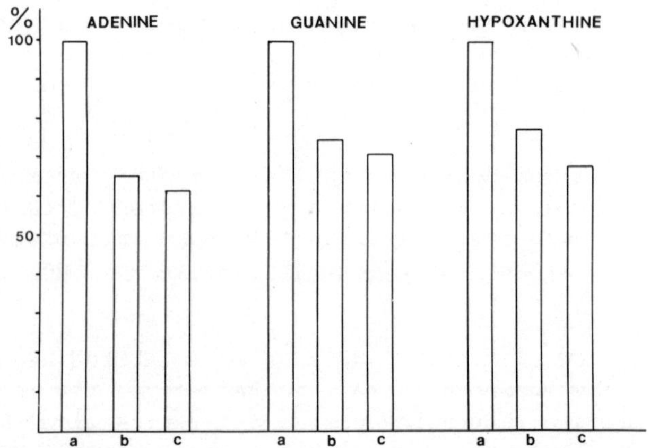

Figure 2. Effect of oxyphenbutazone (b) and flufenamic acid (c) on the purine-PRT from human lymphocytes in % of the control (a).

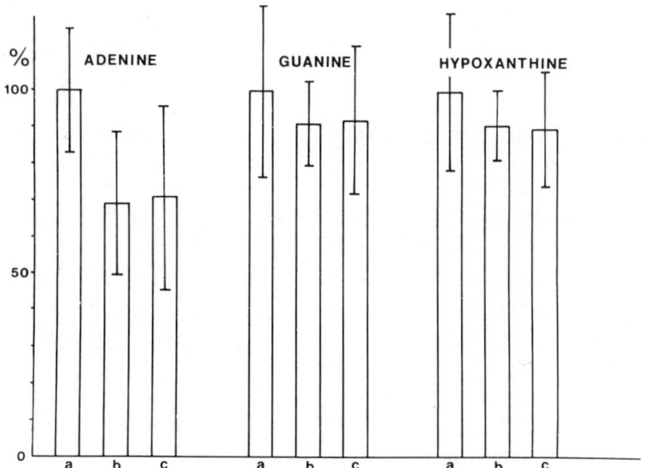

Figure 3. Influence of oxyphenbutazone (b) and flufenamic acid (c) on the purine-PRT of intact human erythrocytes in % of the control (a).

with these enzymes after treatment of whole cells it seems possible that an interaction also occurs in vivo conditions. But we should take into account that the medication of the chronic rheumatic diseases require a long term treatment and that the concentration of these compounds is raised over a long period. There is no evi-

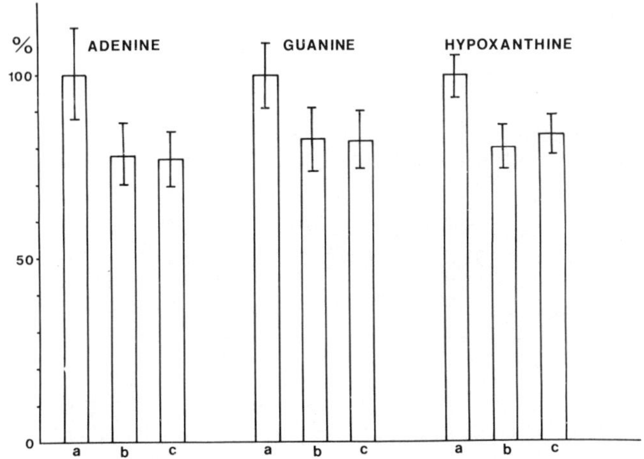

Figure 4. Influence of oxyphenbutazone (b) and flufenamic acid (c) on the purine-PRT of intact human lymphocytes in % of the control (a).

dence that purine-PRT of liver cells were also injured by the application of the drugs but we are able to believe this fact with great probability.

There is an important question for the clinician: Does the reduction of A-PRT and G- H-PRT cause an increase of the uric acid level? Kelley (5) reported that in a case with A-PRT deficiency no overproduction of uric acid could be detected but other papers (6, 7) dealing with the A-PRT deficiency point to partly increased uric acid levels. Because of the reduction of all three enzyme activities by the above compounds uric acid level may be increased. This does not mean that the level has to reach pathological values. But if one normally has a high uric acid level the treatment with oxyphenbutazone or flufenamic acid may lead to hyperuricemia. Results of investigations of Latham (8) done with flufenamic acid correspond with this opinion. Worthington (9) reported about the use of a new anti-inflammatory drug whereby in some cases increased uric acid levels could also be observed depending possibly on an interaction of the substance with the purine-PRT. Even if raised uric acid levels can be observed in some cases the overproduction of uric acid induced by long time medication should be taken into consideration.

REFERENCES

1. Mathies, H.: Klinik und Therapie der Nebenwirkungen. G. Thieme, Stuttgart, 923 (1973).

2. Wottawa, A., Klein, G., Altmann, H.: Wien.klin.Wschr. 86, 161 (1974).

3. Potter, V.R., Elvehjem, C.A.: J.biol.Chem. 114, 495 (1936).

4. Partsch, G., Altmann, H., Eberl, R.: Purine Metabolism in man. Advance in experimental medicine and biology. Plenum Press, New York, Vol. 41A, lo3 (1974).

5. Kelley, W.N., Levy, R.L., Rosenbloom, F.J., Henderson, J.F., Seegmiller, J.E.: J.Clin.Invest. 47, 2281 (1968).

6. Fox, I.H., Kelley, W.N.: Purine Metabolism in man. Advance in experimental medicine and biology. Plenum Press, New York, Vol. 41A, 319 (1974).

7. Emmerson, B.T., Gordon, R.B., Thompson, L.: Purine Metabolism in man. Advance in experimental medicine and biology. Plenum Press, New York, Vol. 41A, 327 (1974).

8. Latham, B.A., Radcliff, R., Robinson, R.G.: Amer.Phys.Med. 9, 242 (1966).

9. Worthington, W.W.: Scand.J.Rheum., Suppl. 8, So2-o4 (1975).

INFLUENCE OF URATE ON CONNECTIVE TISSUE METABOLISM

H.W. Stuhlsatz, W. Enzensberger and H. Greiling

Department of Clinical Chemistry

Technical University Aachen, FRG

Monosodium urate deposits almost exclusively in the connective tissue of patients with gout. Thereby the question arises why the connective tissue is the predilective locus for monosodium urate. Several theories about the localised deposition of monosodium urate crystals in connective tissue demonstrate the interaction of urate with the components of the connective tissue.

1. We have formerly postulated that the urate deposition is increased by elevated hydrogen ion concentration in the synovial fluid of the inflamed joint in connection with cationic exchange by chondroitin sulfate (1).

2. The urate solubility is decreased by the excluded volume effect of the proteoglycans as first demonstrated by LAURENT (2).

3. The solubility of monosodium urate is augmented by native proteoglycans in human cartilage but not by their degraded forms. Increased lysosomal proteoglycan-degrading enzyme activity in the inflamed gouty joint therefore decreases the solubility.

The last theory of KATZ (3) preposes a high concentration of lysosomal enzymes in cartilage. KATZ has shown that hyaluronate glycanohydrolase degrades the proteoglycan and diminishes the solubility of urate in a native proteoglycan solution. But the hyaluronate glycanohydrolase activity in cartilage and in the inflamed cartilage too was not measured and nobody till now has found hyaluronate glycanohydrolase in cartilage. For the degradation of proteoglycans the following steps are necessary:
1. depolymerisation of hyaluronate of the proteoglycan aggregates, 2. degradation of the protein core, 3. degradation of the linkage

region, 4. degradation of the polysaccharide chain. Several lysosomal and other enzymes are involved in the catabolism of the proteoglycans, i. e. lysozyme in the disaggregation of proteoglycan aggregates, cathepsin D in the degradation of the protein core, and a set of other lysosomal enzymes degrading the carbohydrate chain (4). These enzymes are also localised in the leucocytes. We have detected a chondroitin sulfatase activity and also a keratan sulfatase activity in the polymorphnuclear leucocytes. The concentration of these enzymes in the cartilage is very low. In the synovial fluid of chronic joint diseases there is an increase of lysosomal enzymes. The highest activity is found in the synovial fluids of patients with rheumatoid arthritis (table 1).

	N-acetyl-ß-D-glucos-aminidase	ß-glucuro-nidase	α-manno-sidase	arylsul-fatase
rheumatoid arthritis	16.5 (74)	1.48 (77)	1.12 (77)	0.83 (75)
osteo-arthritis	7.4 (9)	0.41 (9)	0.50 (9)	0.26 (9)
gout	6.0 (4)	0.80 (3)	0.64 (4)	0.29 (4)

Table 1: Mean values [mU/ml] (number of cases) of lysosomal enzymes in the synovial fluid.

The lysosomal enzymes can be inhibited by various antiinflammatory drugs, which as prednisolon acetate stabilize the lysosomal membrane or as arteparon inhibits the lysosomal enzymes, i. g. ß-N-acetylglucosaminidase (fig. 1).

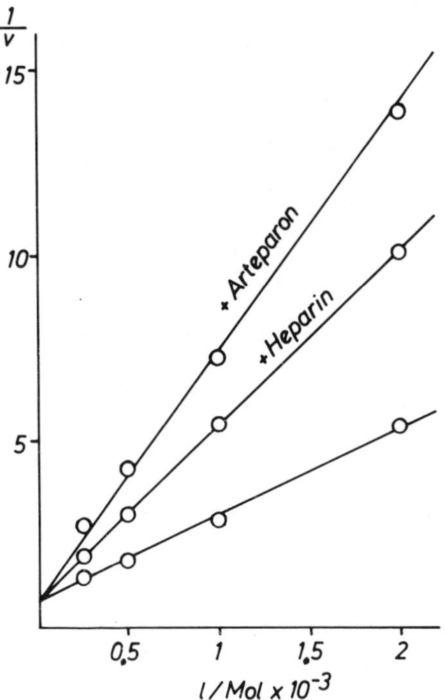

Fig. 1: Competitive inhibition of the ß-N-acetylglucosaminidase from kidney lysosomes by arteparon and heparin. Abscissa: 1/mMol p-nitrophenyl-ß-N-acetylglucosaminide; Ordinate: 1/liberated µMol N-acetylglucosamine groups Concentration of arteparon and heparin: 91 µg/ml. K_M: $3.7 \cdot 10^{-3}$M; K_I (arteparon): $0.7 \cdot 10^{-4}$ M; K_I (heparin): $1.6 \cdot 10^{-4}$M.

After intraarticular injection of prednisolon acetate or arteparon the viscosity and the hyaluronate concentration in the synovial fluid increase. In the inflamed gouty joint, there is a decrease of viscosity and a decrease of the hyaluronate concentration (5), perhaps due to an increased superoxide radical concentration which depolymerises hyaluronate (7). Depolymerized hyaluronate activates the phagocytosis of monosodium urate crystals (6). High polymerised hyaluronate inhibits the phagocytosing phenomenon. We have studied the connective tissue metabolism by tracer experiments concerning the incorporation of $^{35}SO_4$ into the glycosaminoglycans of various connective tissues. The hyperuricemic drug benzbromaron is a competitive inhibitor of the PAPS-chondroitin sulfotransferase (8). In some patients benzbromaron provokes a gouty attack. It is possible that the inhibition of the connective tissue metabolism by benzbromaron may activate the precipitation of monosodium urate in the connective tissue by stimulating the lysosomal degradation. In the cornea and cartilage the in vitro-incorporation of radioactive sulphate into proteokeratan sulfate and proteochondroitin 4- and 6-sulfates is not influenced by an urate concentration even of 10^{-3} M (table 2).

	hyaline cartilage		cornea	
	without urate	with urate	without urate	with urate
cpm/µMol GAG	42 512	44 750	56 342	55 015
n	15	15	4	6
s	2 272	3 125	2 425	3 615
V [%]	5.34	6.98	4.30	6.58

Table 2: Incorporation of $^{35}SO_4$ into the glycosaminoglycans (GAG) of bovine nasal septum cartilage and calf cornea. 10 slices (thickness: 1 mm) of bovine nasal septum cartilage (total wet weight: about 2.5 g) were incubated with 50 µCi $^{35}SO_4$ in 15 ml 0.15 M Krebs-Sörensen buffer pH 7.4 at 37°C for 2 h with and without urate (10^{-3}M) after a preincubation period of 30 min for the urate-containing solutions. Incorporation into calf cornea was performed in the same way with 5 calf corneae (total wet weight: 1.5 g) in 20 ml Krebs-Sörensen buffer.

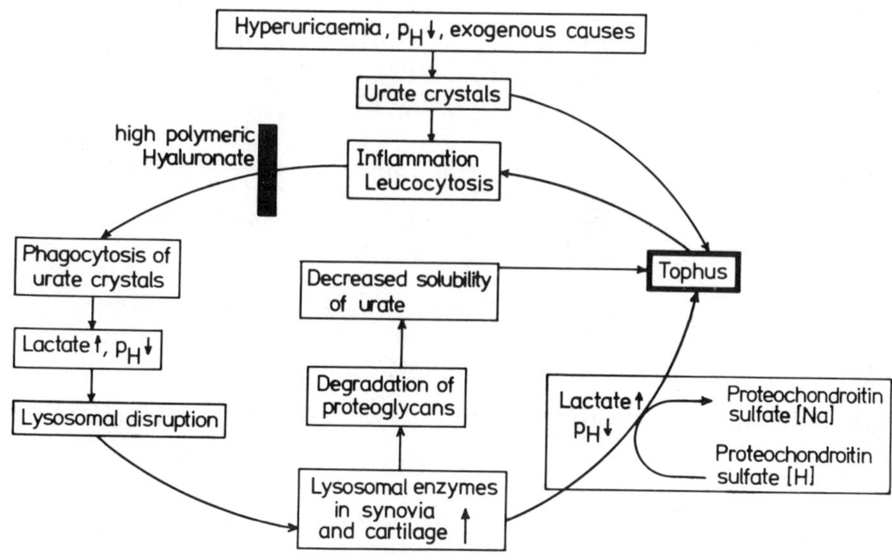

Fig. 2: Proposed mechanism for the crystallisation of monosodium urate and tophus formation

In fig. 2 a mechanism for the crystallisation of monosodium urate in the connective tissue is proposed. In our opinion there are three points essential:

1. The cycle of inflammation is activated by depolymerized hyaluronate which facilitates the phagocytosis of monosodium urate crystals.
2. The lysosomal enzymes degrade the proteoglycan aggregates and decrease the solubility for monosodium urate.
3. The proteochondroitin sulfate acts as cation exchanger performing few amounts of free uric acid from monosodium urate.

References:

1. Greiling, H., Herbertz, T., Schuler, B., and Stuhlsatz, H.W. Z. Rheumaforschung, 21, 50 (1962)

2. Laurent, T. C., Nature (Lond), 202, 1334 (1964)

3. Katz, W. A., Arthritis and Rheumatism, 18, 751 (1975)

4. Greiling, H., in: Vogel (Ed.) Connective Tissue and Ageing, Exerpta Medica, Amsterdam, 168 (1973)

5. Stuhlsatz, H.W., A. Eberhard, H. Kristin, O. Vojtisek, H. Greiling, in: Verhandlungen der Deutschen Ges. für Rheumatologie, Bd. IV, Steinkopff-Verlag Darmstadt, in press

6. Brandt, K.D., Clinica Chimica Acta, 55, 307 - 315 (1974)

7. McCord, J.M., Science, 185, 529 - 531 (1974)

8. Greiling, H., M. Kaneko, in: O. Sperling, H. de Vries and I.B. Wyngarden (Ed.) Purine Metabolism in Man, Plenum Press, New York - London, 693 (1973)

PROTEIN BINDING TO MONOSODIUM URATE CRYSTALS AND ITS EFFECT ON PLATELET DEGRANULATION

Franklin Kozin, Mark Ginsberg, Vladimir Karasek,
John Skosey, Daniel McCarty
The Medical College of Wisconsin and The University of Chicago Medical School
8700 W. Wisconsin Ave., Milwaukee, Wis. 53226 and
950 E. 59th Street, Chicago, Ill. 60037

The membranolytic effects of microcrystalline monosodium urate (MSU) on biologic membranes are well documented for a number of cells and subcellular structures [1-3]. This process has been ascribed to hydrogen-bond formation between the crystal and membrane phospholipids and is blocked by strong hydrogen acceptors such as proteins or anionic polymers [1]. Since the acute inflammatory response of gouty arthritis occurs in protein-rich joint fluid, we have examined the protein adsorptive properties of MSU crystals. The effects of adsorbed protein on polymorphonuclear leukocytes and platelet degranulation were studied and are reported in detail.

MATERIALS AND METHODS

All glassware was acid washed and doubly siliconized prior to use. Doubly deionized, glass-distilled water was used for solutions or crystal synthesis [4]. 0.15 M Sodium Chloride buffered with 0.001 M Sodium Phosphate, pH 7.4 (PBS), was employed routinely. Radioisotopic measurements were performed in Tri-Carb Scintillation Spectrometers, Models 3001 or 3330 (Packard Instrument Co., Inc., Downers Grove, Ill.).

The following were obtained from Sigma Chemical Co., St. Louis, Mo.: Human Cohn Fraction II, bovine serum albumin, egg-white lysozyme, bovine pancreatic ribonuclease, ovalbumin, and beta lactoglobulin. ^{125}I as NaI^{125} in 0.1 M NaOH and ^{14}C-serotonin were purchased from New England Nuclear Co., Boston, Mass. Proteins were radio-iodinated by the Chloramine-T method [5], and free iodine removed by exhaustive dialysis. Prior to use, all protein solutions were passed through Type HAWP, 0.45 micron filters (Millipore Corp., Bedford, Mass.) to remove particulate material. Protein concentrations were measured by standard techniques [6]. Polyvinyl-pyridine-N-oxide (PVPNO) was kindly supplied by Dr. N.H. K. Kiesselbach, Bayer Pharma Forschungszentrum, Aprother Weg, West Germany.

The following method for quantitatively measuring protein binding to crystals was developed. Protein solutions were incubated with crystals for 10 minutes at room temperature in a shaking water bath. Thereafter crystals were separated from the solution by centrifugation at 3000 x g x 10 min, washed x 3 in diluted PBS, and the amount of protein associated with the crystal pellet measured radio-isotopically. This method was reproducible and accurate over a wide range of protein concentrations [7].

All platelet studies were carried out in plastic. Platelets were isolated from unmedicated, normal donors by differential centrifugation, incubated with ^{14}C-serotonin for 30 min at 37°C, washed, and resuspended in 0.03 M TRIS buffer containing 0.12 M NaCl and 0.005 M glucose, pH 7.4. 0.2 cc test solution was added to 0.8 cc platelet suspension, and the mixture incubated for 10 min at 37°C in a shaking water bath. The reaction was stopped by placing the tubes in iced methanol, and the supernatant isolated by centrifugation. The amount of ^{14}C-serotonin, lactic dehydrogenase (LDH) [8], and beta-glucuronidase (BGLU) [9] in the supernatant was measured and compared with the total activity of these substances in the untreated platelet suspension after sonication. Results were expressed as percent release: activity in supernatant/total activity x 100.

EXPERIMENTS

Human Serum Adsorption: MSU crystals were incubated with normal human serum (NHS) diluted 250-, 100-, 50-, and 10-fold. After crystals were separated and washed, the adsorbed protein was eluted from crystals, as described previously [7], concentrated, and analyzed by cellulose acetate and immunoelectrophoresis and by quantitative radial-immunodiffusion against specific antisera to IgG, IgA, and IgM (Hyland Products, Inc., Costa Mesa, Calif.).

Adsorption Isotherms: Cohn Fraction II (CF-II), bovine serum albumin (BSA), and egg-white lysozyme (LYS) at concentrations ranging from 0.2 to 6.0 mg/ml were incubated with MSU crystals at constant concentration, and the amount of binding measured.

Relative Protein Binding: CF-II, BSA, LYS, beta-lactoglobulin (BLG), ovalbumin (OVA), and bovine pancreatic ribonuclease (RIB) at final concentration 0.5 mg/ml were incubated with crystals and the amount of binding compared.

Platelet studies: MSU crystals were incubated with CF-II, 2.0 mg/ml, or PVPNO, 2.0 mg/ml, washed x 3, and resuspended in PBS to a final crystal concentration of 0.8 mg/ml. 0.2 ml of these preparations, uncoated crystals at the same concentration, or buffer were mixed with 0.8 ml platelet suspension and incubated. The release of ^{14}C-serotonin, LDH, and BGLU was measured.

RESULTS

Human Serum Adsorption: At all dilutions of NHS, gamma globulin was found to be adsorbed in greatest quantity (Fig. 1). Immunoelectrophoretic and radial immunodiffusion studies demonstrated that IgG was the predominant species present; at lowest serum dilutions, trace amounts of IgM, but not IgA, were present. Small amounts of albumin and a second unidentified protein also were present.

Adsorption Isotherms: Increased binding of all proteins occurred at higher starting concentrations (Fig. 2). The CF-II isotherm showed a rapid initial slope reaching a plateau at 3.5 - 5.0 mg/ml. The isotherms of BSA and LYS demonstrated a gradual initial slope to a plateau and later rapid rise. The latter phenomenon was not seen with CF-II at the concentrations examined.

Relative Protein Binding: The relative affinities of several proteins for MSU crystals are shown in Table I. These are expressed as percent bound (amount bound/starting amount x 100) as an indication of the number of molecules bound. There was no specific effect of molecular weight or protein charge (as isoelectric point, P_I).

Table I

Protein	P_I [10-11]	Percent Bound
CF-II	6.8 (Approx.)	10.0
LYS	11.0	2.5
RIB	5.8	2.0
BLG	5.2	0.4
BSA	4.8	0.1
OVA	4.6	< 0.1

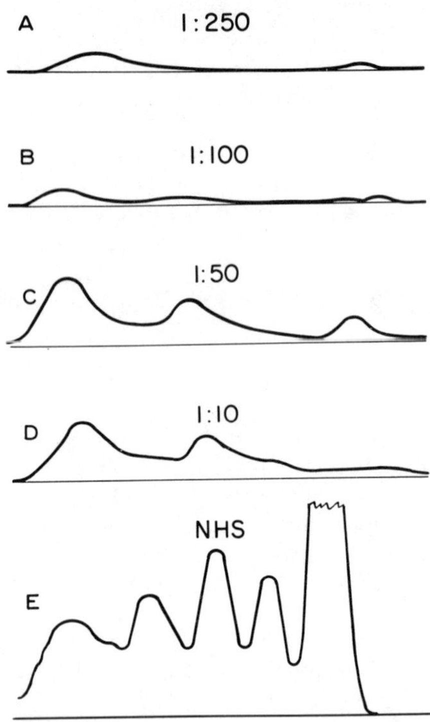

Fig. 1. Cellulose acetate electrophoresis patterns of eluted protein from MSU crystal incubates with serum diluted 1:250 (A), 1:100 (B), 1:50 (C), and 1:10 (D) compared with NHS.

Platelet Studies: MSU crystals themselves stimulated serotonin release in the apparent absence of LDH or BGLU liberation (Fig.3). This effect was found to result from direct binding of LDH and BGLU to the uncoated crystal [12]. When our data were corrected for this phenomenon, significant amounts of LDH and BGLU were liberated into the supernatant, indicating that lysis of the platelet occurred. Coating of MSU crystals with CF-II produced a marked enhancement of serotonin and BGLU release, but not of LDH release, suggesting degranulation in the absence of cytolysis. PVPNO adsorbed to crystals inhibited release of all constituents.

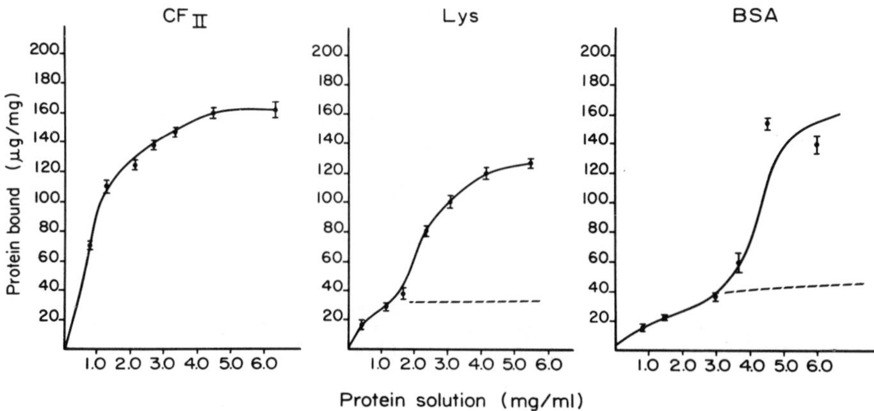

Fig. 2. Isotherms for Cohn Fraction II (CF-II), lysozyme (LYS), and bovine serum albumin (BSA) adsorption to MSU crystals. CF-II isotherm shows a rapid saturation of the crystal, whereas LYS and BSA show a tendency to early saturation (dotted line) and late rapid increase in adsorption.

DISCUSSION

The acute inflammatory response in gouty arthritis depends upon the prior interaction of microcrystalline monosodium urate (MSU) and polymorphonuclear leukocytes (PMN) [3,13,14]. Morphologic and biochemical studies have shown that the PMN readily engulfs the crystal into the primary phagosome [2,3]. Fifteen to thirty minutes later disruption of the phagolysosomal and cellular membrane occurs from within, suggesting that the outer surface of the PMN is protected from the lytic effect of MSU crystals. These observations resemble those of Wallingford and McCarty who found that the addition of plasma proteins to the incubation medium blocked MSU crystal-induced membranolysis of erythrocytes [1].

The phenomenon of protein binding to MSU crystals, demonstrated in this report, may explain their behavior in physiologic fluids. Our studies show that immunoglobulin G (IgG) was adsorbed from normal human serum in greatest amount, although small quantities of IgM, albumin, and another unidentified protein also were found in the eluted material. These results and studies of several isolated proteins suggest that IgG is adsorbed selectively. Binding isotherms of Cohn Fraction II (CF-II), and egg-white

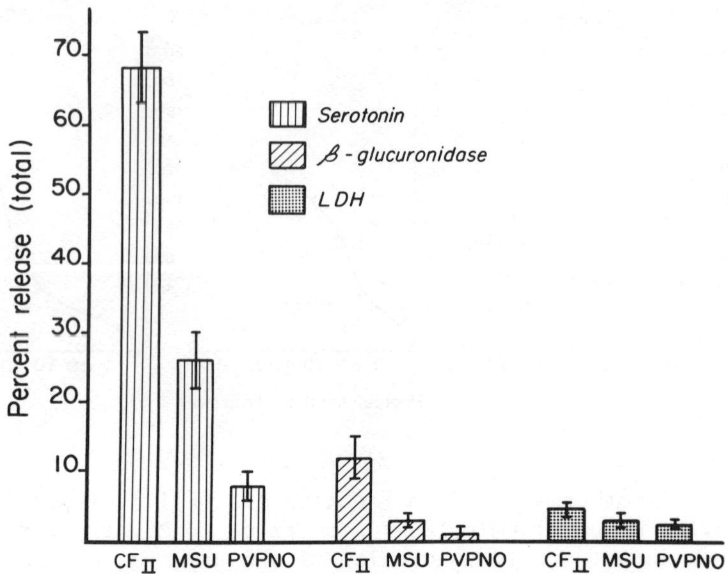

Fig. 3. Percent release of serotonin, β-glucuronidase, and LDH from platelets exposed to MSU crystals (MSU), MSU crystals with adsorbed CF-II (CF-II), or MSU crystals with adsorbed PVPNO (PVPNO). See text for discussion.

lysozyme were examined over a wide range of concentrations. A high affinity configuration, characterized by a rapid initial slope to saturation of "binding sites" [15], was seen with CF-II but not BSA or LYS. The latter isotherms showed a gradual initial slope and later rapid rise, suggesting that protein aggregation occurred at the crystal surface [15].

Binding of several proteins at equal starting concentrations was compared and provided the following order of affinity: CF-II > LYS > RIB > BLG > BSA > OVA. Several variables affecting this reaction were studied and showed that binding increased at lower pH, ionic strength, and temperature [7]. This suggests that the mechanism of binding is predominantly ionic but that other physical forces also may be present.

Previous studies have shown platelet endocytosis of particulate matter including MSU crystals [3,16]. The biochemical correlates of this phenomenon were studied here, using serotonin and beta-glucuronidase release to indicate degranulation and lactic dehydrogenase release to indicate loss of cell membrane integrity or lysis. Adsorbed materials exerted a profound effect on these functions. Uncoated MSU crystals readily induced platelet lysis. Prior coating of crystals with PVPNO blocked this effect as it did in the erythrocyte model [1]. Prior coating with CF-II, on the other hand, produced a marked enhancement of degranulation without parallel disruption of the cell membrane.

We should like to propose the following hypothesis of crystal-induced inflammation based on studies reported here. Once introduced into the joint space, MSU crystals readily adsorb IgG. This serves at once to block the surface forces responsible for membranolysis and provoke endocytosis by stimulating surface F_c receptors on the PMN's. The surface protein is later stripped from the crystal within the phagolysosome by enzymatic digestion or elution, allowing lysis to proceed.

REFERENCES

1. Wallingford WR, McCarty DJ: Differential Membranolytic Effects of Microcrystalline Sodium Urate and Calcium Pyrophosphate Dihydrate. J Exp Med 133:100, 1971.

2. Weissmann G, Rita GH: Molecular Basis of Gouty Inflammation: Interaction of Monosodium Urate Crystals With Lysosomes and Liposomes. Nature (New Biol) 240:167, 1972.

3. Schumacher HR, Phelps P: Sequential Changes in Human Polymorphonuclear Leukocytes After Urate Crystal Phagocytosis: An Electron Microscopic Study. Arthritis Rheum 14:513, 1971.

4. McCarty DJ, Faires JS: Comparison of the Duration of Local Anti-inflammatory Effect of Several Adrenocortical Esters: A Bioassay Technique. Curr Ther Res 5:284, 1963.

5. Hunter R: Standardization of the Chloramine-T Method of Protein Iodination. Proc Soc Exp Med 133:980, 1970.

6. Lowry OH, Rosebrough NJ et al: Protein Measurement with the Folin-Phenol Reagent. J Biol Chem 193:265, 1951.

7. Kozin F, McCarty DJ: Protein Binding to Monosodium Urate Monohydrate, Calcium Pyrophosphate DIhydrate, and Silicon Dioxide Crystals. I. Physical Characteristics. Submitted for publication.

8. Bergmeyer HU, Bernt E, and Hess B: Lactic Dehydrogenase in Methods of Enzymatic Analysis. Bergmeyer HU (Ed.), Academic Press, New York, 1963, p. 736.

9. Skosey JL, Chow D, Domgaard E, Sorenson LB: Effect of Cytochalasin-B on Response of Human Polymorphonuclear Leukocytes to Zymosan. J Cell Biol 57:237, 1973.

10. Saber HA: Handbook of Biochemistry, The Chemical Rubber Co., Cleveland, Ohio, 1968.

11. Alurtz RA: Electrochemical Properties of Proteins and Amino Acids. In The Proteins, H Newrath and K Bailey (Eds.), Academic Press, New York, 1953.

12. Ginsberg MH, Kozin F, Chow D, May J, and Skosey J: The Time Course of Monosodium Urate Crystal-Induced Release of Polymorphonuclear Leukocytes. Submitted for publication.

13. Faires JS and McCarty DJ: Acute Arthritis in Man and Dog After Intra-Synovial Injection of Sodium Urate Crystals. Lancet 2: 682, 1962.

14. Chang Y, Gralla EJ: Suppression of Urate Crystal-Induced Joint Inflammation by Heterologous and Anti-polymorphonuclear Leukocyte Serum. Arthritis Rheum 11:145, 1968.

15. Giles CH and MacEwan TH: Classification of Isotherm Types for Adsorption from Solution. Proc Int Congr of Surf Act II 3:457, 1957.

16. Mustard JF and Packham MA: Platelet Phagocytoses. Ser Haematol 2:168, 1968.

PROTEIN ADSORPTION TO MONOSODIUM URATE CRYSTALS: DIFFERENTIAL RESPONSES OF HUMAN PERIPHERAL BLOOD NEUTROPHILS

John L. Skosey, Frank Kozin, and Mark Ginsberg*

Departments of Medicine, The University of Chicago

Chicago, Illinois, and Medical College of Wisconsin

Milwaukee, Wisconsin

In order for acute gouty arthritis to occur, neutrophils must interact with monosodium urate (MSU) crystals. As a result of this interaction, enzymes, chemotactic factors, and other mediators of the inflammatory response are released from neutrophil lysosomes [1]. We have observed previously that MSU crystals adsorb gamma globulin, albumin, and other proteins found in serum and joint fluid [2]. This study was designed to demonstrate the effects of coating of MSU crystals with proteins on the phlogistic responses of neutrophils to crystals.

For these experiments, leukocyte suspensions containing 85% or more neutrophils were isolated from blood of normal subjects and suspended in Krebs-Ringer phosphate buffer containing 0.01% gelatin. The cells were then incubated with MSU crystals which had either no protein coat or which had been previously coated with various proteins. At the end of the incubation period, cells were separated from the incubation medium by centrifugation. The incubation medium was saved for the assay of cytoplasmic lactate dehydrogenase (LDH) and lysosomal β-glucuronidase, α-mannosidase, and lysozyme. The details of these experimental procedures have been previously given [3].

MSU crystals, when ingested by neutrophils, cause membranolysis which results in release of cytoplasmic lactate dehydrogenase into the incubation medium [4]. This membranolytic effect is proportional to crystal concentration between 2 and 8 mg/ml, and is inhibited by prior coating of crystals with gamma globulin (as Cohn fraction II) (Fig. 1).

*Present address: Department of Immunopathology, Scripps Clinic and Research Foundation, LaJolla, California.

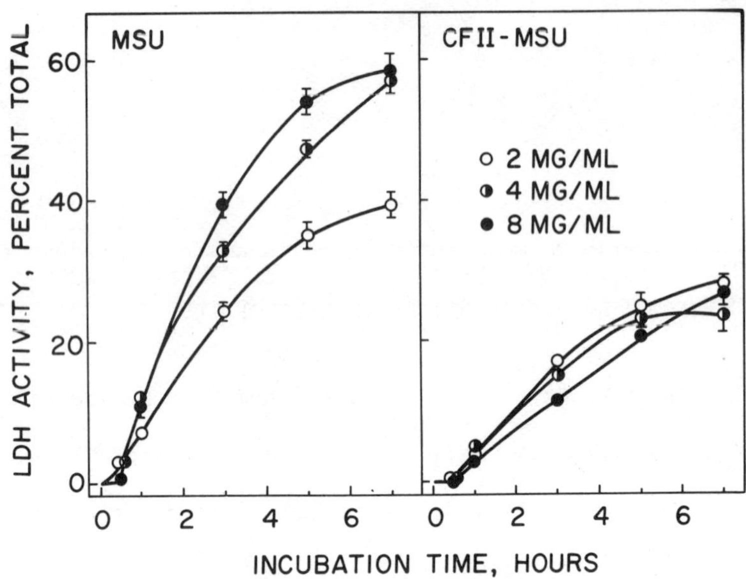

Fig. 1.--Effect of coating of monosodium urate crystals with Cohn fraction II on release of lactate dehydrogenase (LDH) activity from human peripheral blood leukocytes. Cells were incubated with MSU crystals (MSU) or with MSU crystals coated with Cohn fraction II (CF II-MSU). After varying times of incubation, LDH activity in the medium was measured. Results are expressed as the percentage of LDH activity initially present in cells. Each point represents the mean of values obtained from triplicate incubations; the bars represent ±1 standard error of the mean.

Preliminary studies demonstrated that MSU crystals at a lower concentration, 0.4 mg/ml, had only negligible cytolytic effects [5]. Further studies were done using MSU crystals at this concentration in order to study the effect of crystals under conditions where cytolysis was minimal. When MSU crystals were coated with Cohn fracti II, the release of lysosomal α-mannosidase, β-glucuronidase and lysozyme in response to crystals was enhanced. A similar enhancement wa observed when crystals were coated with immunoglobulin G (IgG) (Table

Further studies were done to investigate the specificity of the effects of protein coating of MSU crystals on their effects on isolated leukocytes. The responses of release of lysosomal enzymes and cytolysis (LDH release) were inhibited when crystals were precoated with human serum albumin, fibrinogen, or immunoglobulins A or M (Table II). The disproportionately low recovery of β-glucuronidase from incubation medium can be explained by our previous observation that this enzyme activity is adsorbed to MSU crystals [5].

TABLE I

Effect of Cohn Fraction II- and IgG-Coated MSU Crystals on Release of Lysosomal Enzymes and LDH from Human Peripheral Blood Leukocytes

MSU crystals, mg/ml	Crystal coat	Per cent total enzyme released into incubation medium in 1 h, mean ± standard error			
		LDH	α-mannosidase	β-glucuronidase	lysozyme
Experiment 1					
0	—	1.5 ± 0.6	0.6 ± 0.1	1.1 ± 0.2	1.1 ± 0.3
0.4	—	3.2 ± 0.5	4.3 ± 0.2	2.9 ± 0.4	5.8 ± 0.5
0.4	CF II	3.8 ± 0.2	9.2 ± 0.9	6.8 ± 0	10.4 ± 1.1
Experiment 2					
0	—	1.6 ± 0.3	0.8 ± 0.2	1.0 ± 0.2	3.5 ± 0.6
0.4	—	2.1 ± 0.1	3.6 ± 0	1.2 ± 0.1	7.4 ± 0.2
0.4	IgG	1.9 ± 0.1	5.2 ± 0.1	2.1 ± 0	11.1 ± 0.1

Cells were incubated for one hour with buffer alone, with MSU crystals, or with MSU crystals precoated with Cohn fraction II (CF II) or immunoglobulin G (IgG). Enzyme activities in incubation medium were then measured.

TABLE II

Effect of Human Serum Albumin- , Fibrinogen- , IgA- , or IgM-Coated Monosodium Urate Crystals on Release of Lysosomal Enzymes and LDH from Human Peripheral Blood Leukocytes

MSU crystals, mg/ml	Crystal coat	Per cent total enzyme released into incubation medium in 1 h, mean ± standard error			
		LDH	α-mannosidase	β-glucuronidase	lysozyme
0	—	2.6 ± 0.2	0.7 ± 0.1	0.7 ± 0.1	1.3 ± 0.4
1	—	10.4 ± 1.0	11.7 ± 0.8	0.6 ± 0.4	19.0 ± 1.7
1	albumin	4.9 ± 0.2	3.3 ± 0.2	0 ± 0	7.0 ± 0
1	fibrinogen	6.3 ± 0	4.2 ± 0.3	0.2 ± 0.1	9.5 ± 0.6
1	IgA	7.8 ± 0.4	9.7 ± 0.4	1.4 ± 0.2	15.8 ± 1.7
1	IgM	5.6 ± 0.5	3.0 ± 0.1	0.4 ± 0.2	7.0 ± 1.1

Cells were incubated for one hour with MSU crystals or with crystals which had been precoated with various proteins. Enzyme activities in the incubation medium were then measured. A membranolytic concentration of MSU crystals was chosen for this experiment.

The release of lysosomal enzymes in response to phagocytable stimuli can be dissociated from phagocytosis by cytochalasin B, an antibiotic which inhibits particle ingestion [see ref. 3]. Cytochalasin B had only minimal effects on the release of lysosomal α-mannosidase or on cytolysis (LDH release) which occurred in response to MSU crystals. However, the antibiotic enhanced the release of α-mannosidase and inhibited cytolysis which occurred when cells were incubated with Cohn fraction II-coated MSU crystals (Table III). We interpret these results as indicating that phagocytosis is not required for stimulation of lysosomal enzyme release by MSU crystals coated with IgG. The cytolytic effect, as previously suggested [4], requires particle ingestion.

TABLE III

Effect of Cytochalasin B (CB) on Responses of Human Peripheral Blood Leukocytes to MSU Crystals and IgG-Coated MSU-Crystals (IgG-MSU)

Crystal	CB	Per cent total enzyme released into incubation medium in 1 h, mean ± standard error	
		α-mannosidase	LDH
None	0	1.6 ± 0.4	3.2 ± 0.3
None	+	2.6 ± 0.6	5.0 ± 0.2
MSU	0	3.9 ± 0.2	4.6 ± 0.1
MSU	+	4.0 ± 0.3	5.3 ± 0.3
IgG-MSU	0	5.3 ± 0.2	4.4 ± 0.7
IgG-MSU	+	8.0 ± 0.3	2.5 ± 0.5

Cells were incubated for 1 h. Crystals were present at 0.4 mg/ml, cytochalasin B at 5 μg/ml.

Many proteins present in plasma and joint fluid are avidly adsorbed to MSU crystals. These results demonstrate that the nature and extent of responses of neutrophils to MSU crystals depend upon th species of protein adsorbed to the crystals. Phagocytosis of uncoated MSU crystals leads to cell lysis and subsequent release of both cytoplasmic and lysosomal enzymes. Prior coating of crystals with human serum albumin, fibrinogen, IgA, or IgM results in inhibition of these responses. Coating of crystals with IgG also results in reduced cytolysis, but, in contrast to the other proteins tested, leads to enhancement of secretion of lysosomal enzymes. We propose that the inflammation of acute gout is initiated when neutrophils interact with MSU crystals with adsorbed IgG. The inflammatory response is inhibited when the other proteins tested in these studies are adsorbed to MSU crystals. Experiments designed to elucidate the factors which control these opposing influences are in progress.

References

1. McCarty, D.J. and F. Kozin. 1975. An overview of cellular and molecular mechanisms in crystal-induced inflammation. Arthritis Rheum. 18: 757-764.
2. Kozin, F. and D.J. McCarty. 1976. Protein binding to monosodium urate monohydrate calcium pyrophosphate dihydrate and silicon dioxide crystals. I. Physical characteristics. In press.
3. Skosey, J.L., E. Damgaard, D. Chow and L.B. Sorensen. 1974. Modification of zymosan-induced release of lysosomal enzymes from human polymorphonuclear leukocytes by cytochalasin B. J. Cell Biol. 62: 625-634.
4. Weissmann, G., R.B. Zurier, P.J. Spieler and I.M. Goldstein. 1971. Mechanisms of lysosomal enzyme release from leukocytes exposed to immune complexes and other particles. J. Exp. Med. 134: 149s-165s.
5. Skosey, J.L., F. Kozin, D.C. Chow and J. May. 1976. Differential responses of human neutrophils (PMN) to monosodium urate crystals (MSU) and MSU coated with gamma globulin (Ig-MSU). Clin. Res. 24: 111A.

SERUM URIC ACID LEVELS IN ENGLAND AND SCOTLAND

J.T. Scott,[1] R.A. Sturge,[1] A.C. Kennedy,[2]
D.P. Hart,[3] and W. Watson Buchanan [2]

1. Kennedy Institute of Rheumatology, London
2. Centre for Rheumatic Diseases, Glasgow
3. University of Birmingham

"In Scotland gout is much less frequently met with than in England . . . and when it does occur is generally in the upper classes . . "

A. B. Garrod in Gout and
Rheumatic Gout, 2nd edition, 1876

INTRODUCTION

The prevalences of gout and of hyperuricaemia are know to vary considerably between population groups in different parts of the world (Healey & Hall, 1970). Exactly 100 years ago, A. B. Garrod wrote that gout was far less common in Scotland than in England, a view which continues to be held, though supportive data have never been presented. Most authorities consider the presence of hyperuricaemia, in the absence of recent hypouricaemic therapy, to be mandatory for a diagnosis of gout and it was therefore decided to compare serum uric acid levels between adults in England and Scotland. At the same time the opportunity was taken to investigate other factors which are possibly related to serum uric acid levels, such as physique, blood urea and social class.

METHODS

A total of 1,103 volunteers took part in the survey, comprisir 341 male and 52 female employees of the Atomic Energy Research Establishment, Winfrith, Dorset, England; 171 male and 202 femal blood donors attending the Regional Blood Transfusion Centre, Birmingham, England; and 337 male employees of three light industry establishments in the Glasgow area, Scotland. (The number of female subjects in Glasgow was too small for statistical analysis). Subjects were interviewed at either blood donor sessions (Winfrith and Birmingham) or routine medical examinations (Glasgow) and asked to donate 10 ml of blood. A record was made of each subject's sex, height, weight and occupation, the last being coded into social class according to the Registrar General's Classification of Occupations (1970).

The serum was immediately separated from each blood sample and stored at -20°C. When all collections had been completed the sera were transported in the frozen state to Charing Cross Hospital, London for analysis. Mean storage time was 6.1 months. Serum urea and uric acid were measured colorimetrically on a Technicon auto-analyser in randomized batches of 30 sera (10 from each of the 3 groups) for each run. The analysis sessions took place over 3 weekends and were carried out by the same senior technician and 2 assistants. Quality control checks were carried out at the beginning and at frequent intervals throughout each session.

RESULTS

Comparability of Populations

For the purpose of analysis the subjects were divided by sex and region of domicile. The mean ages, weights and blood ureas for each group are shown in Table I. The ages were similar apart

Table I

	MEN			WOMEN	
	Winfrith	B'ham	Glasgow	Winfrith	B'ham
Number of subjects	341	171	337	52	202
Mean age (years) ± S.D.	41 ± 12	35 ± 12	35 ± 12	35 ± 13	37 ± 13
Mean weight (Kg) ± S.D.	75.9±9.5	77.5±10.9	72.7±10.3	59.9±6.5	63.2±9.6
Mean blood urea (mg%) ± S.D.	31.3±6.7	28.7±6.1	31.4±6.9	25.9±5.8	26.5±6.2

from the Winfrith men who were significantly older than any of the other 4 groups ($p<0.01$). The Winfrith and Birmingham men were heavier than the Glasgow men ($p<0.001$) and the Birmingham women were heavier than the Winfrith women ($p<0.01$). The Birmingham men had a lower mean serum urea than either of the other two male groups ($p<0.001$). Combining the regions the men were of similar age (mean 37.2 years) to the women (mean 36.5 years) but heavier (mean 74.9 Kg and 62.6 Kg respectively, $p<0.01$) and with a higher blood urea (mean 30.8 and 26.4 mg/100 ml respectively, $p<0.001$).

Frequency Distribution of Serum Uric Acid

The serum uric acid values were grouped into 0.5 mg/100 ml classes and plotted as a frequency distribution curve for each group. The curves approximate a normal distribution, those for the men being slightly skewed towards the lower values, those for the women towards the higher values. No individual values were less than 1.3 mg/100 ml or greater than 8.9 mg/100 ml.

Uric acid levels were identical for men in the 3 population groups, with a modal class of 5.5 to 5.9 mg/100 ml and a mean serum uric acid of 5.5 mg/100 ml. The modal class for the Winfrith women was 4.0 to 4.4 mg/100 ml and for the Birmingham women 3.5 to 3.9 mg/100 ml, the mean values being 4.1 and 3.9 mg/100 ml respectively ($0.05<p<0.1$). Combining the two female groups the modal class was 3.5 to 3.9 mg/100 ml and the mean value 3.9 mg/100 ml. Individual values of 7.0 mg/100 ml or over were found in 19 Winfrith men (5.6%), 15 Birmingham men (8.8%) and 27 Glasgow men (8.0%) resulting in an overall prevalence of hyperuricaemia, defined in this way, of 7.2%. Only one women (0.4%), from Birmingham, had a serum uric acid level in excess of 7.0 mg/100 ml.

Age and Serum Uric Acid

The mean serum uric acid values for each 5-year age increment, separated by sex are shown in Table II. In this case the regional groups have been combined as the individual group plots closely overlapped. It can be seen that for adult men there is no significant relation between serum uric acid and age ($r = -0.0248$). In women, however, levels fall slightly at age 30-34 and thereafter gradually rise, more sharply after age 45-49, to reach a new higher level around age 55-59. Overall there is a significantly

Table II Serum uric acid in 5-year age classes

Age group	MEN No.	MEN Mean ± S.D.	WOMEN No.	WOMEN Mean ± S.D.
15 - 19	38	5.53 ± 0.71	18	3.89 ± 0.86
20 - 24	133	5.50 ± 0.97	48	3.95 ± 0.67
25 - 29	122	5.56 ± 1.09	26	3.86 ± 0.94
30 - 34	97	5.46 ± 0.94	33	3.59 ± 0.70
35 - 39	95	5.41 ± 1.09	24	3.65 ± 0.63
40 - 44	84	5.55 ± 0.91	27	3.81 ± 0.56
45 - 49	109	5.41 ± 1.01	22	3.96 ± 0.79
50 - 54	84	5.43 ± 0.92	29	4.41 ± 0.86
55 - 59	57	5.52 ± 0.98	16	4.21 ± 0.01
60 - 64	29	5.42 ± 0.91	11	4.20 ± 1.00
65 - 69	3	6.33 ± 1.10		0
70 - 74	1	5.1 -		0

positive correlation between the individual serum uric acid values in women with age ($t = 0.159$, $p < 0.05$), and there was a highly significant difference ($t = 4.176$, $p < 0.001$) between the mean serum uric acid in those under age 50 (mean 3.8 mg/100 ml) and those aged 50 and over (mean 4.3 mg/100 ml).

Although there was considerable overlap between the individual male and female serum uric acid values at most age groups, the mean values were all significantly different at the 1 per cent level or less, using the "t" test.

Weight and Serum Uric Acid

Levels of serum uric acid rose with increasing weight in each of the 3 male groups and for the individual values the correlation was significantly positive at the 5 per cent level or less. A similar trend was seen in the Birmingham women but not in the smaller group of Winfrith women (Table III).

Serum Urea and Serum Uric Acid

There was a trend for serum uric acid to rise with increasing serum urea, even within the generally accepted range for normal blood urea values (Table IV).

Table III Correlation of serum uric acid with weight

	Weight	
Winfrith men	r = 0.193	p<0.01
Birmingham men	r = 0.171	p<0.05
Glasgow men	r = 0.262	p<0.001
Winfrith women	r = 0.118	N.S.
Birmingham women	r = 0.389	p<0.001
Combined men	r = 0.215	p<0.01
Combined women	r = 0.319	p<0.01

There was significantly positive correlation in two of the three male and one of the two female groups. The correlation was significant in the combined male and combined female groups and increased when the sexes were combined, presumably reflecting the significantly lower (r = 25.66, p<0.001) serum urea in the women (mean 26.4 mg/100 ml) than the men (mean 30.8 mg/100 ml).

Social Class and Serum Uric Acid

No significant variation of serum uric acid between the class groups was found; in these population samples there was under-representation of social class V.

Table IV Correlation of serum uric acid with serum urea

Winfrith men	r = 0.064	N.S.
Birmingham men	r = 0.213	p<0.01
Glasgow men	r = 0.171	p<0.01
Winfrith women	r = 0.345	p<0.02
Birmingham women	r = 0.073	N.S.
Combined men	r = 0.134	p<0.01
Combined women	r = 0.122	p = 0.05
Combined sexes	r = 0.259	p<0.01

DISCUSSION

By separating the blood immediately, storing in the frozen state and analysing the sera in randomized batches over a limited number of sessions in one laboratory, the technical and inter-laboratory errors which can reduce the validity of population comparisons were minimized (Bywaters & Holloway, 1964).

Mean levels in the present study are higher than those in the only previous large-scale United Kingdom survey (Popert & Hewitt, 1962), and cannot be accounted for entirely by methodological differences, but, are similar to a smaller Liverpool survey in men (Finn, Jones, Tweedie, Hall, Dinsdale & Bourdillon, 1966). Our mean levels in men are higher than in a Finnish survey (Isomäki & Takkunen, 1969) but similar to those in a large French survey (Zalokar, Lellouch, Claude & Kuntz, 1972), both using the same method as we have employed. Our findings in men and women are similar to those of two American surveys (Mikkelsen, Dodge & Valkenburg, 1965; Hall, Barry, Dawber & McNamara, 1967) but considerably lower than one other American survey (Acheson & O'Brien, 1966) and three Australasian surveys (Jeremy & Towson, 1971; Garrick, Bauer, Ewan & Neale, 1972; Evans, Prior & Morrison, 1969). Comparison of their findings with those of earlier Australian studies led Jeremy & Towson to suggest that there had been a true increase in serum uric acid levels in the population over the previous 10 to 15 years and the same argument might apply when our findings are compared with those of Popert & Hewitt. This would be in accord with the view that "the associates of a high uric acid are the associates of plenty" (Acheson & Chan, 1969).

The striking finding from this survey was the identical serum uric acid levels of the three male groups, and the closely similar levels between the two female groups. The apparent rarity of gout in Scotland cannot therefore be accounted for by population differences in uric acid levels and the prevalence of hyperuricaemia, defined here as a serum uric acid of 7 mg/100 ml or over, was lower in the combined English male groups (6.6%) than the Scottish group (8.0%).

The well-known sex difference in uric acid levels was confirmed. The slight decrease in mean serum uric acid levels in women between the 3rd and 4th age decades is similar to that in other surveys (Popert & Hewitt, 1962; Mikkelsen et al, 1965; Evans et al, 1969). The only previous investigators to have actually commented on this could not explain it in terms of variations in obesity (Evans et al, 1969). The rise around the menopause has been attributed to hormonal influences (Mikkelsen et al, 1965) though a recent study carefully matching pre- and post-menopausal women of the same age (menopause defined as no menstruation during the previous 6 months) found no difference between the two groups (Bengtsson & Tibblin, 1974). This definition of the menopause may not however exclude a falling off of menstrual blood loss, and hence a rising haemoglobin which is positively correlated with serum uric acid (Acheson & O'Brien, 1966). The known association of serum uric acid with body weight was also confirmed

The association of serum uric acid with serum urea levels, even within the normal range, is unexplained but has been noted before (Kennedy, Brennan, Anderson, Brooks, Buchanan & Dick, 1975) as has an association with serum creatinine (Decker, Healey & Skeith, 1968). Both renal and metabolic mechanisms could account for this and ageing may contribute, as highly significant correlations of serum urea with age were found in both our men (r = 0.201) and women (r = 0.355).

The lack of correlation of serum uric acid with social class is in keeping with the findings of Acheson (1969). The explanation may lie in the lack of true poverty in the population groups studied by Acheson and ourselves.

SUMMARY

A survey of serum uric acid levels in 766 subjects in England and 337 in Glasgow was carried out. There was no difference in the frequency distribution of uric acid or the mean levels in the two countries. A serum uric acid of 7 mg/100 ml or over was found in 7.2 % of the men and 0.4 % of the women. The previously described sex difference and association of serum uric acid with weight were confirmed. No association was found with social class. The suggestion of an increase in uric acid levels in the United Kingdom over the past 14 years is discussed.

REFERENCES

ACHESON, R. M. (1969) Brit. med. J., 4, 65 (Social class gradients in serum uric acid in males and females)

ACHESON, R. M., AND CHAN, Y.-K. (1969) J. chron. Dis., 21, 543 (New Haven survey of joint diseases. The prediction of serum uric acid in a general population)

ACHESON, R. M., AND O'BRIEN, W. M. (1966) Lancet, 2, 777 (Dependence of serum uric acid on haemoglobin and other factors in the general population)

BENGTSSON, C., AND TIBBLIN, E. (1974) Acta med. Scand., 196, 93 (Serum uric acid levels in women)

BYWATERS, E. G. L., AND HOLLOWAY, V. P. (1964) Ann. rheum. Dis., 23, 236 (Measurement of serum uric acid in Great Britain in 1963)

DECKER, J. L., HEALEY, L. A., AND SKEITH, M. D. (1968) In: Population studies of the rheumatic diseases (eds. Bennett, P. H. and Wood, P. H. N.) p. 148 Excerpta med. congr. Int. series No. 148, Excerpta medica foundation, Amsterdam (Ethnic variations in serum uric acid: Filipino hyperuricaemia, the result of hereditary and environmental factors)

EVANS, J. G., PRIOR, I. A. M., AND MORRISON, R. B. I. (1969) N. Z. med. J., 70, 306 (The Caterton study: 5. Serum uric acid levels of a sample of New Zealand European adults.

FINN, R., JONES, P. O., TWEEDIE, M. C. K., HALL, S. M., DINSDALE, O. F., AND BOURDILLON, R. E. (1966) Lancet, 2, 185 (Frequency distribution curve of uric acid in the general population)

GARRICK, R., BAUER, G. E., EWAN, C. E., AND NEALE, F. C. (1972) Aust. N. Z. J. Med., 4, 351 (Serum uric acid in normal and hypertensive Australian subjects)

GARROD, A. B. (1876) A Treatise on Gout and Rheumatic Gout (Rheumatoid Arthritis) 3rd ed. p. 218 Longmans, Green, London

HALL, A. P., BARRY, P. E., DAWBER, T. R., AND McNAMARA, P. (1967) Amer. J. Med., 42, 27 (The epidemiology of gout and hyperuricaemia)

HEALEY, L. A., AND HALL, A. P. (1970) Bull. rheum. Dis., 20, 600 (The epidemiology of hyperuricaemia)

ISOMAKI, H. A., AND TAKKUNEN, H. (1969) Acta rheum. Scand., 15, 112 (Gout and hyperuricaemia in a Finnish rural population)

JEREMY, R., AND TOWSON, J. (1971) Med. J. Aust., 1, 1116 (Serum urate levels and gout in Australian males)

KENNEDY, A.C., BRENNAN, J., ANDERSON, J., BROOKS, P., BUCHANAN, W.W., AND DICK, W.C. (1975) Serum uric acid - its relationship to lean body mass, sex, plasma urea, intracellular potassium and packed cell volume in a normal population group. Paper read to meeting of Heberden Society at Norwich, England. 14th March 1975.

MIKKELSEN, W.M., DODGE, H.J., AND VALKENBURG, H. (1965) Amer. J. med., 39, 242 (The distribution of serum uric acid values in a population unselected as to gout or hyperuricaemia)

POPERT, A.J., AND HEWITT, J.V. (1962) Ann. rheum. Dis., 21, 154 (Gout and hyperuricaemia in rural and urban populations)

REGISTRAR GENERAL (1970) Classification of Occupation. Her Majesty's Stationery Office, London

ZALOKAR, J., LELLOUCH, J., CLAUDE, J.R., AND KUNTZ, D. (1972) J. chron. Dis., 25, 305 (Serum uric acid in 23,923 men and gout in a subsample of 427 men in France)

GOUT DISEASE. ITS NATURAL HISTORY BASED ON 1,000 OBSERVATIONS

A. Rapado and J. M. Castrillo

Unidad Metabólica. Fundación Jiménez Díaz and Universidad

Autónoma. Av. Reyes Católicos, 2. Madrid-3. Spain

Gout disease is the most frequent articular metabolic process. When diagnosed it can be corrected by controlling either the hyperuricemia (1) or the articular inflammation (2). This permits an exact control, the earlier clinical complications and poor prognosis having dramatically improved in the last few years.

The large number of gouty patients presented in this paper can be a good comparison to other series already published (3, 4, 5) and could serve as basis for a long-term follow-up study not only of the evolution of the disease but also to analyze the role of therapy in the natural history of this common process.

MATERIAL AND METHODS

1,000 gouty patients seen from May 1963 to April 1976 were submitted to a clinical protocol including familial history, dietary habits, previous medication and associated diseases. They also were given a complete physical examination, and X-rays of the most commonly affected joints were taken. A biochemical analysis that included the measurement of the glomerular filtration rate by the creatinine clearance and the assessment of uric acid in serum and in 24-hour urine. Where necessary, complementary techniques were used, such as the analysis of the synovial fluid, renal biopsy, lipidrogram, hypoxanthine-guanine phosphoribosil pyrophosphate transferase (HPRT) assay, etc.

Gout disease was defined according to the criteria of the Council of the International Organization of Medical Sciences (6). Normal values of serum uric acid were 6.5 ± 0.3 mg/100 ml for men, and 5.8 ± 0.8 mg/100 ml for women (7). We consider hyperexcretion of uric acid the elimination above 900 mg in 24-hour urine. Clinical results were compared with the findings in 13,146 healthy subjects, matched by sex and age.

RESULTS

From a clinical point of view, we differentiate four forms in the presentation of gout: acute (35.2 per cent of the cases); chroni (47 per cent); tophaceous (13.1 per cent) and polyarticular (4.7 per cent). The latter type has an undoubtable interest because of its sometimes difficult differential diagnosis with other forms of polyarthritis (8).

The great toe was affected in 89.3 per cent of the patients. No joint was free from a severe attack of gout. Therefore pain topography cannot serve as an exclusive index in the diagnosis.

In 33.1 per cent of the cases the acute attack was accompanied by fever. Prodromi were reported in 31.4 per cent while exogenous influences in the acute attack occurred in 71.8 per cent of the cases.

The administration of hyperuricemic drugs was important, and in 56 of our patients it was the cause of the first clinical manifestation of the disease (7, 9). Of these, the most frequently used were the thiazide diuretics. The steroid-dependent effect was important in our series as steroids were frequently employed for the control of the articular inflammation (10).

Gout affects preferently males and in our study the incidence in females was 10.3 per cent. In women, secondary gout is more frequent (Table I). Its primary form was observed in the postmenopausic period. A comparison according to sexes did not show significant changes in the percentage of familial history, incidence of hyperuricemia or uric acid hyperexcretion. However, the percentag of women in whom the disease appeared under 40 years of age was significantly lesser than in men.

The mean age for gout onset rated from 35 to 40 years for men, and from 50 to 55 years for women. This seems to confirm a direct relationship to sexual function and it is parallel to the changes described in serum uric acid values (11, 12).

The most commonly associated diseases were arterial hypertensi (36.9 per cent) (13); renal lithiasis (34.1 per cent) (14) with 16. per cent corresponding to uric acid lithiasis (15) and diabetes mellitus (5.6 per cent) (16). Table II summarizes the relationships between these diseases and gout.

TABLE I. INCIDENCE OF FEMALE SEX IN 1,000 CASES OF GOUT

		no. of Cases
Secondary Gout		42
Primary Gout		
Postmenopausal	36	
Premenopausal	25	61
		103 (10.3%)

	% Female	% Male
Onset before 40 years old	25	37
Family history of gout	30	28
Persistent hyperuricemia	87	90
Uric acid hyperexcretion	20	19

In our series we found the following incidence according to the patient's profession: bartenders, 6.4 per cent; drivers, 10.9 per cent. 2.5 per cent of the cases had had contact with lead.The pathogenic mechanisms of these findings need more complex studies (9, 17).

We found a positive familial history of gout in 27.7 per cent of the cases in comparison with 0.9 per cent in the general population. Familial history of renal lithiasis (25.2 per cent) was significantly higher than among normal subjects (4.5 per cent). There was a direct relationship between a positive familial history of arterial hypertension, renal lithiasis or diabetes, and its incidence in the gouty population (18).

76.9 per cent of our patients presented particular psychological characteristics similar to those described by other authors (19). Overweight was observed in 67.4 per cent and tophi were clinically evident in 37.7 per cent of the cases. No correlation was found between the values of serum uric acid, arterial hypertension or the presence of tophi.The incidence of the latter has decreased over the years (20). This is possibly owed to an earlier diagnosis

TABLE II. GOUT DISEASE IN RELATION TO	ARTERIAL HYPERTENSION	URIC ACID LITHIASIS	DIABETES
Incidence in General Population (%)	15.5	17.2*	6.0
Incidence in Gouty Patients (%)	36.9	16.5	5.6
Female Sex	X		
Age	X		
Duration of Gout	X		
Medication	X	X	
Uric Acid Hyperexcretion		X	
Urinary Excess of Hydrogen Ions		X	
Hyperlipoproteinemia			X
Renal Failure	X		
Family History of Gout	X	X	
Family History of Hypertension	X		
Family History of Renal Lithiasis		X	
Family History of Diabetes			X

*Incidence in renal stone-formers.

and a better treatment of the disease. Thus tophi were observed in 49 of our first 100 patients while they were reported only in 17 among the last 100 cases.

In 60 per cent of our patients we found radiological evidence of tophi in juxta-articular areas. Their insensity was related to the number of gout attacks, visible tophi and duration of the disease. However, chondrocalcinosis was of low incidence in our series (1 per cent of the total) (21).

Our cases were classified (Table III) in secondary gout (to renal disease, to medication or to hematological disorders), and in primary gout. The latter was classified according to a reduced or normal renal function. Of these, we differentiate the cases with uric acid hyperexcretion and the normoexcretors (22).

TABLE III. ETIOLOGICAL CLASSIFICATION OF 1,000 CASES OF GOUT (May 1963 - April 1976)

SECONDARY		CASES
To Renal Disease		87
To Medication		56
To Hematological Disorders		10
		153
PRIMARY		
With Renal Failure		134
With Normal Renal Function		
Urinary Uric Acid Hyperexcretion	91	
Urinary Uric Acid Normoexcretion	622	713

This classification is important not only from a clinical point of view but also from a therapeutic one. The incidence found in our series is similar to that reported by other authors (23, 24).

Hyperuricemia was found in 90.6 per cent of our cases. This percentage was even higher whenever we performed more than one assessment of serum uric acid.

Hyperlipoproteinemia of type IIb and IV (25) was another frequent biochemical finding. And it was not related to obesity or to carbohydrate intolerance.

The increased observed in the Cur/Ccr quotient led us to measurements of HPRT in dried blood. In no case did we find abnormal findings. As reported in other series (26), this defect was of low incidence in adult gouty population.

CONCLUSIONS

Gout disease in our country as both exogenous factors (dietary habits, professions, etc.) and endogenous factors (sex, inheritance, etc.) play a polygenic role in its incidence.

Those conditions resulting in hyperuricemia (diseases, drugs, etc.) facilitate the clinical manifestations. But a genetic determinant in the articular reaction in front of an accumulation of uric acid crystals or in the development of such crystallization is a fundamental factor in the manifestation of the disease.

In our series we were able to confirm the high incidence of arterial hypertension, renal lithiasis and renal participation in gout. The role of lead and of certain professions in favouring the hyperuricemia is suggested.

A classification of gout, from a functional point of view, must include both the clinical form and the analysis of the whole renal function, and the renal handling of uric acid. This is mandatory not only for the diagnosis and prognosis of the disease but to set the basis of its adequate management.

Acknowledgements

To Professor J.E. Seegmiller, who performed the HPRT assessment and to HORMINESA for its financial support.

REFERENCES

1. Kelley, W.N.: Pharmacologic approach to the maintenance of urate homeostasis. Nephron, 14:99 1975
2. Wallace, S.L.: Colchicine and new antiinflammatory drugs for the treatment of gout. Art.&Rheum. 18:847, 1975

3. Kuzell, W.C., Schafferzick, R.W., Naugler, W.E., Koets, P., Mankle, E.A., Brown, B. and Champlin, B.: Some observations on 520 patients of gout. J.chron.Dis. 2:645, 1955

4. Seze, S. and Ryckwaert, A.: La goutte. L'Expansion, ed. Paris 1960, p. 67.

5. Grahame, R. and Scott, J.T.: Clinical survey of 354 patients with gout. Ann.rheum.Dis. 29:461, 1970

6. Kelgre, J.H.: The epidemiology of chronic diseases. Blackwell, Oxford, 1963, p. 104

7. Rapado, A.: Allopurinol in thiazide-induced hyperuricemia. Ann.rheum.Dis. 25:660, 1966

8. Hadler, N.M., Frank, W.A., Bress, N.M. and Robinson, D.R.: Acute polyarticular gout. Am.J.Med. 56:715, 1974

9. Gutman, A.B.: The past four decades of progress in the knowledge of gout, with an assessment of the present status. Art. & Rheum. 16:431, 1973

10. Rapado, A., Herrera, J.L., Jiménez, M. and Rodriguez, J.L.: Cushing iatrogénico y gota corticodependiente. Rev.clin.esp. 98:275, 1965

11. Mikkelsen, W.M., Dodge, H.J. and Valkenburg, H.: The distribution of serum uric acid values in a population unselected as to gout and hyperuricemia. Am.J.Med. 39:242, 1965

12. Healy, L.A.: Epidemiology of hyperuricemia. Art.& Rheum. 18: 709, 1975

13. Rapado, A.: Relationship between gout and arterial hypertension. In: Purine metabolism in man. vol. 41 B, ed. by O. Sperling, A. de Vries and J.B. Wyngaarden. Plenum Pub. Corp. New York 1974, p. 451

14. Cifuentes, L., Rapado,A., Abehsera, A., Traba, M.L. and Cortés, M.: Uric acid lithiasis and gout. In: Urinary Calculi, ed. by L. Cifuentes, A. Rapado and A. Hodgkinson. Karger, Basel 1973, p. 115.

15. Frank, M., Lazebnik, J. and de Vries, A.: Uric acid lithiasis. A study of six hundred and twenty-two patients. Urol.int. 25: 32, 1970

16. Smyth, C.J.: Disorders associated with hyperuricemia. Art.& Rheum. 18:713, 1975

17. Rapado, A.: Gout and saturnism. New Eng.J.Med. 281:851, 1969

18. Seegmiller, J.E.: Genetic considerations of gout. Art.&Rheum. 18:743, 1975

19. Lanese, R.R., Gresham, G.E. and Keller, M.D.: Behavioural and

physiological characteristics in hyperuricemia. J.A.M.A. 207: 1878, 1969

20. O'Duffy, J.D., Hunder, G.G. and Kelly, P.K.: Decreasing prevalence of tophaceous gout. Mayo Clinic Proc. 50:227, 1975

21. Phelps, P., Steele, A.D. and McCarthy, D.J.: Compensated polarized light microscopy: Identification of crystals in synovial fluids from gout and pseudogout. J.A.M.A. 203:166, 1968

22. Wyngaarden, J.B.: Metabolic defects of primary hyperuricemia and gout. Am.J.Med. 56:651, 1974

23. Barlow, K.A. and Beilin, L.J.: Renal disease in primary gout. Quar.J.Med. 37:79, 1968

24. Berger, L. and Yu, T.F.: Renal function in gout. IV. An analysis of 524 gouty patients including long-term follow-up studies. Am.J.Med. 59:605, 1975

25. Wiedeman, E., Rose, H.G. and Schwartz, E.: Plasma lipoproteins, glucose tolerance and insulin response in primary gout. Am.J. Med. 53:299, 1972

26. Yu, T.F., Beilis, E. and Yip, L.C.: Overproduction of uric acid in primary gout. Art.& Rheum. 18:695, 1975

HYPERURICEMIA AS A RISK FACTOR IN CORONARY HEART DISEASE

DUDLEY JACOBS

Discoverer's Memorial Hospital

Florida, Transvaal, South Africa, 1710

SUMMARY

The subjects of this study were 290 hospital patients who were divided into 4 groups: acute myocardial infarction (100), cardiac ischaemia with angina but no infarction (50), white patients without evidence of coronary heart disease (70), and a group of South African black patients without detectable coronary heart disease (70). Mean serum uric acid (SUA) levels were 8,08 $\pm$ 0,32 in the ischaemic group, 7,09 $\pm$ 0,23 in the acute infarction group, 5,78 $\pm$ 0,21 in normal black patients, and 5,75 $\pm$ 0,21 in normal white controls. The differences in SUA levels between the ischaemic and infarction groups as compared with both control groups was highly significant ($p < 0,001$). These differences were most striking in the females aged 60 years and over.

Patients receiving diuretics, uricosuric and uricosuppressive agents were excluded from the study.

Wide racial differences in incidence and mortality from coronary heart disease (CHD) exist, and one of the greatest contrasts is found in the South African multi-racial society. The mortality rate is high in whites and extremely low amongst the black population.

Despite lack of proof that the association is causal, further investigation of SUA levels in CHD, with emphasis upon possible interrelationships between age, sex, and racial factors, appears to be indicated.

There are wide differences in incidence and mortality from CHD among white and non-white populations in different parts of the world. South Africa is one such country, where mortality from CHD among the white population exceeds 40% of all deaths, whereas mortality from this disease amongst the South African black population is extremely low. Neither of these extremes of mortality have been wholly explained.[1]

A recent personal study showed that, as a possible associated risk factor, hyperuricaemia correlated strongly with the occurrence of acute myocardial infarction.[2] The present study was undertaken in an attempt to study further the possible relationship, if any, between high SUA levels, and the development of myocardial infarction. On the assumption that hyperuricaemia is essentially a generalized metabolic disease, and because of the lack of evidence that the relationship was causal, well-defined control groups have been included in the study.

MATERIAL AND METHODS

290 consecutive patients treated in the Medical Unit of Discoverer's Memorial Hospital, Florida, South Africa, were studied. One group of 70 patients were black subjects with no evidence of CHD, a group of white subjects with no evidence of CHD clinically or on E.C.G. was included as a second control group (70). The third group of 50 patients with evidence of ischaemic heart disease, and the fourth of 100 patients with acute myocardial infarction were all white subjects. Criteria for the diagnosis of myocardial infarction were the clinical history and examination with E.C.G. evidence of pathological Q waves, S-T segment changes, later T-wave inversion, a falling R wave in the precordial leads, complete left bundle-branch block with S-T segment changes and, where necessary, significant and transient elevation of serum LDH, SGOT and CPK levels.

All uric acid estimations were performed by the same laboratory and the phosphotungstic acid carbonate method was used. In those with acute myocardial infarction, this was performed at the time of the first prothrombin index, usually the fourth day after admission. The normal range accepted for males was 1 - 7 mg/100 ml and for females 1 - 6 mg/100 ml.

Exclusions from the study were: Evidence of renal failure, manifested by a blood urea over 50 mg.% or serum creatinine over 1,4 mg.%., and concomitant diuretic, allopurinol or uricosuric therapy. Other exogenous factors such as alcoholism were not exclusions.

RESULTS

In the total patient population studied mean SUA for the infarction group was 7,09 ± 0,23 and for the ischaemic group 8,08 ± 0,32, and these levels were higher than in both the black and white control patients (TABLE I.)

TABLE I. SERUM URIC ACID MEANS AND STANDARD ERROR IN mg.% IN 290 PATIENTS

Clinical Group	TOTAL POPULATION		
	TOTAL	MALE	FEMALE
Myocardial Infarction	N = 100 7,09 ± 0,23	N = 50 7,04 ± 0,31	N = 50 7,14 ± 0,35
Cardiac Ischaemia	N = 50 8,08 ± 0,32	N = 25 7,80 ± 0,44	N = 25 8,37 ± 0,47
White Controls	N = 70 5,75 ± 0,21	N = 35 6,19 ± 0,32	N = 35 5,31 ± 0,24
Black Controls	N = 70 5,78 ± 0,21	N = 35 5,37 ± 0,30	N = 35 6,19 ± 0,30

Uric acid levels relative to age and disease state revealed significantly higher levels in patients ⩾ 60, in both the acute myocardial infarction and ischaemic groups (TABLE II.)

The most significant difference was between the myocardial infarction group over and under 60 years of age. In ischaemic subjects similar but lesser degrees of significance were found. In the remaining groups there were no significant differences.

The age differences occurred mainly in females where those ⩾ 60 years in both the acute myocardial infarction, and particularly the ischaemic groups were significantly higher than in the < 60 age group. In the entire series, the highest mean levels occurred in the female ischaemic subjects age ⩾ 60 years (TABLE III.)

TABLE II. SERUM URIC ACID MEANS AND STANDARD ERROR IN BOTH SEXES RELATIVE TO AGE

	Clinical Group		AGE < 60			
			Myocardial Infarction	Cardiac Ischaemia	White Controls	Black Controls
		Means & S.E. in mg.%	$6{,}33 \pm 0{,}26$	$6{,}91 \pm 0{,}47$	$5{,}72 \pm 0{,}25$	$5{,}77 \pm 0{,}23$
AGE ≥ 60	Myocardial Infarction	$7{,}77 \pm 0{,}35$	**	-	-	-
	Cardiac Ischaemia	$8{,}54 \pm 0{,}38$	-	*	-	-
	White Controls	$5{,}79 \pm 0{,}34$	-	-	n.s.	-
	Black Controls	$5{,}77 \pm 0{,}42$	-	-	-	n.s.

** = $p < 0{,}01$
* = $p < 0{,}05$
n.s. = not significant

TABLE III. SERUM URIC ACID MEANS AND STANDARD ERROR IN FEMALES RELATIVE TO AGE

	Clinical Group		AGE ⊲ 60			
			Myocardial Infarction	Cardiac Ischaemia	White Controls	Black Controls
		Means & S.E. in mg.%	6,01 ± 0,43	6,15 ± 0,24	5,44 ± 0,26	6,01 ± 0,30
AGE ⩾ 60	Myocardial Infarction	7,64 ± 0,45	*	-	-	-
	Cardiac Ischaemia	8,79 ± 0,50	-	**	-	-
	White Controls	5,17 ± 0,42	-	-	n.s.	-
	Black Controls	6,63 ± 0,73	-	-	-	n.s.

* = p ⊲ 0,05
** = p ⊲ 0,01
n.s. = not significant

Comparing females under 60 years of age, there were no statistical differences between the disease groups. In the females aged 60 and over, Student's 't' test was applied to the actual values and yielded the following: Cardiac infarction versus white controls t_{49} df = 3,53; $p < 0{,}01$. A comparison between ischaemic subjects and black controls in this female group yielded: t_{29} df = 2,44; $p < 0{,}05$. Comparing ischaemic versus white controls t_{36} df = 5,35; $p < 0{,}001$.

Associated Gout. The incidence rates of associated clinical gout could not be accurately assessed because many of the older patients had painful osteo-arthritis. They were not subjected to therapeutic drug challenge. However, it was clear that the majority of patients had never experienced clinical gout.

DISCUSSION

Several authors have documented an association between hyperuricaemia and coronary heart disease. An earlier study indicated that in acute myocardial infarction, hyperuricaemia is one of the commonly associated risk factors.[3] Unanimity, however, does not exist, and some investigators believe that gouty subjects are no more prone to coronary heart disease than the non-gouty. To add to the problem, we do not know whether the association is causal, effect or coincidental. Finally, there remains controversy as to precisely when a high serum uric acid becomes a pathological hyperuricaemia, and possibly our upper limits of normal should be revised.

It is well known that acute episodes such as myocardial infarction may increase SUA levels. The higher figures found in this ischaemic group suggest that infarction was not the sole reason for the high SUA levels found in the patients with CHD.

SUA is a variable, subject to modification by many interrelated factors. The significance of the present findings requires careful interpretation. It is, however, well known that platelet abnormalities occur with hyperuricaemia. Altered platelet kinetics play a role in arterial thrombosis and possibly in the genesis of atherosclerosis. The interrelationships between platelet abnormalities and hyperuricaemia, as well as common denominator factors such as stress and catecholamine release, are also well known.

On the assumption that the occurrence of high SUA levels in these subjects with CHD is abnormal, and may be causal rather than coincidental, a need for further study, particularly in relation to platelet kinetics, would appear to be indicated.

I should like to thank Dr. P.S. Grobbelaar, Medical Superintendent of Discoverer's Memorial Hospital, Florida, South Africa, for permission to publish; Miss M. Bromfield and staff members of the South African Institute for Medical Research, Discoverer's Memorial Hospital, who performed the uric acid tests for this series; Dr. P.R. Farndell for technical assistance; Mr. R.D. Hutchison and Mrs. L. Salter, Medical Department, Ciba-Geigy (Pty) Limited, for statistical evaluations.

REFERENCES

1. WALKER, A.R.P. (1963) : Amer. Heart J., 66, 293.
2. JACOBS, D. (1972) : S. Afr. Med. J., 46, 367.
3. Ibid (1971) : S. Afr. Med. J., 45, 275.

HYPERURICEMIA AND OTHER CARDIOVASCULAR RISK FACTORS

Heikki Takkunen and Antti Reunanen

The Research Institute for Social Security and the Division of Epidemiology and Preventive Medicine, the Social Insurance Institution, Helsinki, Finland

The prevalence of clinical gout is rather low in Finland. In most studies concerning the epidemiology of gout the prevalence rates have varied from 0.3 to 3.0 % in adult population. Persons having certain chronic diseases are allowed free drug therapy according to the Sickness Insurance Act of Finland. Gout is one of these chronic diseases. According to the latest statistics 0.1 % of men and 0.03 % of women aged 35-64 years are allowed free drug therapy for gout(1). On the other hand, the serum uric acid concentration of Finnish population is not strikingly lower than in other countries (2).

It has been found in many epidemiological studies that serum uric acid has relationship with some cardiovascular risk factors and it has also been claimed that serum uric acid is an independent risk factor for cardiovascular diseases (e.g. 3, 4, 5, 6). In Finland the incidence and mortality for coronary heart disease in the middle-aged men is highest in the world. The prevalence and significance of classical risk factors for cardiovascular diseases in Finland has been studied in some investigations. Until now there has been no study concerning the role of serum uric acid as a risk factor for cardiovascular diseases in Finland.

The mobile clinic of the Social Insurance Institution of Finland has carried out multiphasic health screening examinations in various population groups in Finland (7, 8). The prevalence of hyperuricaemia and the associations of serum uric acid with other cardiovascular risk factors have been studied in the connection of this screening programme in six population groups from western and central parts of Finland. The population studied comprised 8750 men and 7728 women aged 15 years and over. Serum uric acid was deter-

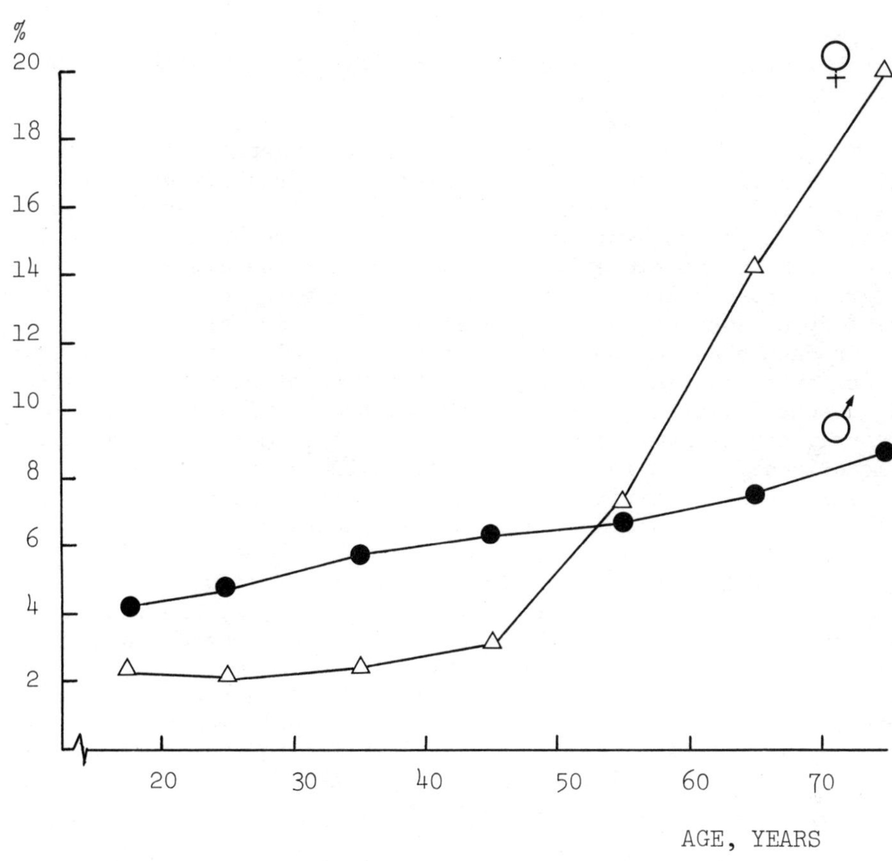

Figure 1. Prevalence of hyperuricaemia in 8570 men (⩾ 7.0 mg/100 ml) and in 7728 women (⩾ 6.0 mg/100 ml) aged 15 years and over.

mined by an autoanalyzer modification of the colorimetric method (9).

The mean concentration of serum uric acid of the men was 5.15 ± 1.08 mg/100 ml, that of the women 4.23 ± 1.08 mg/100 ml, respectively. Serum uric acid was over 7.0 mg/100 ml in 5.9 % of the men and over 6.0 mg/100 ml in 5.3 % of the women. The prevalence rate of hyperuricaemia in men showed a) very slight rising tendency with age but in the women the prevalence rose strikingly after the menopause (figure 1).

The associations of uric acid with other cardiovascular risk factors and the influence on cardiovascular mortality were studied in more detail in a cohort of 2758 men and 2011 women aged 30-59 years. The hyperuricaemic men and women had significantly more ECG changes than normouricaemic men and women but the differences in the prevalences of typical chest pain symptoms (angina pectoris or a history of a severe chest pain attack) were not significant among hyperuricaemic and normouricaemic persons (table 1 and 2). Hyperuricaemic men and women used significantly more often cardiovascular drugs, especially diuretics. The prevalence of high systolic blood pressure, high serum cholesterol of cigarette smoking was not significantly different in hyperuricaemics. On the other hand hyperuricaemics were significantly more obese, had a higher prevalence of glucose intolerance and had a higher packed cell volume.

The mortality of the population has been followed to the end of 1973, with a median follow-up time of six years. The men and women with high serum uric acid level had increased cardiovascular mortality rates (figure 2). When the follow-up time, age, cardiovascular drug therapy, chest pain symptoms, ECG changes and the risk factors mentioned above were adjusted with a multivariate analysis (10) the risk of hyperuricaemic men and women to die in cardiovascular diseases were still significantly higher when compared with the risk of normouricaemic men and women.

The study confirms earlier considerations of the association of serum uric acid to the other cardiovascular risk factors. Preliminary multivariate analysis suggests also that uric acid can have an independent contribution in prediction of cardiovascular mortality. However, it is probable that the main influence of serum uric acid as a cardiovascular risk factor is mediated through other risk factors.

Table 1. Prevalence (%) of symptoms, signs and risk factors of cardiovascular diseases in hyperuricaemic and normouricaemic men aged 30-59 years.

	Serum uric acid (mg/100 ml)		
	⩽ 6.9 (N = 2625)	⩾ 7.0 (N = 133)	p
Chest pains suggesting coronary heart disease	5.0	7.0	N.S.
ECG changes suggesting coronary heart disease	9.6	15.5	<0.05
Drug therapy:			
- digitalis	2.1	6.0	<0.001
- diuretics	0.4	6.0	<0.001
- antihypertensive drugs	1.6	5.3	<0.001
High systolic blood pressure (⩾ 160 mmHg)	12.0	12.4	N.S.
High serum cholesterol (⩾ 310 mg/100 ml)	9.7	7.0	N.S.
Cigarette smoking	49.0	41.4	N.S.
Obesity (Body mass index ⩾ 29.0)	12.5	34.9	<0.001
Glucose intolerance (1 h glucose ⩾ 220 mg/100 ml in OGTT)	8.6	15.5	<0.05
High PCV (⩾ 50 vol. %)	7.3	14.0	<0.01

Table 2. Prevalence (%) of symptoms, signs and risk factors of cardiovascular diseases in hyperuricaemic and normouricaemic women aged 30-59 years.

	Serum uric acid (mg/100 ml)		
	≤ 5.9	≥ 6.0	p
	(N = 1917)	(N = 94)	
Chest pains suggesting coronary heart disease	6.0	6.7	N.S.
ECG changes suggesting coronary heart disease	11.1	20.2	<0.05
Drug therapy:			
- digitalis	3.2	12.8	<0.001
- diuretics	1.8	12.8	<0.001
- antihypertensive drugs	5.3	14.9	<0.001
High systolic blood pressure (≥ 160 mmHg)	14.3	19.1	N.S.
High serum cholesterol (≥ 310 mg/100 ml)	9.1	7.9	N.S.
Cigarette smoking	13.4	7.4	N.S.
Obesity (Body mass index ≥ 32.0)	8.7	28.1	<0.001
Glucose intolerance (1 h glucose ≥ 220 mg/100 ml in OGTT)	9.0	15.7	N.S.
High PCV (≥ 46 vol. %)	8.6	16.9	<0.05

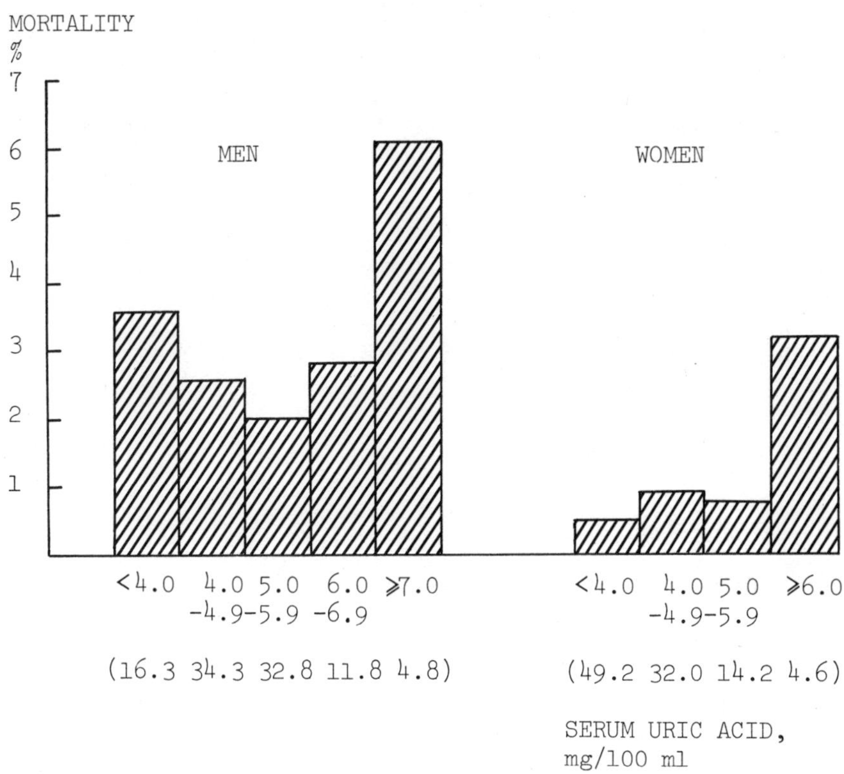

Figure 2. Cardiovascular mortality in six years in 2758 men and 2011 women aged 30-59 years at various serum uric acid levels. Percentage distribution of serum uric acid in parentheses.

References:

1. Statistics on drugs reimbursed in full 30.6.1974. Social Insurance Institution's publication T6:7, Helsinki 1974. (in Finnish)

2. Isomäki, H.: Hyperuricaemia in northern Finland. Ann. clin. Res. suppl. 1. 1969.

3. Hall, A.P.: Correlations among hyperuricaemia, hypercholesterolemia, coronary disease and hypertension. Arthr. Rheum. 8:846, 1965.

4. Klein, R. et al.: Serum uric acid. Its relationship to coronary heart disease risk factors and cardiovascular disease, Evans County, Georgia. Arch. int. med. 32:401, 1973.

5. Pearre, J. et al.: Uric acid and plasma lipids in cerebrovascular disease. Brit. med. J. 4: 78, 1969.

6. Bulpitt, C.J.: Serum uric acid in hypertensive patients. Brit. Heart J. 37:1210, 1975.

7. Pyörälä, K. et al.: Social Insurance Institution's coronary heart disease study. Social Insurance Institution's publications AL 1, Helsinki 1975. (in Finnish with English summary)

8. Takkunen, H.: Iron deficiency in the Finnish adult population. Scand. J. Haemat. suppl. 25, 1976.

9. Uric acid method. In Auto-Analyzer Manual. Technicon Instruments Corp. Chauncey, New York 1965.

10. Andrews, F.M. et al.: "Multiple classification analysis". 2nd ed. Ann Arbor, Institute for Social Research, The University of Michigan, 1973.

RECURRENT GOUTY PHLEBITIS WITHOUT ARTICULAR GOUT

G. Pasero

Istituto di Patologia medica, Università di Pisa

56100 Pisa (Italy)

The ancients said: "Totum corpus est podagra",that is to say: the whole body is subject to gout. The possibility that acute,paroxysmal manifestations,which arise in gouty patients may be ascribed to urate precipitation outside the joints has been certainly overestimated in the past,but now is wrongly disregarded.

Definite criteria are required,of course,to identify such extra-articular gout: a) a chronological one,namely a sudden onset and a self-limited duration with complete recovery; b) a clinical one,which is the severity of the pain; c) a metabolic one,the association with an unequivocally raised blood uric acid level; d) a therapeutical one,the prompt effectiveness of colchicine. An additional therapeutical criterion may be the disappearance of the trouble after urate pool depletion,as obtained with allopurinol or uricosuric agents. Acute extra-articular gout may be associated or alternated with typical gouty arthritis. The ultimate criterion for extra-articular gout - the presence of uric acid crystals or deposits in the affected tissue - cannot be easily demonstrated and therefore it could not be mandatory.

Extra-articular gout,however,should not be confused with the so-called visceral gout: first of all,extra-articular gout may be located in districts not included in the common meaning of the word "viscera"; in the second place,the term "visceral gout" often indicates a vague functional impairment of different parenchymas,unrelated to the metabolic disorder. Gouty phlebitis,gouty pharyngitis (1,2),gouty iritis (3-6),gouty orchitis (7,8) are usually recorded as complications of gout (9); but a complication is a consequence and not a different manifestation of the disease.

Gouty phlebitis is a definite extra-articular localization of acute gout,reported for the first time in 1866 by Sir James Paget (10). It may have no peculiar character or it is described as an acute,painful,segmentary,self-limited,recurrent inflammatory reaction of superficial veins,mainly in the lower extremities (11-14). De Sèze describes gouty phlebitis as very inflamed,very red,very painful (15). Gouty phlebitis,however,is a definitely rare entity. No more than few cases have been reported in the largest series (12); in ours we can recall only two other patients,besides the one described here. As above mentioned,gouty phlebitis,like every other localization of extra-articular gout,is usually associated or alternated with acute gouty arthritis. However we have never found in the literature a detailed report of isolated gouty phlebitis; we believe therefore useful to report this typical observation.

A 50-year-old man come to our clinic because of recurrent attacks of superficial phlebitis in the lower extremities; in the last sommer he was affected at least four times. The attacks were characterized by an extremely acute onset,but spontaneously or after non steroidal anti-inflammatory drugs they subsided in 7-10 days without permanent obstruction of blood flow. The patient noticed that phlebitis arised generally after an unusual physical stress,for example a long walk in the mountains. Family history was negative for gout and he never complained articular or renal symptoms compatible with this disorder.

Physical examination during the attack showed a hard,very painful rope,almost 10 cm long,overlaid with a red and warm skin and surrounded by local,limited oedema. All laboratory tests,including blood clotting tests and platelet count were in the normal limits, except blood uric acid levels,that were repeatedly raised,ranging between 7.0-9.6 mg%. Urate excretion on purine-free diet was 535 mg/day,urate clearance 6.2 ml/min,creatinine clearance 110 ml/min, urinary pH ranging between 5.1-5.2 (a very limited cyrcadian variation,as a rule in gout),urine analysis negative in other respects.

Colchicine (4 mg the first day,3 mg/day subsequently) was administered and resulted in a "dramatic" improvement,with relief of pain within few hours and disappearance of the signs of local inflammation in the next 36 hours. Urate pool depletion with 5-bromephenylindanedione,a uricosuric agent with anti-inflammatory properties (16,17),alternated with allopurinol was started and in the next two years no further attack have occurred. Only recently the patient experienced a transient hardening of a superficial vein in the left leg,promptly regressed with colchicine.

Although it has not been possible to attempt to demonstrate the presence of uric acid deposits in the affected tissue - moreover never reported in gouty phlebitis - there is goood evidence that in this case we are dealing with an isolated,recurrent gouty phlebitis. All the reminded criteria seem to be fulfilled. The effectiveness of colchicine for gouty phlebitis has been reported by Pellet et al. (11) and that of urate pool depletion by Siguier (18).

The peculiarity of this case is the absence of any articular complaint. The matter is just the allowance for identifying as "gouty" a phlebitis never associated with clinical evidence of articular gout. Semantically,gout is a unique term for a double meaning. Gout is a metabolic disorder,which induces an increase of urate pool,but gout is also a clinical entity,with prevailing articular symptoms, due to uric acid precipitation in tissues. The gout-clinical syndrome is usually related to the gout-metabolic disorder and only seldom to other hyperuricaemias; the last condition is called "secondary" gout. Conversely,the gout-metabolic disorder induces in most cases a gout-clinical syndrome,but may be symptomatically silent or unusually associated with kidney disease (19-21),phlebitis or other manifestations. If we bear in mind this double meaning of the word "gout" and we prefer to avoid an unnecessary neologism,it should be legitimated the adjective "gouty" for qualifying clinical patterns, although extra-articular and unassociated with similar joint involvement,when related with fulfilled evidence to gout-metabolic disorder.

SUMMARY

A case of recurrent,superficial phlebitis in the lower extremities,with raised blood uric acid levels,responsive to colchicine and to urate pool depletion,but unassociated with articular complaint is reported. The possibility that such a phlebitis may represent an insolated,critical manifestation of acute gout is stressed.

REFERENCES

1. DEBIDOUR A.: Les manifestations pharyngées de la goutte. Rhumatologie 1951,2,67
2. BERNOT et WETTERWALD: A propos d'un cas de pharyngite goutteuse. Lille méd. 1967,12,599
3. WOOD D.J.: Inflammatory diseases in the eye caused by gout. Brit.J.Ophtalm. 1936,20,510
4. McWILLIAMS J.R.: Ocular findings in gout. Amer.J.Opht. 1952,35, 1778
5. BONAMOUR G.: L'irite goutteuse,à propos de deux observations. Bull.Soc.ophtalm.France 1953,4,413
6. JAYLE G.E.,OURGAUD A.,QUEREILHAC H.: Lésions oculaires au cours des syndromes rhumatismaux et de la goutte. Rev.Rhum. 1954,21,534

7. DECAUX F.: De plusiers cas d'orchite goutteuse et de goutte du scrotum. Rein Foie 1961,2,143
8. LUCHERINI T.,BACCARINI V.: La gotta. Roma,Pensiero scientifico, 1964
9. TALBOTT J.H.: Gout. New York/London,Grune and Stratton,1964
10. PAGET J.: On gouty and other forms of phlebitis. St.Bartholomew's Hosp.Rep. 1866,2,82
11. PELLET C.: Sur trois cas de phlébite goutteuse. Rev.Rhum. 1948, 15,95
12. DIAMOND M.T.: Thrombophlebitis associated with gout. New York J.Med. 1953,53,3011
13. DECAUX F.: A propos d'un certain nombre d'observations de périphlébite goutteuse. Rein Foie 1960,1,159
14. GOSPODINOFF A.,SALVINI S.: Rilievi clinici e considerazioni patogenetiche sulla gotta atipica. La flebite gottosa. Reumatismo 1969,21,300
15. DE SEZE S.,RYCKEWAERT A.: La goutte. Paris,Expansion,1960
16. PASERO G.,RICCIONI N.: Il 2-fenil-5-bromo-1,3-indandione nel trattamento uricurico della gotta e della diatesi urica. Clin. Terap. 1963,26,32
17. LOMBARDINI J.G.,WISEMAN E.H.: Anti-inflammatory 2-aryl-1,3-indandiones. J.Med.Chem. 1968,11,342
18. SIGUIER F.: Maladies vedettes,maladies d'avenir,maladies quotidiennes,maladies d'exception. Paris,Masson,1957
19. DUNCAN H.,DIXON A.St.J.: Gout,familial hyperuricaemia and renal disease. Quart.J.Med.NS 1960,29,127
20. PASERO G.: La nefropatia uratica clinicamente primitiva. Minerva nefrol. 1965,12,113
21. LAGRUE G.,CANLORBE P.,BUSUTTIL R.: Néphropathies apparemment primitives révélatrices d'une dyspurinie latente chez deux adolescents. Sem.Hôp.Paris 1969,45,2367

ALTERATIONS OF HUMAN PURINE METABOLISM IN MEGALOBLASTIC ANEMIA

Irving H. Fox, Dale A. Dotten and Pamela J. Marchant

Purine Research Laboratory, University of Toronto

Rheumatic Disease Unit, Wellesley Hospital, Toronto

Secondary hyperuricemias in hematological diseases are frequently related to an increased turnover of nucleic acid purine in rapidly proliferating cells or in hyperplastic tissue (1). Transient hyperuricemia, hyperuricosuria and acute gouty arthritis have occurred during liver extract induced reticulocytosis and during the therapy for deficiencies of folic acid or iron (1-4). To further explore the acute alterations of purine metabolism, we have investigated 7 patients with severe megaloblastic anemia for changes of uric acid metabolism and 3 important erythrocyte enzymes of purine metabolism (5).

The plasma uric acid, urinary uric acid and urinary oxypurines were measured prior to and during vitamin replacement therapy with B_{12} or folic acid (Figure 1). Before therapy 2 patients were hyperuricemic, while 5 patients were normouricemic. Urinary excretion of uric acid and oxypurines, were normal except in 1 patient who was on allopurinol at the time of the study. Within 3 to 5 days of vitamin therapy, temporally related to the reticulocyte response of a peak value of 21 percent, the plasma uric acid increased in all patients except 1. The mean peak increase was 50 percent over pretherapy values. Urinary uric acid increased by a peak value of 35% of pretherapy values. The urinary oxypurines increased by 243% in the 4 patients in whom these could be measured. In 2 patients no oxypurines could be detected. An increase in the serum uric acid, the urine uric acid and the precursor oxypurines suggested that there was increased uric acid synthesis during the therapy of megaloblastic anemia. This may have resulted in part from degradation of the RNA extruded from normoblasts.

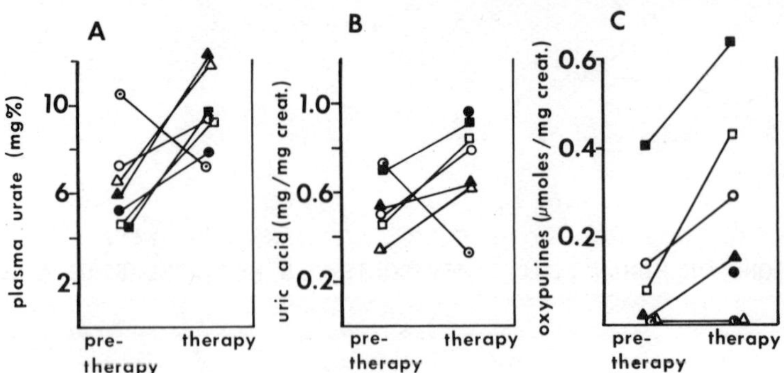

Fig. 1. Uric acid metabolism in megaloblastic anemia. A. Plasma uric acid B. Urine uric acid C. Urine oxypurines. Each symbol represents 1 patient.

Other alterations of human purine metabolism have been observed. Elevation of urinary aminoimidazolecarboxamide, a breakdown product of the de novo pathway, was found in the urine of patients with megaloblastic anemia (6). In addition erythrocytes and leukocytes from these patients had an increase of some enzymes of pyrimidine biosynthesis (7). We investigated 3 erythrocyte enzymes involved with purine metabolism, PP-ribose-P synthetase, adenine phosphoribosyltransferase (APRT) and hypoxanthine-guanine phosphoribosyltransferase (HGPRT) (Figure 2). Four of 6 patients initially had an elevated APRT. The mean value was 35.0 as compared to the normal mean of 23.8 nanomoles/hr/mg, a 47% increase. Following vitamin replacement therapy APRT increased to a peak value of 177% of pretherapy values. Prior to treatment, HGPRT and PP-ribose-P synthetase were in the normal range. The peak values during therapy increased to 121% of pretreatment HGPRT and to 175% of pretreatment PP-ribose-P synthetase levels. The enzyme increases accompanied the elevation of the serum uric acid and reticulocyte percentage (Figure 3).

These elevated erythrocyte enzyme activities could be related to an altered enzyme half life in erythrocytes. This was evaluated by separating peripheral circulating erythrocytes into 4 different fractions of increasing density, known to be related to increasing age. Pretherapy APRT was substantially greater than normal controls (Figure 4). The similar curves indicate that the enzyme half life was normal. One week following the onset of therapy, the youngest cells had markedly increased activity to 57% above pretreatment values. With HGPRT pretreatment values were similar to normal,

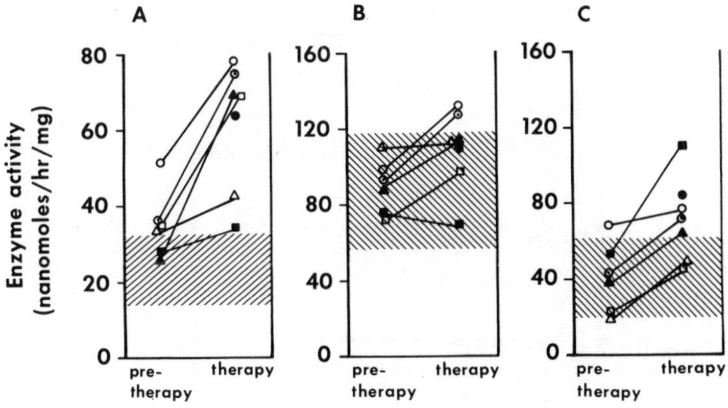

Fig. 2. Erythrocyte purine enzymes in megaloblastic anemia. The base line activity is compared to the peak value observed during the reticulocytosis of treatment. The darkened areas define the mean of normal patients plus or minus 2 standard deviations. A. Adenine phosphoribosyltransferase. B. Hypoxanthine-guanine phosphoribosyltransferase. C. PP-ribose-P synthetase.

while 1 week following therapy the specific activity of the youngest fraction was modestly increased by 24% (Figure 5). Prior to therapy the PP-ribose-P synthetase half life was distinctly different from normal (Figure 6). One week following therapy the youngest cells increased their activity by 397% while the following week the half life was almost normal.

The question arose whether the enzyme changes observed were a non-specific reaction to increased bone marrow activity or an event unique to megaloblastic anemia. The relationship of reticulocyte percentage to erythrocyte APRT from 10 patients with miscellaneous hematologic disorders with reticulocytosis and normal HGPRT was assessed (5). The correlation coefficient between APRT and percent reticulocytosis was 0.81. Thus higher reticulocyte percentages were associated elevated APRT. In contrast in HGPRT deficiency APRT was elevated without reticulocytosis. Similar studies with PP-ribose-P synthetase or HGPRT and reticulocytosis gave correlation of coefficients of only 0.0 and 0.11 respectively suggesting the lack of a relationship with reticulocytosis. Increased activity of HGPRT and PP-ribose-P synthetase was only observed consistently during therapy of megaloblastic anemia, although they were occasionally found in other disorders.

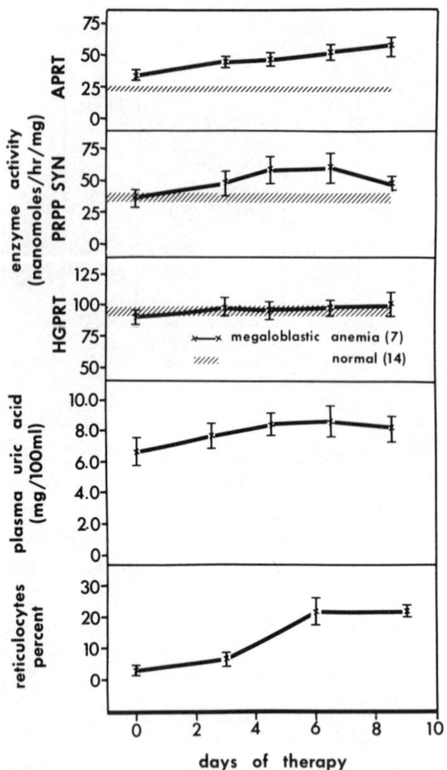

Fig. 3. The temporal relationship of purine erythrocyte enzyme changes. The horizontal axis indicates time in days from the start of vitamin replacement. The normal mean values plus or minus the standard error in 14 patients is indicated by the crosshatched lines.

To further understand the alterations of APRT, its relative half life in peripheral circulating erythrocytes was evaluated (Figure 7). The normal value was 2.3 ± 0.1 in 15 subjects. Two groups of patients were evident. Six individuals had a prolonged half life as suggested by a ratio of 1.2 to 1.6. This phenomenon is recognized in the deficiency of HGPRT. The other 8 patients had normal half lives as did patients with megaloblastic anemia whose mean ratio varied from 1.9 to 2.7.

In conclusion these studies demonstrate that during therapy of megaloblastic anemia there is an elevated activity of APRT, HGPRT and PP-ribose-P synthetase in peripheral circulating

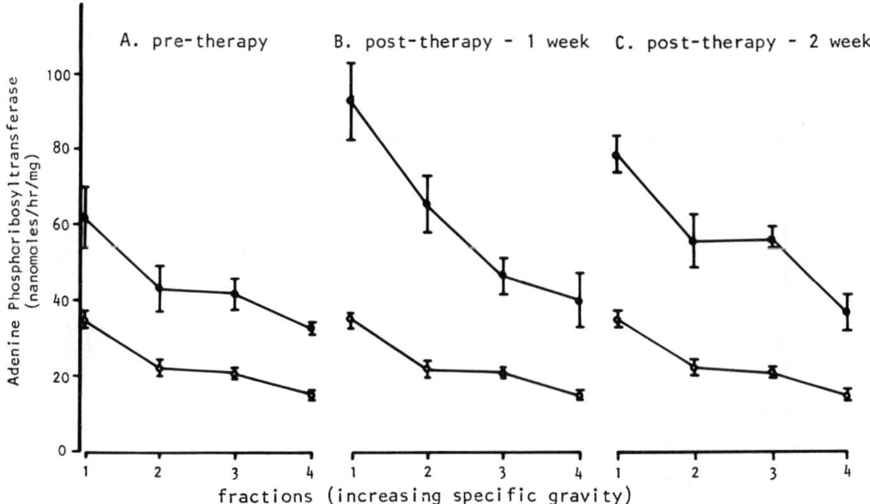

Fig. 4. APRT in density fractionated erythrocytes in megaloblastic anemia. Enzyme values, plus or minus the standard error are indicated pretherapy (A), one (B) and two (C) weeks after the start of therapy (●--●). Comparable values for 15 normal subjects (o--o) are indicated. Two patients were studied pretherapy and 4 patients one and two weeks afterwards.

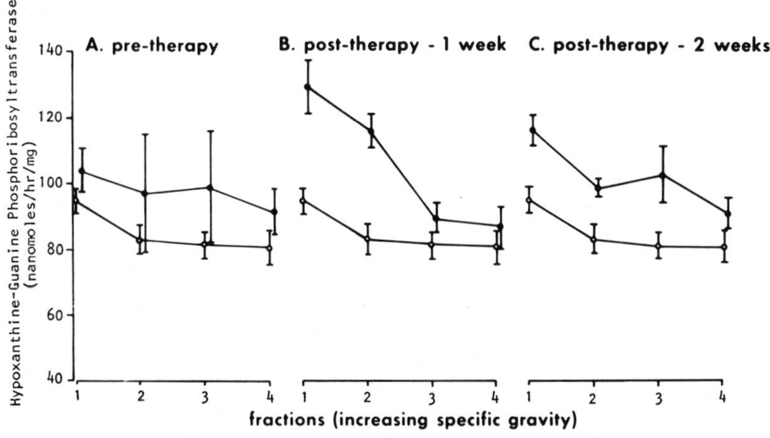

Fig. 5. HGPRT in density fractionated erythrocytes in megaloblastic anemia. Same format as Figure 4.

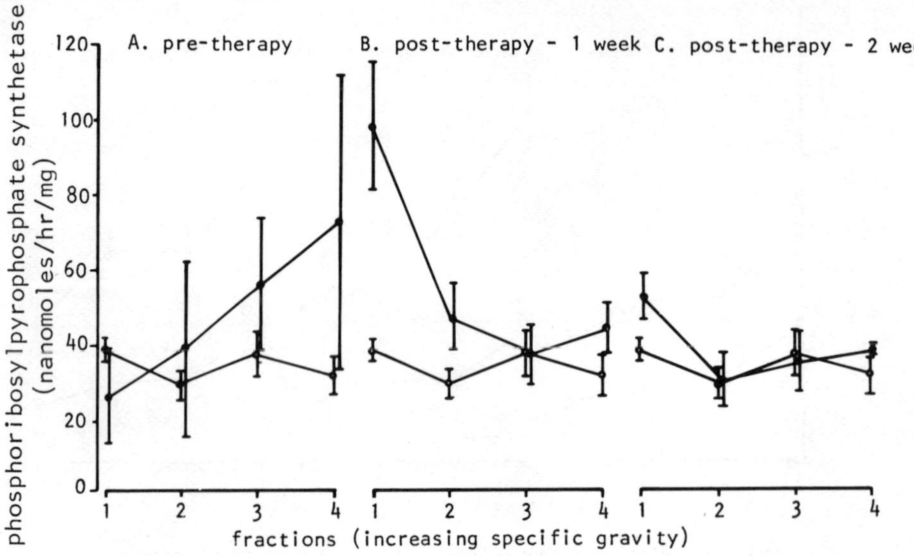

Fig. 6. PP-ribose-P synthetase in density fractionated erythrocytes in megaloblastic anemia. Same format as in Figure 4.

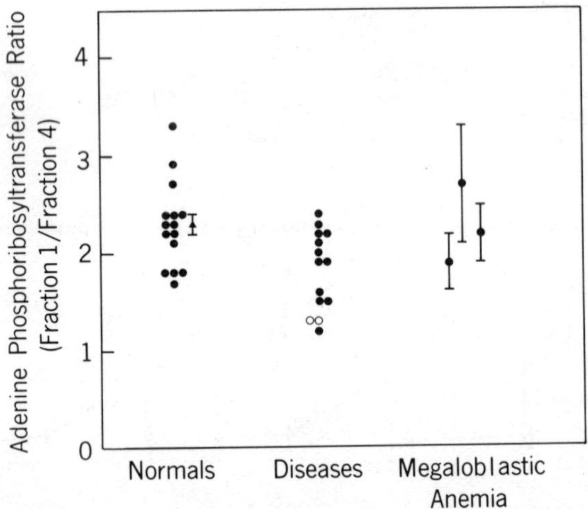

Fig. 7. Relative half life of APRT. The relative half life was estimated by dividing the enzyme values for the youngest cells (fraction 1) by the oldest cells (fraction 4). The mean ratio ± standard error is shown for the normal (subjects) and the patients with megaloblastic anemia prior to and during vitamin replacement therapy. ● normal HGPRT; 0 deficient HGPRT.

erythrocytes and evidence for increased synthesis of uric acid with an elevation of the serum uric acid and urine uric acid and oxypurines. Elevations of erythrocyte APRT, HGPRT and PP-ribose-P synthetase are not unique to megaloblastic anemia and occur in other acquired disorders. APRT elevation is proportional to reticulocytosis with normal HGPRT and may occur with prolonged half life as in HGPRT deficiency or with a normal half life as in megaloblastic anemia.

REFERENCES

1. Talbott, J.H. 1959. Gout and blood dyscrasias. Med. 38:173-205.

2. Riddle, M.C. 1930. The endogenous uric acid metabolism in pernicious anemia. J. Clin. Invest. 8:69-88.

3. Opsahl, R. 1939. Hematopoiesis and endogenous uric acid. Acta Med. Scand. 102:611-628.

4. Krafka, J., Jr. 1929. Endogenous uric acid and hematopoiesis II uric acid, reticulocytosis and erythrocytes after hemolysis by phenylhydrazine hydrochloride. J. Biol. Chem. 83:409-414.

5. Fox, I.H., Dotten, D.A., Marchant, P.J. and Lacroix, S. 1976. Acquired increases of human erythrocyte purine enzymes. Metabolism 25:571-582.

6. Luhby, A.L. and Cooperman, J.M. 1962. Aminoimidazolecarboxamide excretion in Vitamin -B12 and folic acid deficiencies. Lancet 2:1381-1382.

7. Smith, L.H. and Baker, F.A. 1960. Pyrimidine metabolism in man. III Studies on leukocytes and erythrocytes in pernicious anemia. J. Clin. Invest. 39:15-20.

TYPING OF URIC ACID LEVEL IN CEREBROSPINAL FLUID IN NEUROLOGICAL AND PSYCHIATRIC DISEASES

Frieder Láhoda and Dieter Athen

Senior Physician, Reader in Neurology; Research ASS.

Neurological and Psychiatric Clinic Univ.Munich

The increasing number of uric acid metabolism disorders that either are clinically latent or become manifest, plays an important role in the differential diagnosis of neurological and psychiatric disorders. Besides the still largely unsolved problems concerning the clinical significance of uric acid metabolism disorders in this kind of diseases the question arises whether uric acid possibly has a regulative function within the central nervous system under physiological conditions. The basis for this hypothesis is the fact, that famous people suffering from gouty arthitis often were unusually bright minds up to a high age. For example remember Rubens, Leibniz and Frederic 2nd of Prussia. RIGDWAY showed in comparative studies in delphines, which are animals with a more than average "intelligence", that they have a high serum uric acid level. In 1955 ORWAN made the hypothesis that the uricase deficiency and the resulting excess of uric acid might have been one of the main reasons for the dominating development of the intellectual power of the hominides in the course of the evolution. He suspected that uric acid like caffein stimulates the cerebral cortex. The interpretation of statistical data of MUELLER and BROOKS was the following: there might be a certain causal correlation between serum uric acid concentration and the behaviour pattern of a person, especially as far as activity, persistence and guidance qualities are concerned, whereas a correlation between serum uric acid and degree of intelligence still is controversial, last not least because of physiologically lower serum uric acid levels in females.

Material and Methods:

We measured cerebrospinal fluid (CSF) uric acid levels in a group of 173 patients suffering from neurological and psychiatric disorders. The control group consisted of 20 healhty persons of both sexes.
The cerebrospinal fluid was worked up immediately after sampling. Uric acid determinations were done enzymatically using a commercially available kit (biochemica test - combination, Boehringer Mannheim).

Results and Discussion:

In aggreement with the observations of FARSTAD and CARLSSON (1965 and 1973) our results show a correlation of roughly 1:10 between CSF-uric acid and serum uric acid concentration. The mean values were found to be higher in males than in females.
Very low CSF uric acid levels of 0.02 - 0.04 mg% , but normal serum uric acid levels were found in chronic stage of multiple sclerosis, cerebral atrophy and incomplete restoration after severe encephalitis. High CSF uric acid levels of 0.9 - 1.3 mg % with the serum uric acid levels being normal were found in patients with delirium alcoholicum, acute bacterial meningitis and occasionally in acute encephalitis. 35 per cent of patients with parkinsonism who received combined therapy with levodopa and a decarboxylase inhibitor, showed simultaneously rising serum uric acid and CSF uric acid concentration. At the same time an increase in their personality dynamics could be observed.
The discrepancy between the serum uric acid and the CSF uric acid levels in the first two groups can be interpreted in two different ways: it could be due to a selective increase in nucleic acid metabolism as well as due to a disturbance of the blood - CSF - barrier. Since after cerebral convulsions the CSF uric acid concentration was not significantly elevated neither existed a direct correlation between the CSF protein concentration and CSF uric acid concentration, it is not probable that the elevated CSF uric acid level can be explained by cell death.
What we observed in 35% of patients with parkinsonism treated with L-Dopa favours the view that the CSF uric acid may have a regulative function in the central nervous system under physiological conditions.
It appears that in the field of neurology and psychiatry determination of CSF uric acid concentration provides new ways for observing processes that indicate changesof nucleic acid metabolism. Therefore further differentiation of the pathways of uric acid and purine de novo synthesis would be of significant value.

Abstract:

The increasing number of latent and manifest hyperuricemia is important concerning differential diagnosis in neurological and psychiatric diseases. The pathological importance of hyperuricemia in these diseases is particulary unknown. The possibility of a physiological regulative function of uric acid in central nervous system may be discussed. Therefore uric acid typing in cerebrospinal fluid were made of 173 patients with neurological and psychiatric diseases. On an average the uric acid level in cerebrospinal fluid is 1:10 of the blood uric acid level. In advanced age a lower level could be found. High level for example were found after L-Dopa treatment and epileptical seizures. These results and possible pathophysiological correlation to the correspondant diseases are discussed.

References:

Carlsson, C. and Sven Jonas Dencker: Acta Neurol.Scandinav. 49 (1973) 39 - 46

Crone, Ch.: Danish Med.Bull. 4, 1 (1957) 22 - 25

Farstad, M. et.al.: Acta Neurol.Scandinav. 41 (1965) 52 - 58

Láhoda, F. and A. Ross: Münch.med.Wschr. 114, 10 (1972) 441 - 444

Lups, S.: The cerebrospinal Fluid Elsevier, Amsterdam (1954)

RENAL TUBULAR TRANSPORT OF URATE IN FANCONI SYNDROME

Herbert S. Diamond and Allen D. Meisel

Downstate Medical Center

450 Clarkson Avenue, Brooklyn, New York

Recent studies of renal urate handling in man have provided evidence for a model of urate transport which includes the following sequence:

1) uric acid is filtered at the glomerulus; 2) filtered urate is extensively, if not completely, reabsorbed in the proximal tubule; 3) additional urate enters the tubule by secretion; and 4) there is substantial reabsorption of secreted urate coextensive with and/or distal to the urate secretory site (1) (Figure 1). In accordance with this model, hyperuricosuria might result from enhanced secretion or diminished reabsorption of either filtered or secreted urate.

This is a case report of a patient with hyperuricosuria associated with the adult Fanconi syndrome in whom urate excretion was assessed under control conditions, after ribonucleic acid feeding, after pyrazinamide administration and after probenecid administration.

The patient is a 48 year old Caucasian female who was in good health, but on routine urinalysis had glycosuria and albuminuria. The patient was maintained on a constant protein normal purine, isocaloric diet for 3 days prior to and during all clearance studies. Medications known to effect the level of serum and urinary uric acid were withdrawn at least 4 days prior to the studies.

Fasting blood glucose was 88 mg/dL, but random urines showed 1-2+ glycosuria and urinary glucose was 274 mg per 24 hours (Table 1). On a glucose tolerance test, serum glucose peaked at 160 mg/dL at 30 minutes and declined to 88 mg/dL at 2 and 3 hours. Urine

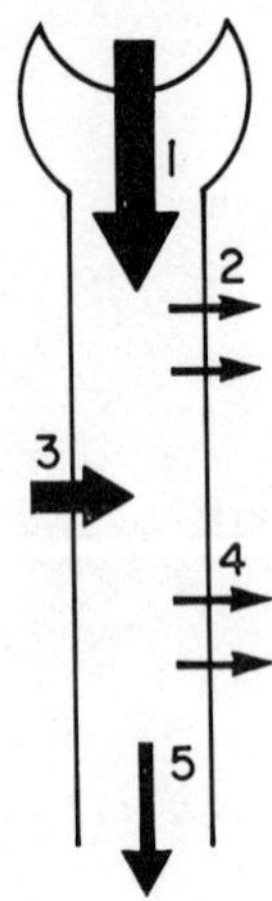

Fig. 1 A theoretical model for urate transport in man. Numbers in the diagram correspond to designations in the text. Reabsorption occurs both proximal and distal to secretion. The two processes may be partially coextensive. Reabsorption may occur at a single reabsorptive site along much of the proximal nephron or at two or more distinct reabsorptive sites.

glucose was trace positive at 30 minutes and 1+ at 2 and 3 hours. Serum phosphate ranged from 2.2 to 2.5 meq/l. Urinary secretion of glycine, histidine, lysine, threonine and serine were 5-10 times normal levels.

Table 1

Evidence for Tubular Dysfunction

Glycosuria	274 mg Glucose/24 hrs. Glycosuria at Plasma Glucose 88 mg/100 ml
Aminoaciduria	Glycine 1.30 μmol/min. Histidine 0.827 μmol/min.
Hypophosphatemia	Serum Phosphate 2.2 - 2.5 meq/L
Renal Tubular Acidosis	A.M. urine pH > 7.0

The patient's first morning urine consistently had a pH of greater than 7. On administration of a 1 g oral dose of ammonium chloride the pH of the patient's urine decreased from 7 to 5.6 within 6 hours. Excretion of ammonia and titratable acidity increased.

Blood urea nitrogen was 15 mg/dL. Mean creatinine clearance was 75 ml/min.

Serum uric acid concentration in this patient ranged from 1.5-1.8 mg/dL (Table 2). Mean urate clearance was 24 ml/min in this patient as compared to a mean urate clearance of 8.2 ml/min in 10 normal subjects. Urate clearance per glomerular filtration rate was 32% in this patient as compared to 7.5% in the controls. When serum uric acid was increased to normal range by feeding yeast RNA, 8 g per day, urate excretion per glomerular filtration rate remained greater than in control subjects.

The Fanconi syndrome is a functional abnormality of proximal renal tubular reabsorption with resultant renal glycosuria, renal hypophosphatemia, hyperaminoaciduria and proximal tubular renal acidosis (2). Thus, this patient can be classified as having idiopathic adult Fanconi syndrome. Hypouricemia and hyperuricosuria have been noted as features of the Fanconi syndrome.

In this patient with multiple other reabsorptive defects, uricosuria probably represents a defect in reabsorption. Maximum urate clearance following probenecid administration was similar in this patient with Fanconi syndrome and control subjects (Table 3). However, the increment in urate clearance in the patient was only 15 ml/min as compared with 31.7 ml/min in the controls. The urico-

Table 2

Renal Function and Uric Acid Excretion

	Fanconi Patient	Normals
Number of subjects	1	10
Serum Uric Acid (mg/100 ml)	1.7	5.3 ± 0.3
Urate clearance (ml/min)	24	8.2 ± 0.6
GFR (ml/min)	75	110 ± 5
Urate clearance/GFR	32	7.5 ± 0.5

Table 3

Effect of Probenecid and Pyrazinamide

	Fanconi Patient	Normals
Peak urate clearance after probenecid (ml/min)	39	40.4 ± 8.0
Change in urate clearance after probenecid (ml/min)	15	31.7 ± 7.9
Pyrazinamide suppressible urate clearance (ml/min)	20.1	6.4 ± 0.8
Pyrazinamide nonsuppressible urate excretion ug/min)	67	57 ± 7
Pyrazinamide suppressible urate clearance at peak response to Probenecid (ml/min)	28.3	30.3

suric effects of probenecid are generally accepted as representing inhibition of urate reabsorption. Therefore, diminished reabsorption of urate might be associated with a diminished uricosuric response to probenecid.

Fractional excretion of urate after pyrazinamide administration is an estimate of the fraction of filtered urate that escapes reabsorption. In normal subjects, reabsorption of filtered urate is almost complete at approximately 98% and remains complete even when filtered load is increased two to threefold by ribonucleic acid feeding (3). At control serum urate levels in this patient, the estimated filtered urate was 1.3 mg/min, of which 1.25 mg/min or 96.5% was reabsorbed. When serum uric acid level was increased by feeding ribonucleic acid, 8 g per day for 3 days, filtered urate increased to 2.2 mg/min, and reabsorption of filtered urate increased to 2.15 mg/min or 98%. Thus, reabsorption of filtered urate was virtually complete and similar to normal controls. This suggests that impaired urate reabsorption in this patient predominantly affects reabsorption of secreted urate.

Oral administration of pyrazinamide in man is associated with a decrease in urate excretion permitting subdivision of excreted urate into pyrazinamide suppressible and nonsuppressible fractions (4). The increased urate clearance in this patient was entirely attributed to increased pyrazinamide suppressible urate excretion. Pyrazinamide suppressible urate clearance in this patient was 20.1 ml/min compared to 6.4 ml/min in controls (Table 3). Pyrazinamide nonsuppressible urate excretion in this patient was similar to that in normal subjects.

Pyrazinamide suppressible urate excretion represents the net effect of urate secretion less reabsorption of secreted urate (1). Although originally interpreted as providing a measure of urate secretion, it is now recognized that an increase in this parameter is consistent with either enhanced urate secretion or diminished reabsorption.

Pyrazinamide suppressible urate clearance at maximal probenecid response was similar in the patient at 28.3 ml/min and the controls at 30.3 ml/min (Table 3). Since this test is influenced by drug interaction, it cannot be considered an accurate measure of urate secretion. However, these results do not suggest increased urate secretion. Thus, uricosuria in this patient appears to result from diminished reabsorption, predominantly affecting secreted urate.

Hyperuricosuria in patients with the Fanconi syndrome secondary to Wilson's disease is associated with increased pyrazinamide suppressible urate excretion (5). Thus, patients with Wilson's disease may have an abnormality of tubular transport of urate similar to the present case. However, hyperuricosuria associated with some other clinical syndromes appears to result from other abnormalities of renal tubular transport of urate. Three patients have been described in whom urate clearance per glomerular filtration rate greatly exceeded filtered urate. There was no evidence of abnormal tubular transport of glucose, phosphate or amino acids. In the patient described by Simkin and associates (6), administration of either pyrazinamide or probenecid suppressed urate excretion but fractional excretion of urate remained greater than 100%. Urate reabsorption in these patients is markedly impaired. Reabsorption of both filtered and secreted urate is probably affected.

In some young adults with sickle cell anemia, renal uricosuria has been attributed to enhanced secretion of urate (7). These patients differ from the present subject in that uricosuric response to probenecid was normal, PAH secretion was increased and maximal uricosuric response to probenecid was increased.

In two patients with idiopathic hypercalcuria without evidence of the Fanconi syndrome, the increase in uric acid clearance was not suppressed by pyrazinamide and was due to increased pyrazinamide nonsuppressible urate clearance (8,9). These patients may have defects in both urate secretion and reabsorption.

Thus, patients with four different clinical syndromes associated with hyperuricosuria; the Fanconi syndrome, isolated renal hypouricemia, sickle cell anemia, and idiopathic hypercalcuria show different patterns of response to pharmacological inhibitors of urate transport. These patterns of response may reflect underlying differences in the defects in renal tubular transport of uric acid in these clinical states.

Acknowledgements

We are indebted to the Staff of the Clinical Research Center for their invaluable aid in these studies. This work is supported by Grant RR 318 from the General Clinical Research Center Program of the Division of Research Resources, and by a grant from the Arthritis Foundation.

Dr. Herbert Diamond is supported, in part, by an Irma T. Hirschl Career Scientist Award.

Dr. Allen Meisel is a Clinical Research Associate of the General Clinical Research Center Program of the Division of Research Resources and is supported, in part, by Grant RR318.

References

1. Diamond, HS, Meisel, AD, Kaplan, D: Renal tubular transport of urate in man. Bull Rheum Dis 26:866-871, 1976.

2. Wallis, LA, Engle, RL: The adult Fanconi syndrome II. Review of eighteen cases. Amer J Med 22:13-23, 1957.

3. Jenkins, P, Rieselbach, RE: Unique characteristics of the mechanism of reabsorption of filtered versus secreted urate. Proc Am Soc Clin Invest 1974 p. 36a.

4. Steele, TH, Rieselbach, RE: The renal mechanism of urate homeostasis in normal man. Amer J Med 43:868-875, 1967.

5. Wilson, DM, Goldstein, NP: Evidence for urate reabsorptive defect in patients with Wilson's disease. In Purine Metabolism in Man, Adv Exp Med & Biol 41B:729-737, 1973.

6. Simkin, PA, Skeith, MD, Healey, LA: Suppression of uric acid secretion in a patient with renal hypouricemia. In Purine Metabolism in Man, Adv Exp Med & Biol 41B:723-728, 1973.

7. Diamond, HS, Meisel, A, Sharon, E, Holden, D, Cacatian, A: Hyperuricosuria and increased tubular secretion of urate in sickle cell anemia. Amer J Med 59:796-802, 1975.

8. Greene, ML, Marcus, R, Aurbach, GD, Kazam, ES, Seegmiller, JE: Hypouricemia due to isolated renal tubular defect: dalmation dog mutation in man. Amer J Med 53:361-367, 1972.

9. Sperling, O, Weinberger, A, Oliver, I, Liberman, UA and deVries, A: Hypouricemia, hypercalciuria and decreased bone density: a new syndrome. Ann Int Med 80:482-487, 1974.

NUTRITIONAL STATE AND PURINE METABOLISM

O.Frank

Special Hospital for Rheumatic Diseases

Baden, Austria

Hyperuricemia and gout are multifarious disorders of purine metabolism with heredity as genetic base and exogenic environmental factors which are prevailing of alimentary nature including overcaloric and fatty nutrition and alcohol consumption. These latter factors may be able to induce the manifestation of clinical gout.

The importance of environmental factors for manifestation of gout is documented by excessive dietary and drinking habits in the 17th century in the upper class in England as reported by G a r r o d producing enormous purine loads as the result of drinking of great amounts of wine and consumtion of meals rich of purine and protein.

The reversal situation has been shown during the last worldwar and the following years. In this time the standard of life was very poor, the people were not obese but underfed and hyperuricemia and gout disappeared for many years. Increasing incidence of hyperuricemia and gout did rise with appearance of prosperity simultaneous with overeating and drinking and development of obesity, a situation which has developed to a steady state. In our patients we have had investigated the distribution of relative weight and age in 3 groups each consisting of 100 patients with gout and hyperuricemia in males and females respectively. 65 % of gouty patients had a relative weight between 110 and 130 %. In males with hyperuricemia the relative weight was somewhat lower, but in females with hyperuricemia the relative weight was between 120 and 140 % in 52 %. The age of gouty patients was in 30% in the 5th decade, all the other patients were elder than 50 years. With regard to the predomi-

DISTRIBUTION OF WEIGHT IN %

% WEIGHT	<100	-110	-120	-130	-140	-150	>150
GOUT	5	2	35	30	18	8	2
HYPERURICEMIA ♂	13	27	40	13	5	–	2
HYPERURICEMIA ♀	–	18	18	29	23	12	–

DISTRIBUTION OF AGE IN %

AGE	<30	-40	-50	-60	-70	-80
GOUT	--	12	30	38	20	--
HYPERURICEMIA ♂	--	18	13	38	27	4
HYPERURICEMIA ♀	6	--	12	35	47	--

nance of males in clinical gout females were not taken into consideration. Obesity is a characteristic feature of gout and hyperuricemia, at which a highly significant correlation in serum uric acid with body weight was found. The difference of mean relative weight between gout and hyperuricemia and on the other hand between gout and hyperuricemia with persons with normal serum uric acid was highly significant.

Undoubtedly overweight is the most essential result of an overcaloric nutrition consisting in a high fat intake and a considerable alcohol consumption. Obesity resulting from overcaloric nutrition is leading to an increase of serum uric acid because of the close correlation with body weight which was verified in our collective. The consumption of alcohol is representing an additional factor leading to a ketose and an interference with the renal excretion of uric acid and thereby to an increase of serum uric acid.
By Lefevre et al. have been shown that alcohol is producing a ketose not in toxic doses only but in habitual consumption in moderate amounts as usually in our patients which are wine grower in a high proportion. The importance of environmental factors, especially in altered nutrition habits, has been shown from colonisation or emigration of polynesian populations leading to hyperuricemia.

Interrogating the patients regarding their eating and drinking habits an abnormal nutrition with partly high intake of calories, a high consumption of fat and alcohol has been shown. The highest daily intake of calories was found in gouty patients. Nearly 50 % of the demand of calories are covered by fat, but especially the intake of protein and carbohydrates is proportional reduced. The mean share of alcohol in the daily consumed calories amounts from 10 to 30 %.

	n	SUA $\bar{x}$	Age	% Weight
GOUT	40	7.6	51.8 ± 9.2	126.5 ± 17.4
CONTROLS	40	5.6	52.4 ± 8.7	126.3 ± 14.0
HYPERURICEMIA	45	7.7	53.5 ± 10.1	113.7 ± 14.3
CONTROLS ♂	45	5.0	54.0 ± 10.3	113.4 ± 12.9
HYPERURICEMIA	17	6.3	56.9 ± 11.1	124.4 ± 12.7
CONTROLS ♀	17	4.7	58.8 ± 7.2	126.0 ± 12.8

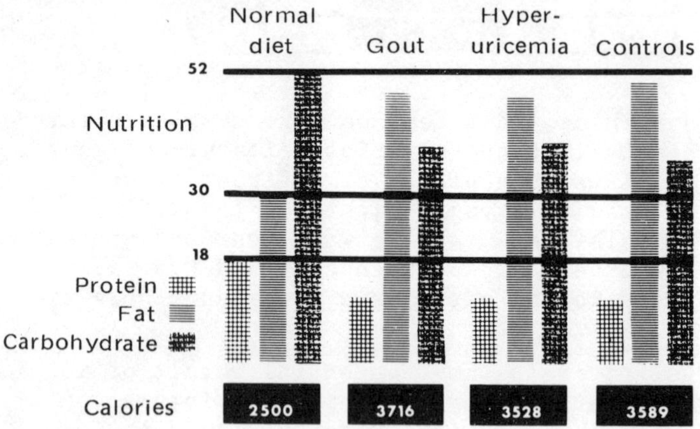

Primary gout is associated with a variety of disorders such as a manifest or more frequently a subclinical diabetes, a frequent incidence of hyperlipoproteinemia of type IV and hypertension and finally obesity. All these disorders represent an accumulation of risk factors for coronary heart disease.

	n	HYPERLIPOPROT. type IIa	IIb	IV	DIABETES overt	lat.	FHD	HYPER-TENSION	CHD
GOUT	40	1	1	21	4	14	3	9	13
CONTROLS	40	1	1	4	3	14	9	10	6
HYPERURICEMIA ♂	45	2	3	7	1	17	4	9	8
CONTROLS	45	1	1	6		17	5	12	5
HYPERURICEMIA ♀	17	2	1	1	3	7	3	6	6
CONTROLS	17	1	1	2		8	3	8	4

FHD = FAMILY HISTORY OF DIABETES
CHD = CORONARY HEART DISEASE

THE EFFECT OF DIFFERENT PURINES AND PYRIMIDINES ON HUMAN PYRIMIDINE BIOSYNTHESIS

N. Zöllner, W. Gröbner, A. Rauch-Janssen
Medizinische Poliklinik der Universität München
Pettenkoferstrasse 8a
D-8000 München 2

Hereditary orotaciduria is caused by the patient's inability to convert orotic acid to uridine monophosphate, due to an absence of the activity of orotidine phosphate decarboxylase in all cases and orotate phosphoribosyltransferase in all cases but one. All or nearly all of the orotic acid formed by biosynthesis must be excreted or degraded. Therefore, urinary excretion permits an estimation of pyrimidine biosynthesis. Children with the disease excrete 1.0 to 1.5 grammes of orotic acid per day while the normal excretion of orotic acid is only 1.4 mg/d.

In 1970 FOX et al. as well as KELLEY and BEARDMORE observed orotaciduria - consisting of orotic acid and orotidine - in patients under therapy with allopurinol. Later it could be shown (BEARDMORE and KELLEY 1971) that mononucleotides of allopurinol and oxipurinol inhibit orotidine phosphate decarboxylase in vitro. This inhibition is accompanied (in vitro) by changes in the molecular configuration (GRÖBNER and KELLEY 1975).

Allopurinol induced orotaciduria may be considered an incomplete phenocopy of the genetic disease, incomplete in so far as the excretion of total orotic acid amounts only to 15 - 100 mg/d., i.e. 1 - 5 per cent of the amount excreted by the patients with hereditary orotaciduria. Nevertheless, it can be used as a model to estimate pyrimidine biosynthesis or changes of it.

Several years ago, we (ZÖLLNER and GRÖBNER 1971) found that oral administration of RNA decreases allopurinol induced orotaciduria considerably, e.g. from 21 mg/d to 7 mg/d. Of course our results reminded us of the earlier observations that administration of uridylic acid and/or cytidylic acid decreases orotaciduria in patients with the genetic disease. One reasonable explanation of our experimental findings as well as of the clinical observation is an inhibition of orotic acid biosynthesis by exogenous nucleotides. However, it remained to be established which nucleotide is responsible. It is the purpose of this report, to report and discuss further findings.

The basic design of our experiments was to administer a diet free of purines and pyrimidines to which allopurinol and/or measured amounts of chemically defined sources of purines or pyrimidines were added. Such, experimental diets must be isoenergetic, i. e. provide the same amount of energy the subject has turned over before the experiment. For our purposes it was also necessary that every experimental period was continued until a steady state of the parameters in question was reached. Experiments fulfilling these requirements last thirty days or longer. Over that extended period of time, subjects prefer the neutral taste of liquid formula diets to the monotony of a no purine and/or pyrimidine diet from conventional foods (in our laboratory slogan "spaghetti-egg-diet").

Nearly all of our experiments have been done with liquid formula diets. The technical details have been described repeatedly, theyare all based on the work of AHRENS. Our own formulas were changed several times, mainly due to improvements in the commercial products for protein and carbohydrate. However, the relation between protein, carbohydrate and fat was always the same. The protein content of our diets always was 15 energypercent and the proportion of fat and carbohydrates was roughly 1:2. The total amount of energy to be administered was determined by a three day recall as well as by the subjects' sex, body surface and activities.

When an alkali hydrolysate of RNA (4 g/d) (consisting of all mononucleotides contained in RNA) was administered, the results with regard to diminution of orotaciduria were identical with those obtained with RNA itself.

In the next set of experiments 1 g/d of each of the nucleotides was used. None of these was as potent with respect to reducing orotaciduria as 4 g/d of the hydrolysate, demonstrating that more than one nucleotide is responsible for the effect. When the nucleotides were compared,not only the pyrimidine nucleotides were found to be effective but also guanosine monophosphate. Obviously, the influence of RNA on allopurinol induced orotaciduria is due to the combined effects of several of its constituents.

In order to estimate equivalent doses, larger amounts of mononucleotides had to be given. As far as these experiments are completed (table 1 summarizes the results) they demonstrate that 3 g/d of uridylic acid are equivalent to 4 g/d of RNA. Lesser effects were observed for cytidylic acid and adenylic acid. (No data for guanylic acid a available at present). These data show that the effect of RNA is not the effect of any single one of its nucleotides.

In addition to orotic acid excretion, orotidine phosphate decarboxylase activity was measured in erythrocytes. In experiments with RNA hydrolysate and one gramme doses of nucleotides, the enzyme activity was not changed from its allopurinol induced level.

We interpret our findings to indicate that dietary nucleotides or metabolites derived from them inhibit pyrimidine biosynthesis. About the site of this inhibition there are only speculations. In mammalian systems it appears that the most important site of feedback inhibition of pyrimidine synthesis is at the level of the soluble carbamyl phosphate synthetase (for a review see BLAKELY and VITOLS, 1968). Another speculation involves dihydroorotase which is inhibited by a number of purines and pyrimidines (BRESNICK and BLATCHFORD 1964).

Two further explanations for our findings come to mind, inhibition of the renal excretion of orotic acid and an influence of the dietary nucleotides or their metabolites on the effects of allopurinol on orotidine phosphate decarboxylase.

The renal effect is very unlikely since the physiological orotaciduria is not influenced by reducing the normal dietary administration of RNA to zero during the first phase of formula diet experiments.

Table 1

Orotic acid excretion of healthy subjects on a formula diet free of purines and pyrimidines and treated with 400 mg allopurinol daily (column a, basal experiment). Column b (nucleotide supplement) shows the excretion when nucleotides were supplemented. The values for orotic acid excretion are averages of the total experimental periods of ten days each.

Nucleotide	Dose	Orotic Acid Excretion (mg/d)			
		(a) basal experiment	(b) nucleotide supplement	diffe-rence	per-cent
RNA	4	19	6	13	68
RNA-Hydrolysate	4	18.8	5.5	13.3	69
Adenine	3	34	31	3	8.8
AMP	1	10.9	10.2	-	-
	3	16.9	8.8	8.1	48
GMP	1	15.2	7.5	7.7	50
Guanine	3	12	13.5	-	-
IMP	1	22.5	27.5	-	-
CMP	1	18.9	12.2	6.7	35
	3	21.6	11.9	9.7	45
UMP	1	15.8	11.3	4.5	28
	3	16.1	5.7	10.4	65
	3	18.2	6.8	12.4	68

Our observation that RNA does not change increased in vitro activity of orotidine phosphate decarboxylase induced by allopurinol suggests that the allopurinol and oxipurinol ribonucleotides responsible for the inhibition of the enzyme are not displaced by larger amounts of oral ribonucleotides. Otherwise, the enzyme would have returned towards normal.

References:

Beardmore, T. D. and Kelley, W. N.: Mechanism of allopurinol-mediated inhibition of pyrimidine biosynthesis. J.Lab.Clin.Med. 78, 696 (1971)

Blakely, R. L. and Vitols, E.: The control of nucleotide biosynthesis. Ann. Rev. Biochem. 37, 201 (1968)

Bresnick, E. and Blatchford, K.: Inhibition of dihydro-orotase by purines and pyrimidines. Biochim. Biophys. Acta 81, 150 (1964)

Fox, R. M., Royse-Smith, D. and O'Sullivan, W. J.: Orotidinuria induced by allopurinol. Science 168, 861 (1970)

Gröbner, W. and Kelley, W. N.: Effect of allopurinol and its metabolic derivatives on the configuration of human orotate phosphoribosyltransferase and orotidine-5-phosphate decarboxylase. Biochem. Pharmac. 24, 379 (1975)

Kelley, W. N. and Beardmore, T. D.: Allopurinol: alteration in pyrimidine metabolism in man. Science 169, 388 (1970)

Zöllner, N. and Gröbner, W.: Influence of oral ribonucleic acid on orotaciduria due to allopurinol administration. Z.ges.exper.Med. 156, 317 (1971)

THE EFFECT OF WEIGHT LOSS ON PLASMA AND URINARY URIC ACID AND LIPID LEVELS

J. T. Scott and R.A. Sturge

Kennedy Institute of Rheumatology

Bute Gardens, London W6 7DW

INTRODUCTION

There is considerable evidence from epidemiological and clinical studies of a relation between body weight and the plasma level of uric acid. An association exists between serum uric acid and triglyceride levels, while a relationship between serum uric acid and cholesterol has been claimed by some but disputed by others (Darlington and Scott, 1972). The serum uric acid appears to fall with weight loss (Nicholls and Scott, 1972; Emmerson, 1973).

The overall relationships between weight loss, serum uric acid, urinary uric acid and blood lipids have not previously been examined and preliminary findings of such a study are presented.

METHODS

The plan of the study was as previously described, when fifteen overweight subjects were examined before and after a period of weight loss (Nicholls and Scott, 1972). Twenty further gouty and non-gouty obese subjects (defined as over 10% of ideal body weight), 14 men and 6 women, participated. At the onset of the study they took a low-purine, alcohol-free diet (containing less than 200 mg purine daily) for one week, during the last 3 days of which blood and 24-hour urine collections were obtained for uric acid, cholesterol, triglyceride and creatinine estimations. No drugs were allowed.

The subjects then lost weight on a low carbohydrate diet. After adequate weight reduction had been achieved they again took a low-purine diet for a week, blood and urine collections being obtained during the last three days as before. Plasma and urinary uric acid were estimated by an enzymic spectrophotometric method, cholesterol and triglycerides (under fasting conditions) on the Technicon Autoanalyser. Results are expressed as a mean of the three estimations before and after weight reduction.

RESULTS

Weight-loss in the 20 subjects ranged from 1.6 kg to 12.3 kg, mean 7.0 kg.

Plasma uric acid levels fell in 15 of the 20 subjects, the mean value before weight loss being 6.9 mg/100 ml and that afterwards being 6.4 mg/100 ml. This fall, though not pronounced, is statistically significant (t test for paired differences, $p = 0.007$).

Levels of urinary uric acid also fell in 16 of the 20 subjects. The mean level of 658 mg/24 hours before weight loss fell to 567 mg/24 hours ($p = 0.005$). As in the previous study, creatinine and urate clearances were virtually unchanged.

Changes in plasma and urinary uric acid from the total of 35 subjects in this and the previous study are shown in Table I.

Table I

Uric acid data before and after weight-loss

	No. of subjects	Initial	Final	Mean fall	Significance
Plasma uric acid mg/100 ml	35	6.96	6.27	0.69	p 0.000025
Urinary uric acid mg/24 hour	35	603	516	87	p 0.0006

The mean fasting triglyceride level in the 20 subjects in the present study fell from 182 mg/100 ml to 137 mg/100 ml ($p = 0.03$) and that of cholesterol from 244 mg/100 ml to 232 mg/100 ml ($p = 0.03$). The majority of subjects showed a fall in both lipids and in plasma uric acid. There was some association between the degree of change in cholesterol and that of uric acid ($r = 0.62$, $p = 0.004$) but not between that of triglycerides and uric acid ($r = 0.01$, n. s.).

DISCUSSION

A number of epidemiological investigations have demonstrated an association between plasma levels of uric acid and body-weight or body-bulk (Healey and Hall, 1970) and clinical surveys also indicate that gouty patients tend to be overweight (Grahame and Scott, 1970).

The present study confirms our previous findings that weight reduction usually leads to a fall in the plasma level of uric acid, the mean fall in 35 subjects being 0.69 mg/100 ml. An individual's plasma uric acid therefore appears to be influenced to some extent by his weight.

The fall in urinary uric acid found in our previous study was of marginal significance, but with greater numbers the mean fall, though not large (87 mg/24 hours), is now significant, indicating that there is an alteration in urate production with weight loss. This is consistent with the findings of Emmerson (1973), who from isotope kinetic studies in an obese patient before and after weight loss concluded that urate production, elevated during obesity, fell to within the normal range; in this patient, however, there was also a rise in renal clearance of urate.

The nature of the relation between uric acid and lipids remains to be determined. There is general agreement of an association with triglycerides, less in the case of cholesterol. Some have emphasized the importance of obesity and alcohol consumption as contributory factors in the relationship (Gibson and Grahame, 1974), but if indeed the higher triglyceride levels found in gout patients are independent of body weight (Wiedmann, Rose and Schwartz, 1972) other factors must be involved. In the present study there has been an overall fall in cholesterol and triglyceride levels with weight loss: to date a correlation appears to exist between the degree of fall in uric acid and that of cholesterol, but not that of triglycerides. However, numbers are small and further data are needed.

Metabolic inter-relationships between caloric intake, lipogenesis, obesity and hyperuricaemia, involving perhaps factors such as conversion of excess carbohydrate intake to fatty acids, production of additional pentose phosphate and consequently of PRPP and purine ribonucleotides (Holman, Kelley and Wyngaarden, 1975-76) also remain to be explored.

SUMMARY

Plasma uric acid levels in 35 obese gouty and non-gouty subjects showed a highly significant mean fall of 0.69 mg/100 ml after weight reduction on a low-calorie diet. Urinary urate also fell by a mean value of 87 mg/100 ml. Serum cholesterol and triglyceride levels have been studied in 20 of the subjects. Both have shown some overall fall with weight loss. So far the degree of change of plasma urate has shown a correlation with that of cholesterol but not that of triglyceride, but further data are needed on these inter-relationships.

REFERENCES

DARLINGTON, L. G., AND SCOTT, J. T. (1972) Ann. rheum. Dis., 31, 487 (Plasma lipid levels in gout)

EMMERSON, B. T. (1973) Aust. N. Z. J. Med., 3, 410 (Alteration of urate metabolism by weight reduction)

GIBSON, T., AND GRAHAME, R. (1974) Ann. rheum. Dis., 33, 298 (Gout and hyperlipidaemia)

GRAHAME, R., AND SCOTT, J. T. (1970) Ann. rheum. Dis., 29, 461 (Clinical survey of 354 patients with gout)

HEALEY, L.A., AND HALL, A.P. (1970) Bull. rheum. Dis., 20, 600 (The epidemiology of hyperuricaemia)

HOLMES, E.W., KELLEY, W.N., AND WYNGAARDEN, J.B. (1975-76) Bull. rheum. Dis., 26, 848 (Control of purine biosynthesis in normal and pathologic states)

NICHOLLS, A., AND SCOTT, J. T. (1972) Lancet, 2, 1223 (Effect of weight-loss on plasma and urinary levels of uric acid)

WIEDEMANN, E., ROSE, H. G., AND SCHWARTZ, E. (1972) Amer. J. Med., 53, 299 (Plasma lipoproteins, glucose tolerance and insulin response in primary gout)

PATHOGENESIS OF "FASTING HYPERURICEMIA" AND ITS PROPHYLAXIS

P. SCHRÄPLER, E. SCHULZ and A. KLEINSCHMIDT

Medical University of Lübeck
Ratzeburger Allee 160
D - 2400 Lübeck

Since the initial descriptions by BENEDICT (4) and BLOOM (5), several other reports on fasting as an introduction to the treatment of severe obesity have been made (8,22,28,29). Examples of acute gout (9,29), urate nephrolithiasis (9,14) and renal insufficiency (14 are reported. In order to prevent such complications during fasting, medicamentous treatment seems to be necessary to reduce urate synthesis or - on the other hand - to increase the renal clearance of uric acid.

This study was carried out to elucidate the question whether daily applications of allopurinol, benzbromarone and probenecid respectively in normal doses are able so to lower the uric acid level during prolonged fasting that uricopathic complications can be avoided. Informations on the pathogenesis of "fasting hyperuricemia" and on the mode of action of the drugs just mentioned is to be hoped for from special methods of controlling renal elimination mechanisms and their partial functions.

Patients, experimental structure, methods

The collective to be investigated consisted of 77 overweight probands (46 women, 31 men), who voluntarily subjected themselves to a strict (total) fasting regime lasting about 3 weeks (20,3 ± 3,9 days).

From the second day onwards, 20 unselected patients received 300 mg allopurinol/day, 11 received probenecid (2g daily) and 26 patients got benzbromarone (100 or 300 mg daily). Untreated probands served as controls. Uric acid[+)] (serum/24- hr - urine), lactate[+)], pyruvate[+)], blood- (micro-Astrup method) and urine-pH (special indicator paper from Merck), and also the glomerular filtration rate (inulin, creatinine[+)]) were determined and the uric acid clearance rate (24-hr) and a rough estimate of the net-reabsorption were calculated as well as the proportion of the uric acid reabsorbed from the quantity filtered (24).

+) = Biochemica Test Packs , Boehringer GmbH, Mannheim/GFR

Statistics and results

The results observed are reproduced in 5 figures and 1 table. In addition to the mean values ($\bar{x}$) with the standard deviation (s) and the confidence limits ($s\bar{x}$), the significances (Student's- test) between the corresponding daily values of the collectives are determined. A significant difference in the mean values is assumed for $p < 0.05$.

Fig. 1 shows the serum level, excretion and clearance rates of the control collective which received no medication. After an approximately linear increase in concentration, already statistically confirmed from the second day onwards ($p < 0.01$) with an upper average level of 13.9 mg/100 ml (+ 178,0 %/10th day), the plasma levels fall again in the second fasting period (11th day), but remain distinctly above ($p < 0.01$) the basic level at the beginning of the fasting-cycle. Excretion and clearance values fall rapidly at first and become stabilised at a significantly lower niveau.

In Fig. 2 the uric acid values of the untreated group and the individual test groups with therapeutic doses of allopurinol, probenecid and/or high doses of benzbromarone are given. Taking into account a common initial level, the plasma levels on the corresponding test days lie distinctly ($p < 0.01$) below the values for the drug-free control group. 100 mg benzbromarone/day not only prevents the natural increase of uric acid but even lowers it significantly below the basic value ($p1/14 < 0.05$; $p1/21 < 0.01$). The reduction after 7, 14 and 21 days was 12%, 19.8% and 35.2%.

Fig. 3 shows the comparison of serum uric acid, excretion and clearance rates of the control and the allopurinol group. The plasma levels differ from each other significantly from the third day onwards ($p < 0.05$). Excretion and clearance rates fall rather more rapidly at first with allopurinol, the differences amounting to 13.3 and 4.9%. From the 10th day on, they lie slightly above the values of the drug-free control collective. But if the common basic level is taken into consideration, this difference is only about 10.2 and 8.8%. The corresponding parameters do not differ to any extent worth speaking of in relation to the 14-day medication period, as the percentage figures for the elimination and clearance rates of 3.8 and 0.5% show.

In Fig. 4, the mean level profile of plasma uric acid and uric acid excretion during therapeutic treatment with probenecid and high-dosed therapy with benzbromarone are shown. The serum levels already differ significantly ($p < 0.01$) after only one day of medication, considering that the antigout drugs were applied on the morning of the 2nd day of fasting. Whereas with high doses of benzbromarone (3 x 100 mg/day), the uric acid becomes stabilised at a significantly lower level from the 6th fasting day onwards, the corresponding daily levels with probenecid show a distinct increase with an upper mean value of 9.8 mg/100 ml (+ 67.3%, 14th day). The mean plasma levels for comparable medication periods (2nd - 15 th day) are 8.4 mg/100 ml (probenecid) and 2.6 mg/100 ml (benzbromarone). In both collectives there was a distinct and persistent increase of excretion, as the comparison with the elimination values of the control group (Fig. 1) and the drug-free period of the probenecid collective shows. While the excretion rates with probenecid up to the 9th day are altogether rather lower with an average of 428.1 mg, the difference being about 42.0%, the mean excretion rates from the 10th day onwards are almost the same at 407 or 435 mg.

In Table 1, the proportion of the amount of filtered uric acid which is reabsorbed through the tubules is given for 4 collectives; the uric acid clearance on corresponding days of the test being related to the creatinine clearance of the first fasting day (C_{Cr} 1), supported by our own findings (32). The initial value of 0.944 for all groups says that at the beginning of fasting 94.4% of the filtered amount of uric acid is reabsorbed through the tubules. Without therapy and with allopurinol this proportion increases during the starvation-period, with probenecid it remains at about the same niveau as the initial value, while with fractional dosage of benzbromarone (3 x 100 mg/day) it is distinctly and persistently lowered. The differences from the untreated control and the probenecid group averaged 15.2 and 11.6%.

Fig. 5 shows the correlation between withdrawal of calories and plasma uric acid in the initial fasting-period (total-fasting).

Discussion

LENNOX (16) had already pointed out the connection between withdrawal of calories and hyperuricemia in 1924. In agreement with the findings of other teams (1,20,22,29) there is an excessive increase in serum uric acid. We observed an upper mean level of 13.9 mg/100 ml (+ 178%, 10th day) and a significant restriction of excretion and clearance from the 3rd and 2nd day onwards (Fig. 1). Other pathogenetic factors are increased breakdown of nuclear-protein (21) and increased reabsorption (6,30,31) as a result of ketoacidic secretion blockade in

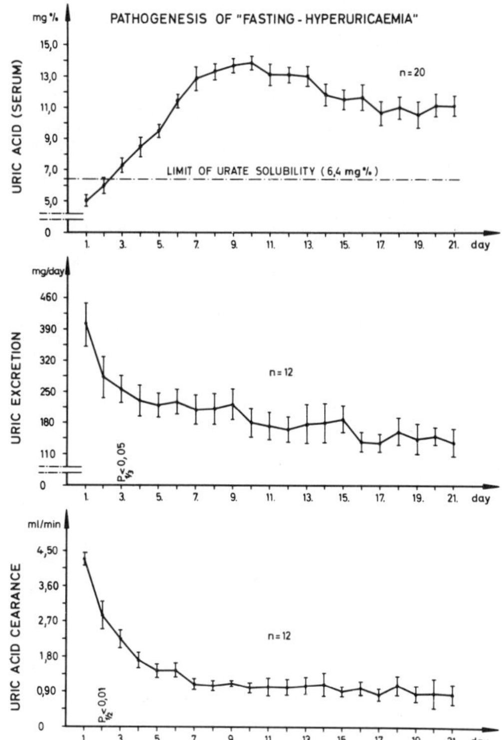

Fig.: 1 Reaction of uric acid (serum level, excretion and clearance/24-hr) during prolonged fasting

the distal tubule section. Whereas CHRISTOFORI et al.(6) saw the cause of the increased reabsorption in the reduced concentrations of glucose and aminoacids , the significant shift of the lactate-pyruvate ratio in favour of lactate from the second day onwards seems us to be more important, the more so as HANDLER (13), REEM et al.(25) and YÜ et al.(33) demonstrated experimentally the uric acid retaining effect of lactate. The significance of ketosis for the changes, we have shown summarily in Fig. 1 is supported by the findings of CHEIFETZ (7), LENNOX (16) and PADOVA et al.(23).

Since the plasma levels and excretion rates return to the original initial levels again within a few days after ending the starvation-period, we see no contraindication to strict fasting with consistent continuation - possibly with increased dosage and alkalisation of the urine - of a uric acid lowering drug in the rather frequent overweight gouty patients.

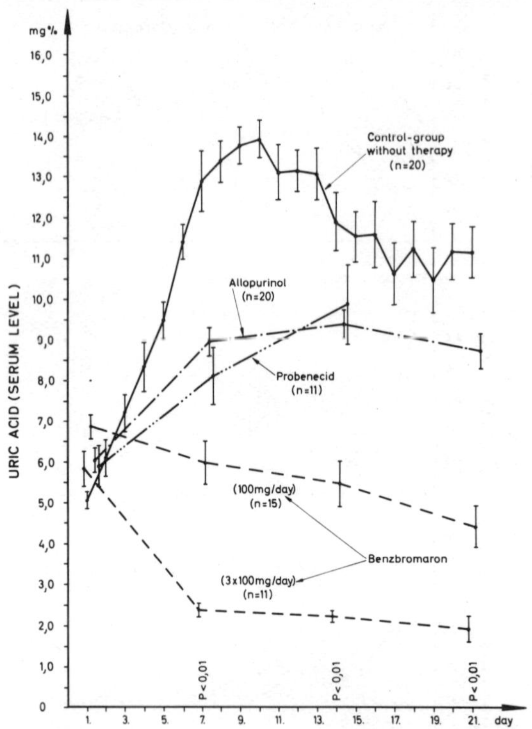

Fig.: 2 Influence of therapeutic daily doses of allopurinol (300 mg daily), probenecid (2g daily) and/or high doses of benzbromarone (100 mg or 300 mg daily) on fasting hyperuricemia

Fig. 2 shows pharmacodynamic aspects. The usual therapeutic daily doses of allopurinol and probenecid lower the serum uric acid significantly without being able to compensate the "fasting hyperuricemia" (Fig. 2). The high rate of intolerance in the probenecid group was striking. The medication had to be discontinued prematurely in 4 patients because of extensive allergic reactions. It is possible that a compensation of the "fasting hyperuricemia" might have been achieved by the daily administration of 500 - 600 mg allopurinol, referri to DRENICK et al.(10), LLOYD-MOSTYN et al.(17) and SCHATONOFF et al. (27) among others. In agreement with SCHMAHL et al.(28), on the othe hand, 100 mg benzbromarone/day lowers the plasma level significantly below the initial fasting level. In relation to the basic level, the decrease on the corresponding test days was 12.0, 19.8 and 35.2%. With the divided dosage of 300 mg benzbromarone the uric acid loweri effect is clearly greater, as can be seen also from Fig. 2.

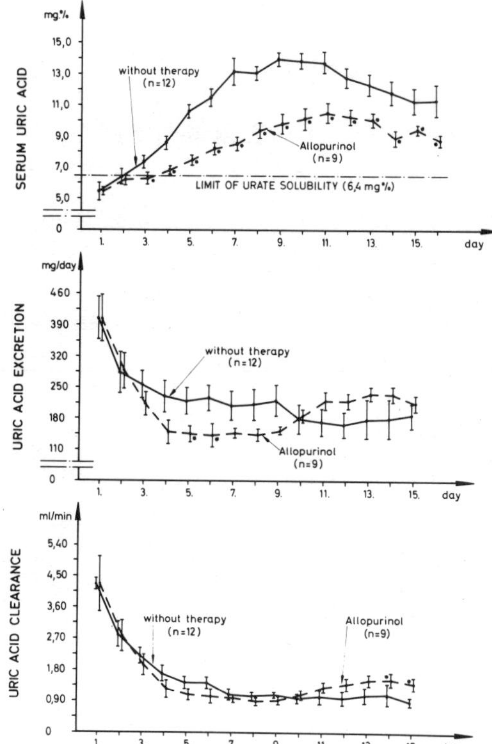

Fig.: 3 The influence of therapeutic daily doses of allopurinol (300 mg daily) on serum uric acid, excretion and clearance of uric acid during prolonged strict fasting

The correlation between withdrawal of calories and plasma uric acid is shown in Fig. 5. The limit of urate solubility is transmitted after a few days as can be seen also in Fig. 1. In order to prevent uricopathic complications (9,14,29) medicamentous treatment seems to be necessary.

Since urate calculus formation and renal colics have been observed within 3-4 weeks during uricosuric therapy despite the usual precautions (alkalisation, increased diuresis, 9,12,18), it may not be without significance that the excretion rates with probenecid and benzbromarone are more than 2 - 3 times that of the allopurinol- and the drug-free collectives, even within prolonged fasting, the more so since the acid urine also favours potential concrement formation. From this point of view it seems advisable to us to render the urine alkaline to improve the solubility of uric acid when giving an uricosuric

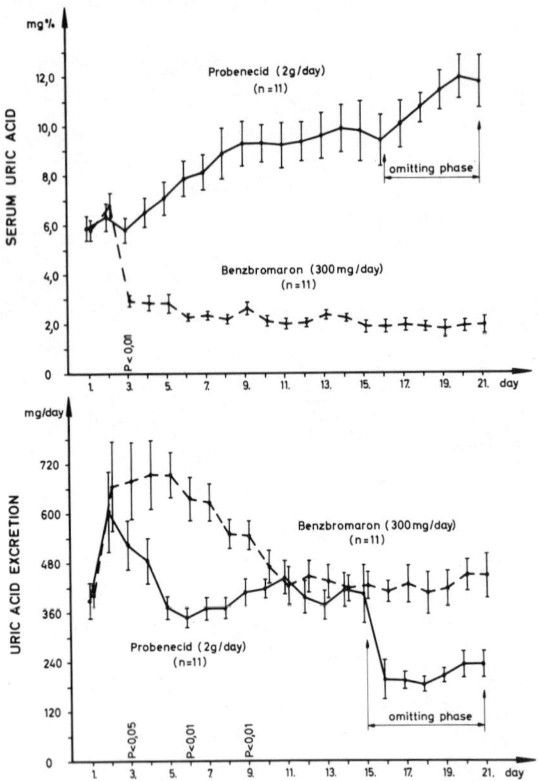

Fig.: 4 The influence of probenecid (2g daily) and benzbromarone (300mg daily) on serum uric acid and uric acid excretion during fasting

agent. An adequate diuresis is provided for in the conditions for the strict fasting regime.

Significantly reduced serum levels with approximately equal excretion and clearance-rates confirm the extrarenal sites of action of the xanthine oxidase inhibitor "allopurinol"(Fig. 3). The slightly raised excretion- and clearance-rates in prolonged fasting may lie within the range of physiological dispersion. An uricosuric effect can also be excluded from the same date and also from the portion of uric acid filtrate reabsorbed by the tubules, as shown in Table 1.

With probenecid (2g/day) and benzbromarone (100 or 300 mg/day) there is an immediate and persistent rise in the clearance. During the comparable observation period (2nd - 16th days of fast) the serum levels and excretion rates with benzbromarone lie 95% lower and 27% higher (Fig. 4). Apparently the amount of the increased uric acid

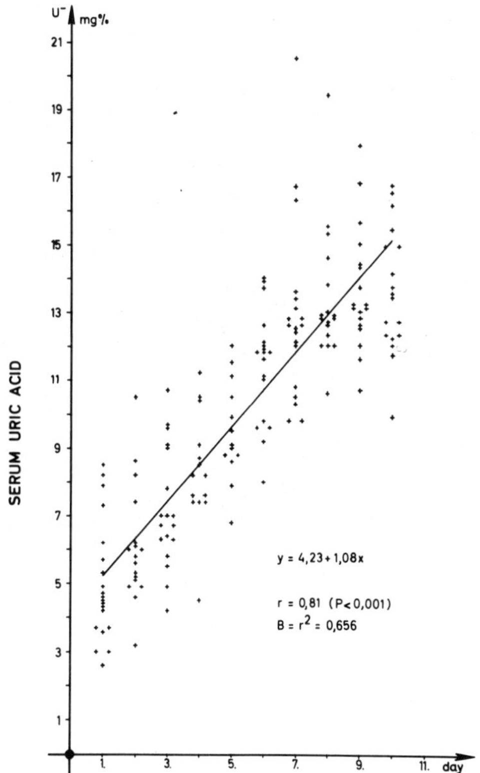

Fig.: 5 Hyperuricemia as function [y=f(x)] of starvation in the initial fasting-period (total fasting)

excretion consequently does not alone explain the great reduction in the plasma levels as the significantly different plasma levels with approximately equal excretion rates would lead one to suspect.

Since probenecid has an exclusively uricosuric action (2,26), our findings may also point to an extrarenal site of action for the benzofuran derivatives (11,15,19,20). A further differentiation, in particular a comment on the enteral point of attack discussed by MÜLLER et al. (20), we expect from the determination of uricolysis products, with which we are at present concerned. Our previous allantoin determinations did not indicate any signs of an increase in intestinal uricolysis. Clearance, net-reabsorption, and the amount of filtered uric acid reabsorbed (Tab. 1) also indicate a predominantly uricosuric effect which is greater than that of probenecid. But this action does not explain the uric acid lowering effect observed with benzbromarone

Table 1. For the influence of Allopurinol (300mg/day), Probenecid (2g/day) and Benzbromaron (300mg/day) on the resorbed portion of the filtered quantity of uric acid

	Periods of starvation (fasting-day)	1.	2.	3.	4.	5.	6.	7.	8.	9.	10.	11.	12.	13.	14.	15.
$1-\frac{C_{U^-}}{C_{Cr_1}}$	Without therapy (n=12)	0,944 ±0,02	0,959 ±0,03	0,969 ±0,01	0,977 ±0,01	0,979 ±0,01	0,979 ±0,01	0,986 ±0,01	0,984 ±0,01	0,984 ±0,01	0,984 ±0,04	0,982 ±0,02	0,984 ±0,01	0,985 ±0,01	0,982 ±0,02	0,985 ±0,01
	Allopurinol (n=9)	0,944 ±0,02	0,964 ±0,05	0,974 ±0,05	0,985 ±0,01	0,994 ±0,01	0,994 ±0,01	0,994 ±0,01	0,994 ±0,07	0,984 ±0,01	0,985 ±0,01	0,984 ±0,01	0,985 ±0,01	0,985 ±0,01	0,985 ±0,01	0,985 ±0,01
	Probenecid (n=11)	0,944 ±0,01	0,917 ±0,05	0,918 ±0,07	0,925 ±0,06	0,944 ±0,04	0,954 ±0,09	0,951 ±0,04	0,952 ±0,03	0,951 ±0,03	0,952 ±0,08	0,951 ±0,03	0,952 ±0,04	0,958 ±0,03	0,957 ±0,03	0,958 ±0,03
	Benzbromaron (n=11)	0,944 ±0,03	0,889 ±0,08	0,837 ±0,14	0,814 ±0,10	0,794 ±0,08	0,791 ±0,12	0,817 ±0,06	0,848 ±0,05	0,820 ±0,08	0,824 ±0,07	0,841 ±0,19	0,862 ±0,11	0,868 ±0,05	0,853 ±0,19	0,844 ±0,18

medication in anephric patients (15) or in patients with severe renal insufficiency up to creatinine and glomerular filtration values of 5.0 mg/100 ml and 20 ml/min (3,15). The problems of an uricosuric therapy for patients with this kind of serious disorder of renal function cannot be entered into here.

SUMMARY

The essence of "fasting hyperuricemia" is a positive uric acid balance (limitation of renal clearance, increase of net-reabsorption and nucleoprotein breakdown).

Without or with increasing the normal renal uric acid excretion, the usual therapeutic daily doses of allopurinol and probenecid lower the serum uric acid during starvation significantly, without, however being able to compensate the "fasting hyperuricemia".

Administration of 100 mg benzbromarone/day lowers the uric acid significantly below the basic level at the beginning of the fasting. The increased amount of uric acid excreted does not of itself explain the greatly reduced plasma level. However a distinctly stronger uricosuri effect compared with probenecid is a primary feature. Based on supplementary studies we calculate the extrarenal effect of benzbromarone with less than 15% of the predominantly uricosuric action (85%).

Serum levels, excretion, clearance and reabsorption rates show the extrarenal point of attack for allopurinol and the exclusively or predominantly uricosuric effect of probenecid and benzbromarone.

Allopurinol and benzbromarone were well tolerated under fasting conditions. In the probenecid group (n=11), the medication had to be prematurely discontinued in 4 patients because of allergic reactions.

References

1. Aldermann,M.H. et al.:Proc. Soc. exp. Biol. (N.Y.) 118, 790 (1965), 2. Bartels, E.C.: Ann. intern. Med. 42, 1 (1955), 3. Begemann, H. et al.: Therapiewoche 25, 16, 2184 (1975), 4. Benedict, F.C.: Carnegie Inst. (Washington) 203, 27, 42, 182 (1915), 5. Bloom, W.L.: Metabolism 8, 214 (1959), 6. Christofori, F.C. et al.: Metabolism 13, 303 (1964), 7. Cheifetz, P.N.: Metabolism 14, 1267 (1965), 8. Ditschuneit, H. et al.: Internist 11, 176 (1970), 9. Drenick, E.J.: Arthritis Rheum. 8, 988 (1965), 10. Drenick, E.J. et al.: Clin. Pharm. Therap. 12, 1, 68 (1971), 11. Greiling, H.: Dtsch. med. J. 10, 336 (1969), 12. Griebsch, A. et al.: Medizin heute 13, 5 (1972), 13. Handler, J.: J. Clin. Invest. 39, 1526 (1960), 14. Karcher, G.P. et al.: Zschr. Urol. 11, 827 (1972), 15. Köthe, E. et al.: Therapiewoche 23, 35, 2927 (1973), 16. Lennox, W.G.: J. Amer. Ass. 82, 602 (1924), 17. Lloyd-Mostyn, R.H. et al.: Ann. rheum. Dis. 29, 553 (1970), 18. Mertz, D.P.: Ther. d. Gegenw. 9, 1378 (1975), 19. Müller, M.M. et al.: Therapiewoche 25, 5, 514 (1975), 20. Müller, M.M. et al.: Dtsch. med. Wschr. 100, 198 (1975), 21. Murphy, R. et al.: Arch. intern. Med. 112, 954 (1963), 22. Pabico, R.C. et al.: Clin. Res. 13, 45 (1965), 23. Padova, J. et al.: New Engl. J. Med. 267, 530 (1962), 24. Pitts, R.F.: F.K. Schattauer Verlag, 2. Auflage, Stuttgart (1972), 25. Reem, G.H. et al.: Amer. J. Physiol. 207, 113 (1964), 26. Sirota, J.H. et al.: J. Clin. Invest. 31, 692 (1952), 27. Schatonoff, J. et al.:Metabolism 19, 84 (1970), 28. Schmahl, K., P. Schräpler : Therapiewoche 25, 47, 7193 (1975), 29. Schmahl, K., P. Schräpler : Wehrmed. Mschr. 20, 39 (1976), 30. Schräpler, P., E. Schulz : Med. Welt (N.F.) 27, 575 (1976), 31. Schräpler, P., E. Schulz : Verh. Dtsch. Ges. innere Medizin, Bd. 82, Hrsg.: B. Schlegel, J.F. Bergmann Verlag, München (1976), 31a. Schräpler, P., E. Schulz, A. Kleinschmidt : Hyperuricemia - Risk Factor - Therapy, 4th Hungarian Arteriosclerosis Conference, Budapest, 10th - 12th Nov. 1976, 32. Schulz, E., P. Schräpler : Verh. Dtsch. Ges. innere Medizin, Bd. 82, Hrsg.: B. Schlegel, J.F. Bergmann Verlag, München (1976), 33. Yü, T.F. et al.: Proc. exp. Biol. (N.Y.) 96, 809 (1957)

THE TREATMENT OF ACUTE GOUTY ARTHRITIS

O.Frank

Special Hospital for Rheumatic Diseases

A 2500 Baden, Renngasse 2, Austria

With increasing incidence of clinical gout the treatment of acute gouty arthritis has gained much interest especially because Colchicine is getting to the background more and more at least in Europe. Undoubtedly the treatment of acute gouty arthritis with Colchicine is effective. However, usually the clinical effect is associated with toxic reactions from the gastrointestinal tract, sometimes occurring before the clinical effect.

The substrate of the gouty attack is an inflammation. Therefore it seems to be obvious that the acute gouty attack may be influenced by antiphlogistic drugs.

We have treated 11 gouty attacks, 1 patient had 2 attacks, with Azapropazone (Prolixan 300) in an open trial. The oral dose of Azapropazone on 3 following days was 2400 mg on the first, 1800 mg on the second and 1200 on the third day.

In a second group of 6 patients 7 gouty attacks were treated randomized with intramuscular injektions of 600 mg Phenylbutazone and 75 mg Voltaren for 3 days.

The medication was restricted for 3 days in both groups because a medication of more than 3 days could overlap the drug effect with the tendence to self limitation of the attack which can be explained by removal of monosodium urate from the joint cavity by diffusion aided by increased blood flow, slow breakdown of urate to allantoin by a lysosomal myeloperoxidase and finally by release of corticosteroids from the adrenal cortex as a general stress response to the gouty attack.

The first figure shows the effect of Azapropazone, an antiphlogistic drug, well known from the treatment of rheumatoid arthritis, in acute gouty arthritis. In 7 cases the gouty attack was terminated within 3 days, in 3 cases the duration of the attack was one day, in further 2 patients 2 days and 3 days in 3 cases and in the remained 3 cases the duration was 4, 5 and 6 days respectively.

In 4 patients treated with Phenylbutazone the duration of the acute attack was in 1 case 5 hours, in 3 cases 3 days. In the patients treated with Voltaren the attack was terminated in 6 hours and in 2 days and 8 hours respectively.

Acute gouty arthritis could be terminated in all the cases, in 3 cases treated with Azapropazone the duration of the attack was between 4 and 6 days. Despite of the relatively small number of patients we have the impression that Phenylbutazone and Voltaren are more effective than Azapropazone.

The duration of the attacks has been shown to be different despite of the uniform treatment in the particular groups. These results show that the effect of an antiphlogistic drug for the treatment of acute gouty arthritis is not depending from the antiphlogistic properties only, but is influenced from a variety of factors such as the intensity of the inflammatory reaction in the joint cavity which is dependent from the extent of the crystallization. Urate crystals are deposited primarily in connective tissue. Proteoglycanes are connective tissue components that have been shown to enhance the solubility of sodium urate and to prevent it from crystallizing from a supersaturated medium, a property that might explain the accumulation of urate in connective tissue. Increased catabolism has been postulated to explain the crystallization of urates in the joint cavity. Repeated microtrauma to joints with consequent removal of proteoglycanes by enzymatic degradation, thereby altering the ration of proteoglycan to urate is another mechanism for crystallization in the joint cavity and the initiation of the acute gouty attack.

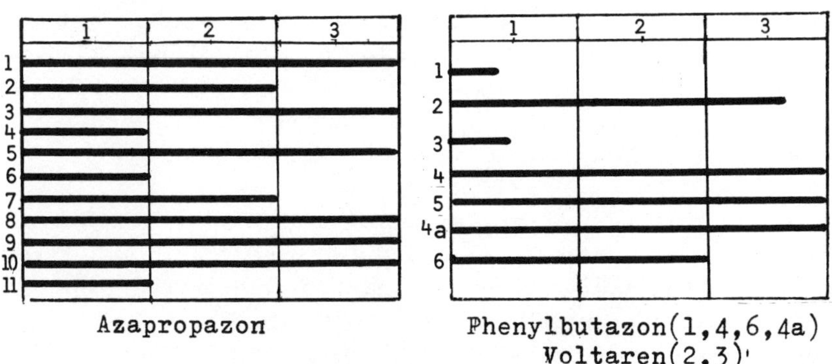

Azapropazon

Phenylbutazon(1,4,6,4a)
Voltaren(2,3)

MULTI-CENTRE TRIAL OF NAPROXEN AND PHENYLBUTAZONE IN ACUTE GOUT

R.A. Sturge,[1] J.T. Scott,[1] E.B.D. Hamilton,[2] S.P. Liyanage,[3] A. St. J. Dixon,[4] and C. Engler[5]

1. Charing Cross Hospital and Kennedy Institute of Rheumatology, London. 2. King's College Hospital, London. 3. The London Hospital. 4. Royal National Hospital for Rheumatic Diseases, Bath. 5. Syntex Pharmaceuticals, Maidenhead

INTRODUCTION

There are three drugs in general use in the treatment of acute gout. Colchicine has been in intermittent use since Byzantine times but requires frequent administration and its gastro-intestinal side effects too often terminate the treatment before the attack. Phenylbutazone is as effective as colchicine (Freyberg, 1962; Gutman, 1965) and untoward reactions occurring during the treatment of acute gout are rare (Smyth and Percy, 1973), although gastro-intestinal intolerance and fluid retention are predominant among the side effects that occasionally limit its usefulness. Indomethacin in adequate dosage has been shown to be as effective as phenylbutazone (Smyth and Percy, 1973) but unpleasant side effects are not infrequent (Boardman and Hart, 1965). There is therefore need for further exploration of safe and rapid therapy in acute gout, particularly as the sufferers are often relatively young and active members of the community.

Naproxen is an anti-inflammatory and analgesic agent which has proved to be useful in rheumatoid arthritis with fewer reported side effects, particularly those of gastro-intestinal intolerance, than aspirin or indomethacin (Hill, Hill, Mowat, Ansell, Mathews, Seifert, Gumpel and Christie, 1973; Kogstad, 1973). It has been suggested that naproxen is effective in the treatment of acute gout (Willkens, Case and Huix, 1975; Cuq, 1973) and it was accordingly decided to compare the efficacy of this agent with that of phenylbutazone.

METHODS

41 patients with acute gout were seen at one of four centres comprising King's College Hospital (5 patients), the London Hospital (4 patients), the Royal National Hospital for Rheumatic Diseases, Bath (6 patients) and Charing Cross Hospital (Kennedy Institute of Rheumatology), London (26 patients). The diagnosis was made by the investigating physician on generally acceptable clinical grounds, and all but two patients, who had recently started allopurinol, were hyperuricaemic.

The trial was of open design and the patients were assigned to either naproxen 750 mg as a single dose followed by 250 mg three times daily (22 patients) or to phenylbutazone 200 mg four times daily for 48 hours followed by 200 mg three times daily (23 patients), four patients receiving both drugs in different attacks. Treatment was continued until the affected joint was pain-free. For each individual the age, sex and duration of attack before starting treatment was recorded. At follow-up the patient was asked to assess the length of time the attack had taken to settle down in terms of absence of pain, swelling and tenderness and ability to walk in shoes without a limp. He was also asked if the treatment had upset him in any way.

Results were analysed for statistical significance by the Mann-Whitney and Spearman's rank correlation tests.

Table I

Drug	No. of patients Male	Female	Mean age in yrs (Range in parentheses)
Naproxen	20	2	58.8 (34 - 84)
Phenylbutazone	23	0	50.4 (30 - 73)

Table II

Time in days from onset of attack to start of treatment							
	1 or less	2	3	4	5	6	7 or more
Naproxen	9	1	3	2	1	1	5
Phenylbutazone	15	1	5	0	0	0	2

RESULTS

Comparability of the two treatment groups is shown in Tables I and II. The phenylbutazone-treated patients were found to be rather younger than the naproxen group (p=0.03) and duration of attack before starting treatment was significantly shorter (p=0.03) in the phenylbutazone group (mean less than 1 day) than the naproxen group (mean 2.0 days).

There was, however, no significant difference in the duration of the attack after starting treatment between the phenylbutazone-treated patients (mean 3.4 days) and the naproxen-treated patients (mean 2.9 days) (Table III). In both groups there were weak but positive correlations between age and duration of attack after starting treatment and between delay in starting treatment and duration of attack after treatment as shown in Table IV.

An attack persisting for more than 7 days after institution of therapy might be taken as indication of treatment failure. Three patients in the naproxen group, two of whom started treatment late in the attack, fulfilled this criterion, as did three patients in the phenylbutazone group, one of whom started treatment late in the attack.

Table III

Time in days from start of treatment to end of attack							
	1 or less	1-2	2-3	3-4	4-5	5-6	7 or more
Naproxen	5	2	5	3	3	1	3
Phenylbutazone	2	3	5	6	3	1	3

Table IV

Correlation of duration of attack after starting treatment with a) age and b) duration before starting treatment

Patient group	Age	Duration before starting treatment
Naproxen	+ 0.410 (p=0.06)	+ 0.296 (N.S.)
Phenylbutazone	+ 0.112 (N.S.)	+ 0.365 (p=0.09)
Combined patients	+ 0.220 (N.S.)	+ 0.281 (p=0.07)

One elderly patient receiving naproxen for 7 days developed ankle oedema which resolved on completion of therapy. One patient on phenylbutazone developed mild diarrhoea and one complained of "wind and palpitations".

DISCUSSION

There are indications that the worldwide prevalence of gout is increasing (Talbott, 1976), an impression which is in accord with the postulate that gout is related to the "associates of plenty" (Acheson and Chan, 1969). However, there is also an impression among rheumatologists in this country that gout is becoming less common in hospital practice, presumably because of the ease and effectiveness of contemporary methods of therapy applicable in general practice. Hence a multi-centre trial was planned and to simplify the procedure the open method was used comparing the trial drug with an agent known to be effective in acute gout. A placebo comparison was not used as it was thought unlikely that patients suffering from gout would consent to this type of investigation. Assessment was entirely subjective but it has been demonstrated that patients' assessment of pain relief and drug preference correlate well with changes in more complicated clinical and investigative indices of joint inflammation, at least in rheumatoid arthritis (Deodhar, Dick, Hodgkinson and Buchanan, 1973) and the aim of treatment in acute gout is largely to obtain symptomatic relief.

The ages of the patients and the delay in starting treatment were significantly greater in the naproxen group than in the phenylbutazone group. Both these variables tended to be positively correlated with duration of attack after starting treatment and may therefore have introduced a bias against naproxen. Nevertheless, the present study shows that naproxen is equally as effective as phenylbutazone in the treatment of acute gout, the success rate (87% phenylbutazone, 83% naproxen), being similar to that previously reported for phenylbutazone (Boardman and Hart, 1965; Wilson, Huffman and Smyth, 1956).

Side effects were mild and infrequent occurring in only three of the 41 patients in the trial. The single side effect recorded with naproxen was the development of ankle oedema in one patient. Fluid retention has been an infrequently reported side effect of this drug but was probably the explanation for weight gain of 4 Kgs in a patient with known cardio-vascular disease receiving this drug for acute gout in a similar dose regime to that used in our study (Willkens, Case and Huix, 1975) and has been noted in a patient with rheumatic heart disease by one of the present authors (E. B. D. H.).

This study demonstrates that naproxen is a safe and effective addition to the small group of drugs in current use for the treatment of acute gout. A previous study has suggested that the present dose regime is close to optimal, lower doses being less effective and higher doses providing no increased benefit (Cuq, 1973). The long plasma half-life of the drug (14 hours) suggests that more frequent administration is unnecessary.

There is now identified a group of drugs - naproxen, phenylbutazone, indomethacin and possibly other anti-inflammatory agents - whose often dramatic efficacy in acute gout is in striking contrast to the variable and relatively mediocre response which they produce in more chronic forms of inflammatory arthritis, such as rheumatoid disease. Is it justifiable to conclude from this that a single inflammatory pathway is predictably interrupted by these drugs in gouty arthritis whereas in rheumatoid arthritis a number of alternative mechanisms may be operating, varying from patient to patient? The mechanism of the intensely painful but eventually self-limiting inflammation of the acute gouty joint is unknown. Activation of Hagemann factor, complement and the kinin system have been implicated in the past, but their importance doubted in subsequent work (Spilberg, 1974; Phelps and McCarty, 1969; Phelps, Prockop and McCarty, 1966) and even the role of the polymorphonuclear leucocyte has been thrown in doubt (Ortel and Newcombe, 1974). The drugs in question are all potent inhibitors of prostaglandin synthetase (Tacheguchi and Sih, 1972; Flower, Cheung and Cushman, 1973), suggesting that prostaglandins play a major part in the inflammation of acute gout. Animal studies lend support to this hypothesis (Denko, 1974). It is conceded that colchicine does not inhibit prostaglandin synthesis, but this drug is not an anti-inflammatory agent in the usual sense of the term and presumably acts by a mechanism different from that of the other drugs.

SUMMARY

Naproxen 750 mg as a single dose followed by 250 mg three times daily has been compared with phenylbutazone 200 mg four times daily for 48 hours followed by 200 mg three times daily for the treatment of acute gout in an open study on 41 patients. The drugs were equally effective with few and relatively mild side effects. Naproxen is a useful alternative agent for the treatment of acute gout.

ACKNOWLEDGEMENT

We are indebted to Dr. P. Freeman, Department of Statistics, University College, London for the statistical analysis.

REFERENCES

ACHESON, R. M., AND CHAN, Y.-K. (1969) J. chron. Dis., 21, 543 (New Haven survey of joint diseases. The prediction of serum uric acid in a general population)

BOARDMAN, P. L., AND HART, F. DUDLEY (1965) Practitioner, 194, 560 (Indomethacin in the treatment of acute gout)

CUQ, P. (1973) Scand. J. Rheum., Suppl. 2, 64 (Experience francaise du traitement de la crise de goutte aigüe par le naproxen - C1674)

DENKO, C. W. (1974) J. Rheumatol., 1, 222 (A phlogistic function of Prostaglandin E_1 in urate crystal inflammation)

DEODHAR, S. D., DICK, W. C., HODGKINSON, R., AND BUCHANAN, W. W. (1973) Quart. J. Med., 42, 387 (Measurement of clinical response to anti-inflammatory drug therapy in rheumatoid arthritis)

FLOWER, R. J., CHEUNG, H. S., AND CUSHMAN, D. W. (1973) Prostaglandins, 4, 325 (Quantitative determination of prostaglandins and malandialdehyde formed by the arachidonate oxygenase (prostaglandin synthetase) system of bovine seminal vesicle)

FREYBERG, R.H. (1962) Arthr. & Rheum., 4, 624 (Gout)

GUTMAN, A.B. (1965) Arthr. & Rheum., 8, 911 (The treatment of primary gout: the present status)

HILL, F.H., HILL, A.G.S., MOWAT, A.G., ANSELL, B.M., MATHEWS, J.A., SEIFERT, M.H., GUMPEL, J.M., AND CHRISTIE, G.A. (1973) Scand. J. Rheum., Suppl. 2, 176 (Multi-centre double-blind cross-over trial comparing naproxen and aspirin in rheumatoid arthritis)

KOGSTAD, O. (1973) Scand. J. Rheum., Suppl. 2, 159 (A double-blind cross-over study of naproxen and indomethacin in patients with rheumatoid arthritis)

ORTEL, R.W., AND NEWCOMBE, D.S. (1974) New Engl. J. Med., 29, 1363 (Acute gouty arthritis and response to colchicine in the virtual absence of synovial fluid leucocytes)

PHELPS, P., AND McCARTY, D.J., Jr. (1969) Postgrad. Med., 45, 87 (Crystal-induced arthritis)

PHELPS, P., PROCKOP, D.J., AND McCARTY, D.J. (1966) J. Lab. Clin. Med., 68, 433 (Crystal induced inflammation in canine joints. III. Evidence against bradykinin as a mediator of inflammation)

SMYTH, C.J., AND PERCY, J.S. (1973) Ann. rheum. Dis., 32, 351 (Comparison of indomethacin and phenylbutazone in acute gout)

SPILBERG, I. (1974) Arthr. & Rheum., 17, 143 (Urate crystal arthritis in animals lacking Hageman factor)

TACHEGUCHI, C., AND SIH, C.J. (1972) Prostaglandins, 2, 169 (A rapid spectrophotometric assay for prostaglandin synthetase: application to the study of non-steroidal anti-inflammatory agents)

TALBOTT, J.H. (1976) Arthr. & Rheum., 18, 6 (Suppl.) (It happened on the way to the XIII International Congress on Rheumatology in Kyoto and after I had arrived)

WILLKENS, R.F., CASE, J.B., AND HUIX, F.J. (1975) J. Clin. Pharmacol., 15, 363 (The treatment of acute gout with naproxen)

WILSON, G.M., HUFFMAN, E.R., AND SMYTH, C.J. (1956) Amer. J. Med., 21, 232 (Oral phenylbutazone in the treatment of acute gouty arthritis)

THE INFLUENCE OF ALLOPURINOL IN CUSTOMARY AND IN SLOW RELEASE PREPARATION ON DIFFERENT PARAMETERS OF PURINE AND PYRIMIDINE METABOLISM

W. Gröbner, I. Walter, A. Rauch-Janssen,
N. Zöllner
Medical Polyclinic, University of Munich,
Pettenkoferstrasse 8a, Munich

Allopurinol is widely used as a hypouricemic agent in man. This compound serves as both a substrate and an inhibitor of xanthine oxidase the enzyme which catalyses the oxidation of hypoxanthine to xanthine and xanthine to uric acid. Allopurinol itself is rapidly oxidized in vivo by xanthine oxidase to oxipurinol, which is also an inhibitor of xanthine oxidase.

In addition to their inhibitory effect on purine catabolism, both allopurinol and oxipurinol influence pyrimidine biosynthesis as evidenced by an increase in the urinary excretion of orotic acid and orotidine. This effect is attributed to inhibition of orotidine-5-phosphate decarboxylase (FOX et al. 1970, KELLEY and BEARDMORE 1970).

In the present study we compared the effect of a customary preparation of allopurinol (Zyloric 300) and a slow release preparation (Allopurinol retard 300) on the concentration of plasma oxipurinol, serum uric acid and renal excretion of uric acid and total orotic acid (orotic acid plus orotidine). The study was designed as a randomized cross over trial, in which each subject was its own control.

Five young healthy volunteers received an isocaloric formula diet without purines which contained 55 energy-percent carbohydrate, 30 energypercent fat and 15 energypercent protein. Part of the subjects took daily

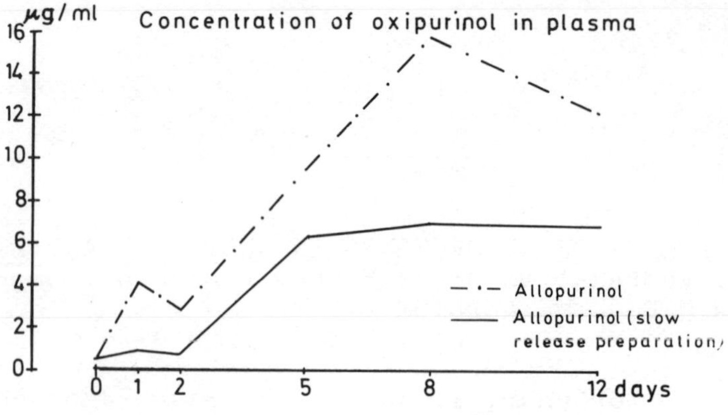

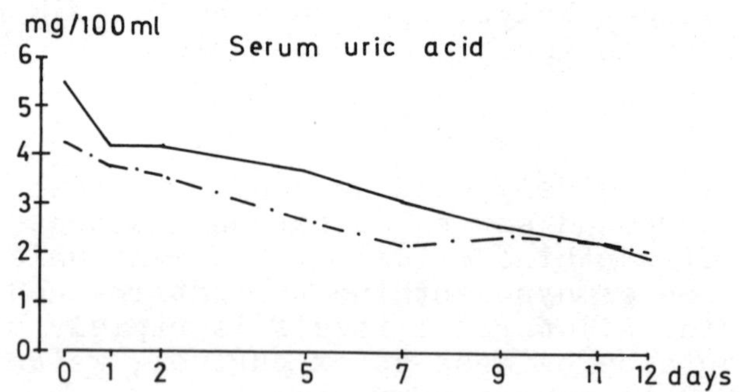

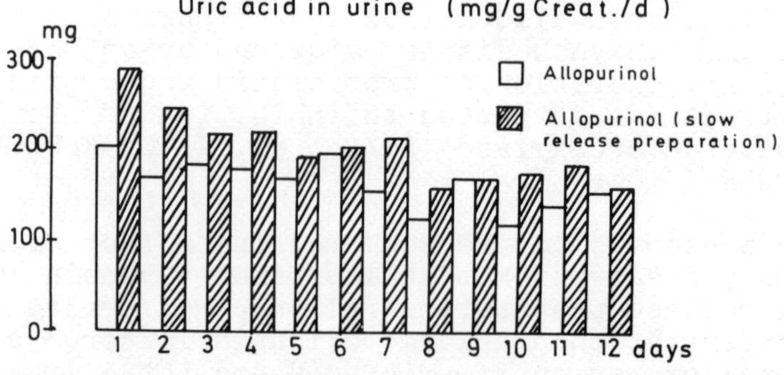

Fig. 1. The effect of 300 mg allopurinol in customary and slow release preparation on plasma oxipurinol, serum uric acid and renal excretion of uric acid and total orotic acid in one subject. A single dose of allopurinol was given daily for 12 days except for the second day.

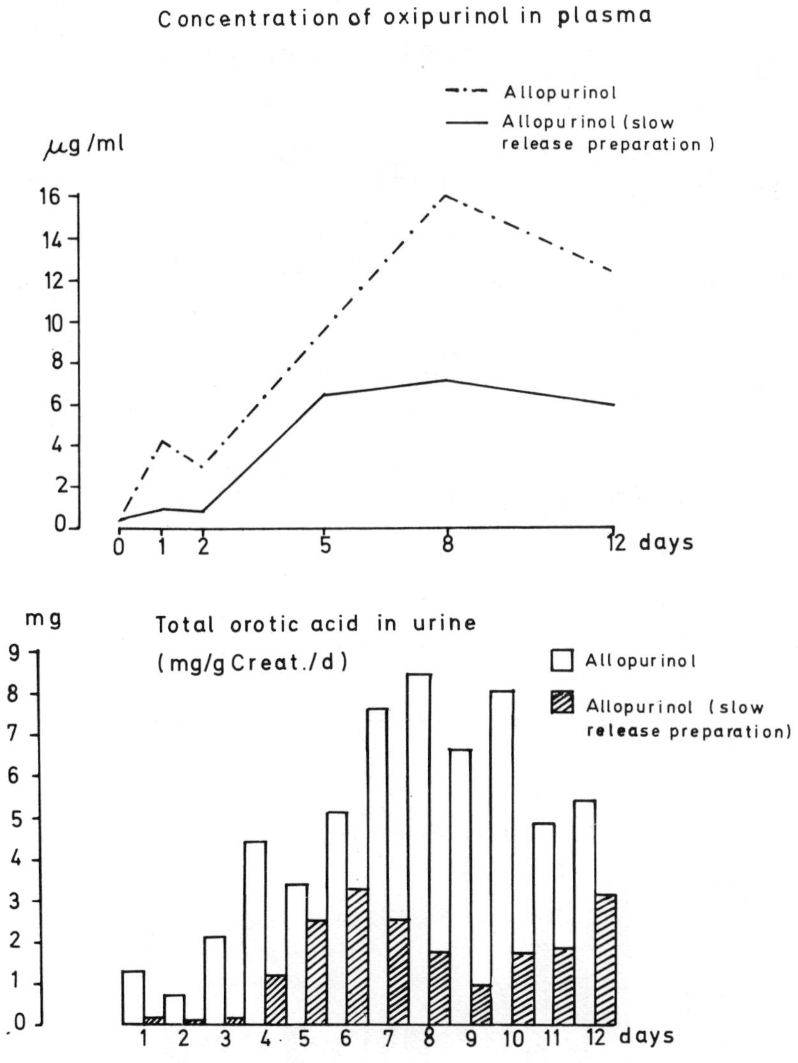

Fig.2. The effect of 300 mg allopurinol in customary and slow release preparation on plasma oxipurinol, serum uric acid and renal excretion of uric acid and total orotic acid in one subject. A single dose of allopurinol was given daily for 12 days except for the second day.

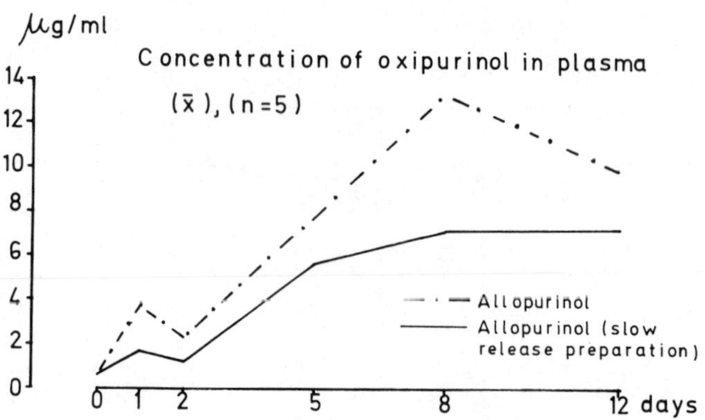

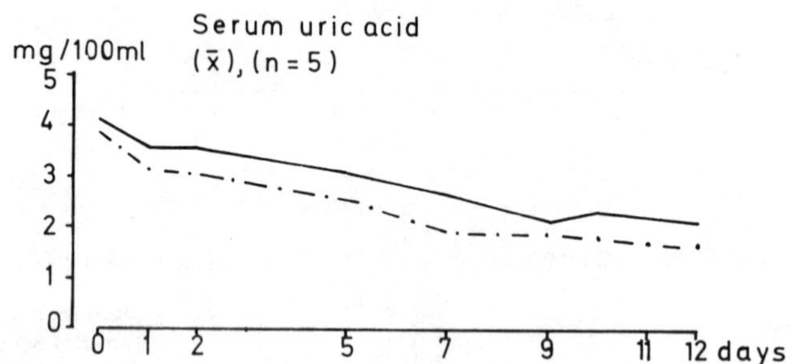

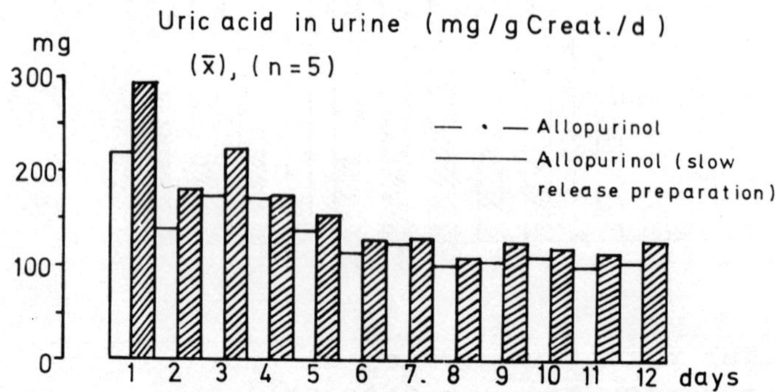

Fig.3. Show the mean values of five subjects. The data are illustrated according to fig. 1 and 2.

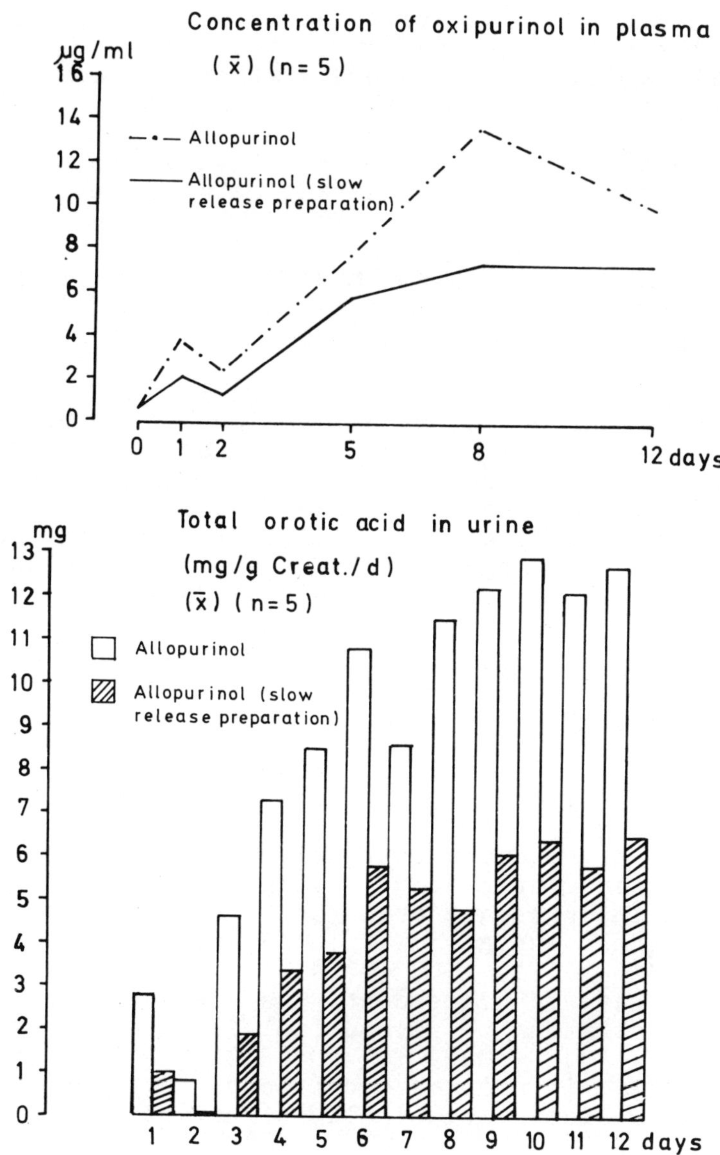

Fig.4. Show the mean values of five subjects. The data are illustrated according to fig. 1 and 2.

300 mg allopurinol in customary preparation as a single dose in the morning. After 12 days of allopurinol administration the drug was discontinued for 12 days (control period). Then the administration of 300 mg allopurinol in slow release preparation was started under the same conditions like mentioned above. In the rest of the subjects the sequence of allopurinol and allopurinol in slow release preparation was reversed. During the control periods the subjects took no formula diet.

Uric acid was determined enzymatically, total orotic acid colorimetrically according to ROGERS and PORTER (1968). Oxipurinol was determined by high pressure liquid chromatography.

An example of the effect of allopurinol in customary preparation and in slow release preparation on the parameters mentioned above is shown in figure 1 and 2. The mean values of the five subjects are illustrated in figure 3 and 4. During the administration of allopurinol in customary preparation the average concentration of plasma oxipurinol increased within 12 days from less than 0.5 µg/ml to 9.98 ± 2.23 (s) µg/ml. The administration of allopurinol in slow release preparation gave an increase of plasma oxipurinol from less than 0.5 ug/ml to 7.14 ± 1.48 (s) µg/ml within the same time. The difference between the steady state plasma levels resulting from different allopurinol preparations is statistically significant ($p < 0.025$).

The increase of plasma oxipurinol is accompanied by a decrease of serum uric acid. During administration of allopurinol in customary preparation the average serum uric acid level decreased within 12 days from 3.96 ± 0.29 (s) mg/100 ml to 1.69 ± 0.41 (s) mg/100 ml; this means a decrease of 57 percent. The corresponding values under administration of allopurinol in slow release preparation were 4.12 ± 0.89 (s) and 2.12 ± 0.27 (s)mg/100 ml respectively; this means a decrease of 49 percent. This difference is also statistically significant ($p = < 0.05$).

The average excretion of total orotic acid after administration of allopurinol in customary preparation for 12 days was 12.7 mg/g creat. and 6.45 mg/g creat. under administration of allopurinol in slow release preparation respectively. Despite of these values this difference was not statistically significant because of a high interindividual variation ($p < 0.10$).

The effect of the two preparations on renal excretion of uric acid was only slightly different.

Our observations show, that administration of allopurinol in customary preparation results in higher plasma levels of oxipurinol than administration of allopurinol in slow release preparation. This is accompanied by a stronger effect of the customary preparation on serum uric acid and renal orotic acid excretion than that of the slow release preparation. However, the effect of both drugs on renal excretion of uric acid was only slightly different. The difference in the effect of both preparations examined is probably due to an incomplete absorption of allopurinol in slow release preparation.

References:

Fox, R. M.; Royse-Smith, D.; O'Sullivan, W. J.: Science 168, 861 (1970)

Kelley, W. N.; Beardmore, T. D.: Science 169, 388 (1970)

Rogers, L. E.; Porter, F. St.: Pediatrics 42, 423 (1968)

PURINE EXCRETION IN COMPLETE ADENINE PHOSPHORIBOSYLTRANSFERASE DEFICIENCY: EFFECT OF DIET AND ALLOPURINOL THERAPY

*H.A. Simmonds, **K.J. Van Acker, *J.S. Cameron, *A. McBurney

*Department of Medicine, Guy's Hospital, London

**Department of Paediatrics, University of Antwerp, Belgium

Partial adenine phosphoribosyltransferase (APRTase) deficiency has been considered hitherto as relatively benign in that the only detectable abnormality of purine metabolism has been hyperuricaemia varyingly accompanied by hyperuricosuria in some, but not all, cases (3). Elsewhere in this symposium we present clinical details of a child apparently homozygous for APRTase deficiency with urolithiasis due to 2,8-dihydroxyadenine stone formation (11). An identical case of complete APRTase deficiency with 2,8-dihydroxyadenine stone formation has been described by Cartier and Hamet (2). In both cases the stones were originally misdiagnosed as uric acid because of the non-specificity of the methods routinely used (2,8). The metabolic basis for the stone formation results directly from the enzyme deficiency. In the absence of APRTase adenine is oxidised to the extremely insoluble purine 2,8-dihydroxyadenine. The nephrotoxicity of 2,8-dihydroxyadenine has been known since 1898 and is discussed by Cameron et al. However, free adenine is not normally readily detectable in man, and the pathway for its production via adenosine and nucleoside phosphorylase has very low activity in mammalian tissues (9,5). On this basis, endogenous adenine production would appear unlikely in normal circumstances.

In this paper we report studies to assess the contribution of diet and the effect of therapy on the abnormal metabolism of adenine in this APRTase-deficient child. These studies were performed before beginning the long-term allopurinol therapy which has successfully eliminated his renal complication. A healthy child of comparable age was investigated using a similar protocol for comparison. An apparently healthy elder brother and the mother were also studied.

METHODS

Both the propositus and control child were males aged 3 years. Clinical details and enzyme studies in the kindred have been given in detail by Van Acker et al. Both subjects were first investigated after a three-day run-in on a low purine (caffeine-free) diet. All subsequent regimes were also caffeine-free, and similar periods of equilibration were used following the change to a high purine diet or the addition of allopurinol (125 mg/day) to either dietary regime. The mother and the brother were studied on a normal caffeine-free diet. The experimental methods used have been reported previously (6).

RESULTS

Purine Metabolism

Plasma and urinary uric acid concentrations in the propositus were within normal limits (0.19-0.23 mmol/l and 0.7-1.3 mmol/24hr) and did not differ greatly from those of the control child (0.24-0.26 mmol/l and 0.87-1.09 mmol/24hr) with the exception that urinary uric acid excretion was almost doubled on a high purine diet in the propositus. Levels in the mother and brother were also normal (Table 1).

		Control*	APRT*	Mother**	Brother**
Plasma Uric Acid (mmol/l)		0.26	0.20	0.24	0.19
Urine (mmol/24hr)	Uric Acid	0.87	0.71	3.37	2.04
	Xanthine	0.02	0.03	0.06	0.02
	Hypoxanthine	0.02	0.03	0.07	0.03
	Adenine	0	0.10	0	0.20
	8-OH adenine	0	0.02	0	0.01
	2,8-OH adenine	0	0.14	0	0.08
	Total	0.91	1.03	3.50	2.38

* Low Purine Diet ** Normal Diet

Table I. Urinary purine excretion in a family with APRTase deficiency and a control child.

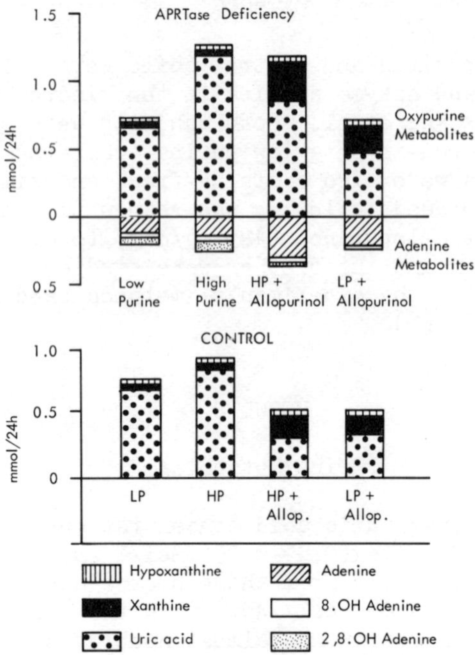

Fig. 1. Effect of allopurinol and diet on urinary purine excretion in the propositus compared with a control child.

The level of the minor purine bases hypoxanthine and xanthine was comparable in all three children and levels in the mother were also identical with those for a healthy adult by this method (6). The one abnormality noted in the propositus (Table I) (Fig. 1) was the excretion of abnormal amounts of adenine and the adenine metabolites 8-OH adenine and 2,8-dihydroxyadenine. These compounds accounted for up to 25% of the total urinary purine metabolites and comprised approximately 50% 2,8-dihydroxyadenine. No adenine or other adenine metabolites could be detected in the urine of the control child or of the mother, although she is a heterozygote with 47% of normal APRTase activity. All three metabolites were excreted in the urine of the apparently healthy brother (12% of total urinary purine metabolites; 28% 2,8-dihydroxyadenine) (Table I). (APRTase activity in haemolysed erythrocytes was likewise later found to be almost undetectable (11)).

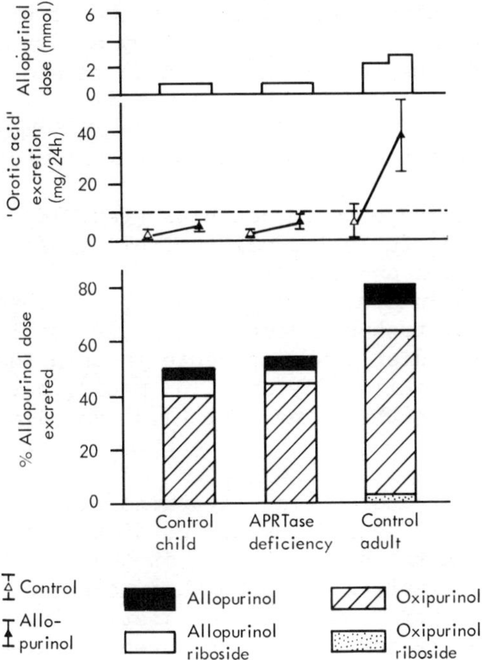

Fig. 2. Allopurinol metabolites expressed as a per cent of the dose and effect of allopurinol on pyrimidine ('orotic acid') excretion in controls and and propositus.

Allopurinol Metabolism

Approximately 50% of the dose of allopurinol (Fig. 2) was excreted in the urine of both the propositus and the control child (as oxipurinol: allopurinol riboside: allopurinol, in the approximate ratio 4:0.6:0.4). This represents a slightly lower absorption than is found in adults (80%) (6) but is probably related to the much lower dosage.

Effect of Allopurinol on Pyrimidine Metabolism

Although allopurinol increased orotidine/orotic acid levels two to threefold in both the propositus and the control child

(Fig. 2), the results were still below the upper adult limit of normal (7) (the mean increase in healthy adults being approximately tenfold). Again the results may be dose related.

Effect of Diet and Allopurinol on Urinary Purine Excretion

The greater increase in xanthine (tenfold) over hypoxanthine excretion (Fig. 1) following allopurinol (twofold) was comparable in both the control and the APRTase deficient child and is similar to that reported in adults (6), although the absolute magnitude is smaller; (again, probably dose related). Total purine excretion on the low purine diet was comparable in both the control and APRTase deficient child (0.9 and 1.03 mmol), but was greater in the propositus on the high purine diet (1.13 and 1.60 mmol, respectively (Fig. 1).

Allopurinol reduced total oxypurine excretion in the control child both on the high and low purine diet, the effect being more marked (43.4%) on the high purine diet than on the low purine diet (15.6%). By contrast, allopurinol had no effect on total oxypurine or total purine excretion on either diet in the propositus. However allopurinol therapy resulted in the eventual disappearance of 2,8-dihydroxyadenine from the urine with a concomitant increase in adenine excretion which was greater on the high than the low purine diet (Fig. 1).

DISCUSSION

From these studies it appears that excessive amounts of adenine and/or its oxidation products are not normally excreted in childhood nor from a single study would they appear to be excreted in the heterozygous state for APRTase deficiency. However, in homozygotes for APRTase deficiency all three compounds, adenine, 8-hydroxyadenine and 2,8-dihydroxyadenine are excreted and may account for up to 25% of total urinary purine metabolites. Their presence in the urine of the symptomless brother, subsequently found to be homozygous for APRTase deficiency, indicates that their excretion is diagnostic of the homozygous state. No other abnormality in excretion of the minor purine bases was apparent and uric acid levels in plasma and urine, at least in childhood, appear entirely normal. The absence of clinical manifestations in the brother is unexplained. Although comparable amounts of adenine metabolites, on a millimolar basis, were excreted by both siblings, 2,8-dihydroxyadenine represented a much lower proportion of the total in the apparently unaffected brother. Nevertheless, in both cases the approximate concentration of 2,8-dihydroxyadenine (20 mg per litre as compared with 70 mg/litre in the propositus) was well in excess of the reported solubility for 2,8-dihydroxyadenine (1-3 mg/litre) (1).

The elimination of 2,8-dihydroxyadenine from the urine following allopurinol therapy corresponds with the clinical findings that the affected child has been free of stones since beginning allopurinol therapy. The reduction in 2,8-dihydroxyadenine excretion during allopurinol therapy coincided with an increase in urinary adenine levels which was greater on the high as compared with the low purine diet; which suggests that some at least of the urinary adenine was derived from the diet.

The metabolism of allopurinol is obviously normal in the propositus since the excretion of all three principal metabolites of allopurinol (unchanged allopurinol, allopurinol riboside and oxipurinol) were comparable in both children and similar to that in adults. By contrast HGPRTase deficient children on allopurinol do not excrete allopurinol riboside (10). This finding has been attributed to the increased hypoxanthine levels in children with this defect; hypoxanthine having a competitive advantage over allopurinol for nucleoside phosphorylase for which it is a much better substrate (5). The excretion of normal levels of allopurinol riboside in the APRTase deficient child excreting high levels of adenine is consistent with the low activity reported for nucleoside phosphorylase with adenine as substrate (9).

Several other points of interest relating to the effect of diet and allopurinol emerge from this study but because of the limited amount of data must be only speculative at present.

The high purine diet produced only a moderate increase in total purine excretion in the control child, but the effect in the APRTase deficient child was much more marked, the purine excretion being almost doubled. These results would suggest that the APRTase deficient child was much more sensitive to dietary purine.

In the control child on allopurinol total urinary purine excretion was reduced to the same level on both the high and low purine diet and was therefore more marked on the high purine diet. This reduction in total oxypurine excretion by allopurinol has been reported in normal adults (6) and in heterozygotes for APRTase deficiency (3). The much greater reduction on the high purine diet has not previously been noted in man, but it corresponds with an effect of allopurinol in reducing the increment in urinary purine excretion which follows either the feeding of RNA in man or guanine in pigs (7).

By contrast, allopurinol had no effect on total purine excretion (or oxypurine excretion) in the APRTase deficient child, irrespective of diet. A similar lack of effect of allopurinol on total purine excretion is seen in the Lesch-Nyhan syndrome (4). If it is considered that dietary adenine normally contributes to urinary purine excretion via the salvage pathway, then this lack of effect may possibly be explained in both cases. In the APRTase deficient child the dietary adenine would not be available for conversion to hypoxanthine via AMP and hence for reutilisation, either normally or following xanthine oxidase inhibition by allopurinol; in the HGPRTas deficient child dietary adenine converted to hypoxanthine via AMP could not be reutilised either.

SUMMARY

1. Abnormal amounts of adenine, 8-hydroxyadenine and 2,8-dihydroxyadenine are found in the urine of homozygotes for APRTase deficiency and are diagnostic of this condition.

2. The renal complication is due to the excessive amounts of 2,8-dihydroxyadenine excreted since it is removed by allopurinol which blocks 2,8-dihydroxyadenine formation.

3. Uric acid metabolism and the excretion of the other minor purine bases is normal, at least in childhood, in homozygotes for APRTase deficiency.

4. Patients with the defect appear to be very sensitive to dietary purine.
At least some of the adenine metabolites may have a dietary origin.

COMMENT

1. Allopurinol therapy is effective in this condition but since it results in the accumulation of free adenine, the toxicity of which is unknown, patients should be followed carefully both clinically and biochemically.

2. The original misdiagnosis of the condition as "uric acid urolithiasis" (because it is impossible to distinguish between uric acid and 2,8-dihydroxyadenine routinely) suggests that stones should, where possible, be submitted for specialist investigation.

REFERENCES

1. Cameron, J.S., Simmonds, H.A., Cadenhead, A. and Farebrother, D. Metabolism of intravenous adenine in the pig. (This Symposium).

2. Cartier, P. and Hamet, M. Une nouvelle maladie metabolique: le deficit complet en adenine phosphoribosyltransferase avec lithiasis de 2,8-dihydroxyadenine. C.R. Acad. Sci., 279, 883-886 (1974).

3. Emmerson, B.T., Gordon, R.B. and Thompson, L. Adenine phosphoribosyltransferase deficiency: Its inheritance and occurrence in a female with gout and renal disease. Aust. N.Z. J. Med., 5, 440-446 (1975).

4. Jones, C.E., Smith, E.E., Hicks, W. and Crowell, J.W. Determination of urinary purines in hyperuricosuric children by thin-layer chromatography. J. Lab. Clin. Med., 76, 163-170 (1970).

5. Krenitsky, T.A., Elion, G.B., Henderson, A.M. and Hitchings, G.N. Inhibition of human purine nucleoside phosphorylase. Studies with intact erythrocytes and the purified enzyme. J. Biol. Chem., 243, 2876-2881 (1968).

6. Simmonds, H.A. Urinary excretion of purines, pyrimidines and pyrazolopyrimidines in patients treated with allopurinol and oxipurinol. Clin. Chim. Acta, 23, 353-364 (1969).

7. Simmonds, H.A., Hatfield, P.J., Cameron, J.S., Jones, A.S. and Cadenhead, A. Metabolic studies on purine metabolism in the pig during the oral administration of guanine and allopurinol. Biochem. Pharmacol., 22, 2537-2551 (1973).

8. Simmonds, H.A., Van Acker, K.J., Cameron, J.S. and Snedden, W. The identification of 2,8-dihydroxyadenine: A new component of urinary stones. Biochem. J., 157, (1976) (in press).

9. Snyder, F.F. and Henderson, J.F. Alternative pathways of deoxyadenosine and adenosine metabolism. J. Biol. Chem., 248, 5899-5904 (1973).

10. Sweetman, L. Urinary and CSF oxypurine levels and allopurinol metabolism in the Lesch-Nyhan syndrome. Fed. Proc., 27, 1055-1059 (1968).

11. Van Acker, K.J., Simmonds, H.A. and Cameron, J.S. Complete deficiency of adenine phosphoribosyltransferase (APRTase): report of a family. (This Symposium).

12. Wyngaarden, J.B. and Dunn, J.T. 8-hydroxyadenine as the intermediate in the oxidation of adenine to 2,8-dihydroxyadenine by xanthine oxidase. Arch. Biochem. Biophys., 70, 150-156 (1957).

THIOPURINOL: DOSE-RELATED EFFECT ON URINARY OXYPURINE EXCRETION

R.Grahame, H.A.Simmonds, J.S.Cameron, A.Cadenhea

Arthritis Research Unit and Department of Medici

Guy's Hospital Medical School, London SE1 9RT

INTRODUCTION

Thiopurinol (mercapto-4-pyrazolo-(3,4-d)pyrimidine) has been shown to effectively reduce both plasma and urinary uric acid levels in the treatment of gout (2,4,5) These effects were observed without showing a concomitant increase in urinary hypoxanthine and xanthine excretion, (2,4,5). As an inhibitor of xanthine oxidase, thiopurinc has been shown to be only one tenth as active as allopur-inol *in vitro* (3). Thiopurinol is ineffective in lowerir uric acid levels in gout associated with a deficiency of the enzyme HGPRTase. This observation has led to the suggestion that the drug acts mainly by feedback inhibiti of de novo purine synthesis; either by causing an alteration in the balance of available purine nucleotides or by the formation of thiopurinol ribotide (2,5). Extensive metabolic studies have recently been carried out in pigs using ^{14}C allopurinol and ^{14}C thiopurinol (7) These studies revealed no measurable tissue incorporatior of either drug and showed that thiopurinol was less well absorbed than allopurinol when administered orally.

These studies have been extended by investigating the effect of thiopurinol in the pig at dosages in excess of human therapeutic levels in experiments comparable to those with allopurinol in the pig (6). The effect of thiopurinol on purine excretion following a dietary purine supplement in the form of guanine has also been investigated.

MATERIALS AND METHODS

The animals used were littermates of a breed of large white/landraces cross pigs weighing approximately 30-45 Kg. The animals were kept in metabolic cages throughout the experiment and fed a diet of skim milk and barley, low in purines. All treatments were given as supplements mixed with the food, the animals being fed twice daily. Details of sample collection and storage are essentially those reported previously (6). Analytical methods employed in these studies for the identification of both purine, pyrimidine and drug metabolites and the investigation of renal function have likewise been reported in a previous communication (6).

DESIGN OF THE EXPERIMENT

The animals were studied on three different regimes as indicated in Tables 1 and 2, representing three different sets of experiments. All results are the mean

TABLE 1

URINARY PURINE AND PYRIMIDINE EXCRETION DURING THIOPURINOL THERAPY

Thiopurinol Dosage (mmol/kg)	C	0·13	C	0·92	C	2·3
Allantoin	4·15	4·03	4·40	2·29	4·60	1·30
Xanthine	0·16	0·16	0·24	1·39	0·29	2·15
Hypoxanthine	0·08	0·07	0·13	0·42	0·15	0·33
Uric Acid	0·42	0·30	0·59	0·50	0·60	0·39
Total Oxypurine Excretion	4·81	4·56	5·36	4·60	5·64	4·17
Orotic Acid and Orotidine	0·04	0·04	0·08	0·16	0·13	0·21

TABLE 2

EFFECT OF THIOPURINOL ON URINARY PURINE EXCRETION DURING GUANINE THERAPY

Dosage mmol/kg								
	Guanine	Nil	0·75	0·75	Nil	0·75	0·75	0·75
	Thiopurinol	Nil		0·33	Nil	Nil	0·92	2·3
	Allantoin	4·69	25·23	12·58	4·40	22·86	9·94	0·71
	Xanthine	0·14	0·34	3·95	0·23	0·30	7·17	5·41
	Hypoxanthine	0·12	0·31	0·4	0·16	0·15	0·82	0·96
	Uric acid	0·59	0·67	0·41	0·52	0·56	0·39	0·21
	Total	5·54	26·03	17·34	5·31	23·87	18·32	7·29
	Orotic Acid and Orotidine	0·13	0·15	0·26	0·08	0·09	0·161	0·145

of three consecutive mid-week specimens as outlined below. The first study involved four animals, two controls and two test animals and the data are given in the first two columns of Table 1. The next two experiments involved three control and three test animals. Data from the second experiment are given in the last four columns of tables 1 and 2 and from the third experiment, the first two columns of Table 2.

RESULTS

Effect of Increasing Thiopurinol Dosage on Urinary Purine Excretion

Thiopurinol at levels within the human therapeutic range (Table 1) produced no increase in the levels of the precursor purines xanthine and hypoxanthine and total urinary purine excretion was reduced only slightly. However, at both higher dosages a substantial reduction in allantoin excretion was noted with a concomitant increase in xanthine and hypoxanthine excretion which was not stoichiometric so that total oxipurine excretion was reduced up to 30% on this regime. At the highest dosages the animals refused their food, necessitating reduction of the drug dosage so that the high dosage results are the mean only for two days of therapy.

Effect of Thiopurinol and Dietary Purine on Urinary Purine Levels

The guanine dosages employed have been used in previous studies and the magnitude of the increase in urinary purine excretion, predominantly an increase in urinary allantoin is similar to that found previously (6). Addition of thiopurinol to the guanine supplement diet at levels approximately three times human therapeutic levels produced a marked increase in xanthine excretion but the increase in hypoxanthine excretion was not so marked. Further increase in the amount of thiopurinol administered with the same guanine dosage produced an even more marked increase in xanthine and hypoxanthine levels as shown in Table 2. At the same time, the increase in total urinary purine excretion resulting from the guanine supplement was substantially reduced by the thiopurinol at all dosages. This effect was dose-related and was most marked at the highest dosage when total urinary purine levels were reduced to near control values.

However, results at the high dosage must be viewed in the light of food refusals at this time.

Pyrimidine Excretion During Thiopurinol Therapy

Thiopurinol had no effect on pyrimidine excretion in the human dosage but orotic acid and orotidine levels were increased on both the higher thiopurinol regimes.

Drug Metabolites

The only metabolite of thiopurinol found in the urine at the low dosage was oxithiopurinol. At the higher dosages in addition to oxithiopurinol, a substantial proportion of the dose was excreted as unchanged thiopurinol (approximatley 50%) in the urine. This finding coincided with a reduction in GFR and PCV at the highest dosage levels but may also be complicated by the food refusals at the time.

DISCUSSION

In contrast to allopurinol, the reduction in plasma and urine uric acid levels during treatment of gout by thiopurinol is not accompanied by increased levels of the precursor oxypurines, xanthine and hypoxanthine (2,4,5). The effect of thiopurinol has therefore been attributed to nucleotide feedback inhibition of de novo purine synthesis. We have previously found no significant effect on hypoxanthine and xanthine excretion in the pig, when given thiopurinol in doses equivalent to the usual human therapeutic range (7) with only a slight reduction in total purine excretion.

However, in the present studies at dosages well in excess of the therapeutic range, definite evidence of xanthine oxidase inhibitory activity was indicated by the considerable increase in xanthine excretion (7-10 fold) and hypoxanthine excretion (2-4 fold). The quantities of xanthine excretion were, in fact, comparable with the maximal xanthine excretion, achieved at the highest dosages in the previous study with allopurinol (6), which was considered evidence for enzyme saturation; since no further increase could be achieved by trebling the dose in that study. Further evidence for enzyme saturation at the above levels was the finding of unchanged thiopurinol in approximately equal proportion to oxithiopurinol in drug metabolites excreted at the highest dosages. At therapeutic levels, oxithiopurinol is normally the only urinary metabolite and our previous studies using ^{14}C thiopurinol have substantiated that this oxidation is rapid and complete at low levels.

The above dosage related increase in xanthine and hypoxanthine coincided with a considerable reduction in urinary allantoin excretion which was not stoichiometric and total urinary oxypurine excretion was reduced by up to thirty percent. This result is consistent with the mode of action previously proposed for thiopurinol (2,5), namely inhibition of _de novo_ purine synthesis. The present studies at doses in excess of the therapeutic range, indicate a dual mode of action for thiopurinol: xanthine oxidase inhibition and inhibition of _de novo_ purine synthesis.

The study involving thiopurinol administration following dietary purine (guanine) supplemenation demonstrated that thiopurinol is also capable of inhibiting the conversion of exogenous purine to allantoin; the increase in xanthine excretion being three times noted in the study without guanine supplementation. Furthermore when thiopurinol was added to the guanine regime, a considerable reduction in total urinary purine excretion also resulted even at doses close to the human therapeutic range. The reduction was most marked at the higher thiopurinol dosages and was of much greater magnitude than could be accounted for solely by inhibition of _de novo_ purine synthesis alone. This effect has also been noted previously with guanine and allopurinol in the pig (6) and indicates that, like its analogue allopurinol, thiopurinol is also capable of exerting an inhibitory effect on dietary purine absorption in this animal.

An inhibitory effect of thiopurinol, comparable to that noted with allopurinol on pyrimidine excretion was also noted at the higher thiopurinol dosages. This could obviously be due equally to a metabolite of thiopurinol or a metabolite of oxithiopurinol formed at the higher oxithiopurinol levels. An inhibitory effect with allopurinol on pyrimidine metabolism has been noted in both man and the pig even at dosages in the therapeutic range, however, this effect is not manifest with thiopurinol in the therapeutic range in either man or pig (4,6,7). No conclusions can be drawn at present from the adverse effects at the highest drug dosages on renal function and haematological parameters because the animals were not on this dosage for a sufficiently long period and further experiments are required. We have however, related these effects to the presence of unchanged thiopurinol which would suggest caution in the use of thiopurinol in patients with reduced xanthine oxidase activity until these results may be clarified.

SUMMARY

1. The important fact which emerges from these studies both without and with dietary purine supplementation is that thiopurinol does possess apparent *in vivo* xanthine oxidase inhibitory activity but the dose must be increased well above therapeutic levels for this effect to manifest.

2. The considerable reduction in total endogenous urinary purine excretion at these high doses substantiates an additional inhibitory effect of thiopurinol on *de novo* purine synthesis.

3. Thiopurinol, like its analogue allopurinol, is also capable of reducing the absorption of dietary purine administered in the form of guanine in the pig.

REFERENCES

1. Auscher, C., Mercier, N., Pasquier, C., and Delbarre, F., (1973), "Allopurinol and thiopurinol; effect on oxypurine excretion and on rate of in vitro synthesis of ribunucleotides" Purine Metabolism in Man, 41B, Plenum Press, 657-667

2. Delbarre, F., Auscher, C., De Gery, A., Brouilhet, H., and Olivier, L., (1968) "Le traitement de la dyspurinurie goutteuse par la mercapto-pyrazolo-pyrimidine.(M.P.P. Thiopurinol)" Presse Med 76 2329-2332

3. Elion, G.B., Benezra, F.M., Canellas, I., Carrington L.O. and Hitchings, G.H. (1968) "Effects of xanthine oxidase inhibitors on purine catabolism" Israel J. Chem., 6, 787-796

4. Grahame, R., Simmonds, H.A., Cadenhead, A., and Dean, B.M. (1974) "Metabolic Studies of Thiopurinol in man

and the pig" Purine Metabolism in Man, 41B, Plenum Press, 597-605

5. Serre, H., Simon, L., and Claustre, J., (1970), "Les Urico-frenateurs dans le traitement de la goutte: A propos de 126 cas" Sem. Hop. Paris. 46, 3295-3301.

6. Simmonds, H.A., Rising, T.J., Cadenhead, A. Hatfield, P.J., Jones, A.S. and Cameron, J.S. (1973) "Radioisotope studies of Purine Metabolism during administration of Guanine and Allopurinol in the Pig" Biochemical Pharmacology, 22, 2553-2563

DOSE-RESPONSE RELATIONSHIP OF A URICOSURIC DIURETIC

B.T. EMMERSON, L. THOMPSON and K. MITCHELL

University of Queensland Department of Medicine

Princess Alexandra Hospital, Brisbane, Australia 4102

Hyperuricaemia is a common complication of diuretic use in man and its simple elimination as a side effect would be an advantage. The finding that the indanone diuretic, MK-196, was actually uricosuric in the chimpanzee (Fanelli et alii, 1974), was of potential importance to human therapeutics. Initial studies had confirmed that a 10 mg dose was diuretic in healthy volunteers but had very little effect upon the serum urate concentration. An important question therefore was to decide whether larger doses of the drug would have a greater uricosuric effect without a corresponding increase in the natriuretic or kaluretic effects.

Protocol

The study was undertaken according to the protocol shown below to include five healthy male subjects who were studied on the same three consecutive days of the week for five consecutive weeks. In order to maintain comparability between weeks, as many variables as possible were eliminated. The same diet was taken for four days of the week starting 24 hours before the first urine collection and fluids were taken at regular intervals throughout. Of the three study days, the first was a control day, the second was the treatment day and the third was a recovery day. For the next three days, the patient was given an unrestricted diet and whatever sodium and potassium supplements were required to restore any deficit. Each study day began at 8.00 a.m. and urine was collected in two-hourly aliquots for the first 8 hours, then in a 4-hour and 12-hour aliquot. Blood was collected at hours 0, 4, 8 and 12 on the first two days of each study period and at hours 0 and 8 on the third day of the study period. The drug was administered at weekly intervals in rising doses with frusemide 40 mg inserted in random order into this sequence as illustrated in the allocation schedule.

PROTOCOL

	Monday	Tuesday Control Day	Wednesday Treatment Day	Thursday Recovery Day	Friday
Constant diet and fluid	X	X	X	X	
Medication			R_x		
Electrolyte supplement					X
Blood		0 4 8	0 4 8 12	0 8	0
Urine		0-2 hr 2-4 4-6 6-8 8-12 12-24	0-2 2-4 4-6 6-8 8-12 12-24	0-2 2-4 4-6 6-8 8-12 12-24	

WEEKLY FOR FIVE WEEKS

ALLOCATION SCHEDULE

Subject No.	Treatment Period				
	I	II	III	IV	V
1	A	B	C	D	E
2	A	E	B	C	D
3	A	B	E	C	D
4	A	B	C	E	D
5	E	A	B	C	D

Medication:

A – 20 mg MK-196 at 8.00 a.m.
B – 40 mg MK-196 at 8.00 a.m.
C – 60 mg MK-196 at 8.00 a.m.
D – 80 mg MK-196 at 8.00 a.m.
E – 40 mg Frusemide

Results

The urine flow rate increased significantly during the first 12 hours after administration of MK-196 in comparison with the appropriate flow rate during the control day (Figure 1). Increasing doses of MK-196 produced a steadily increasing urine flow rate even when viewed over a 24 hour interval. Its action appeared to be more prolonged than that of 40 mg of frusemide.

During the first 12 hours, natriuresis also was greater than during the corresponding control day. Urinary sodium loss also increased with increasing dose of the indanone diuretic and lasted for 12 hours (Figure 2). In these normal subjects, it was followed by sodium retention.

Significant kaluresis occurred during the first 24 hour period, lasting longer than the diuresis or natriuresis (Figure 3) At no stage however, was the serum potassium concentration observe

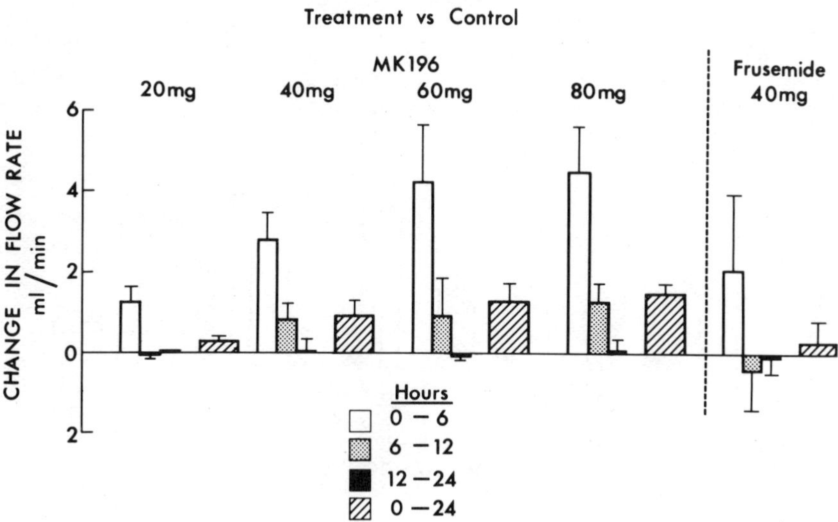

Fig. 1. Mean changes in urine flow rate on treatment day in comparison with control day (+ S.D.) with increasing doses of the indanone diuretic MK-196 and frusemide.

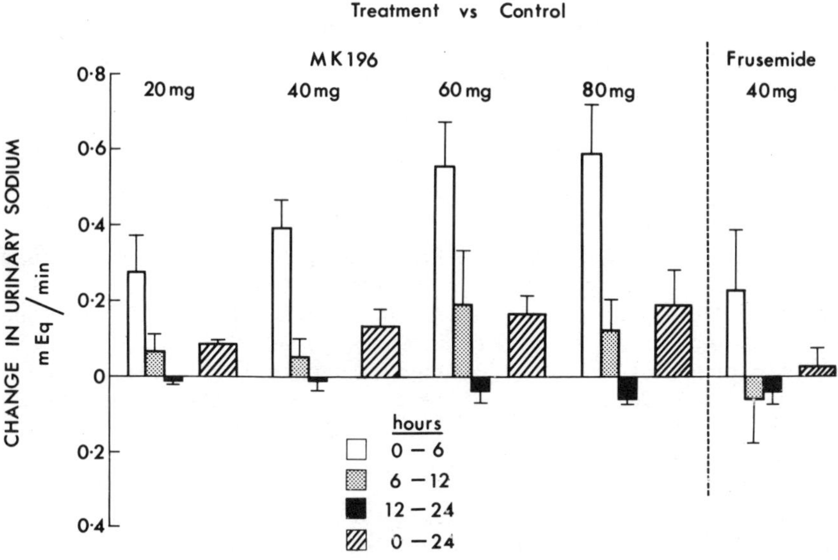

Fig. 2. Mean changes in urinary sodium excretion on treatment day in comparison with control day (+ S.D.) with increasing doses of MK-196 and frusemide.

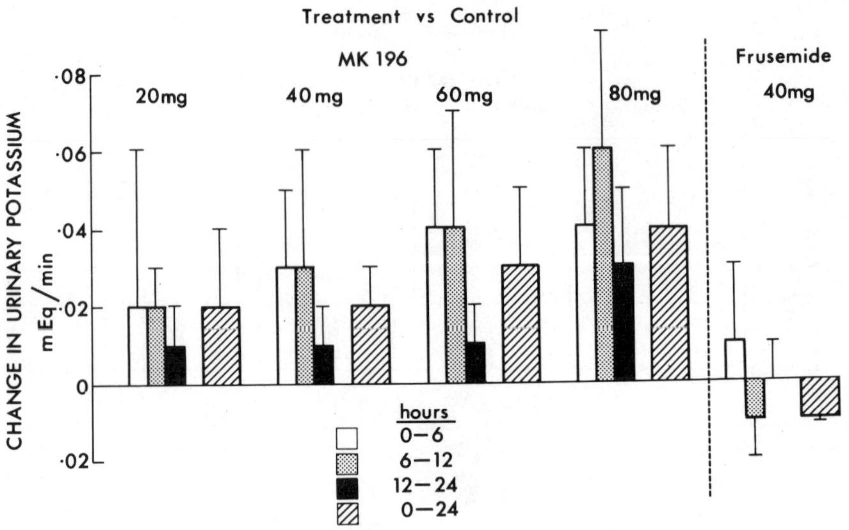

Fig. 3. Mean changes in urinary potassium excretion on treatment day in comparison with control day (+ S.D.) with increasing doses of MK-196 and frusemide.

to fall below 3.5 mEq/litre. The kaluresis also increased with increasing dose of the diuretic. The diuresis, natriuresis and kaluresis were generally more marked and more prolonged with MK-196 than with frusemide, but this was probably because only the 20 mg dose of the indanone was comparable with the 40 mg dose of frusemide which was employed.

The urinary excretion of urate increased during the first 6 hours after administration of MK-196 and its extent increased with increasing doses (Figure 4). Frusemide, for comparison, caused a reduction in urinary excretion of urate at all time intervals. This uricosuric effect did not last beyond 6 hours at any dose level and had finished before completion of either the diuresis and natriuresis (12 hours) or the kaluresis (24 hours). The uricosuria was followed by urate retention so that over the 24 hour period, there was little over-all change in renal excretion of urate.

Although a purine-containing meal had been taken on the control day, the serum urate concentration was slightly lower on the treatment day (before administration of the drug) than on the control day. Particularly with the larger doses of MK-196, the serum urate fell by about 0.2 mg/100 ml during the first 8 hours (Figure 5). Thereafter, it rose, often by about 0.4 mg/100 ml

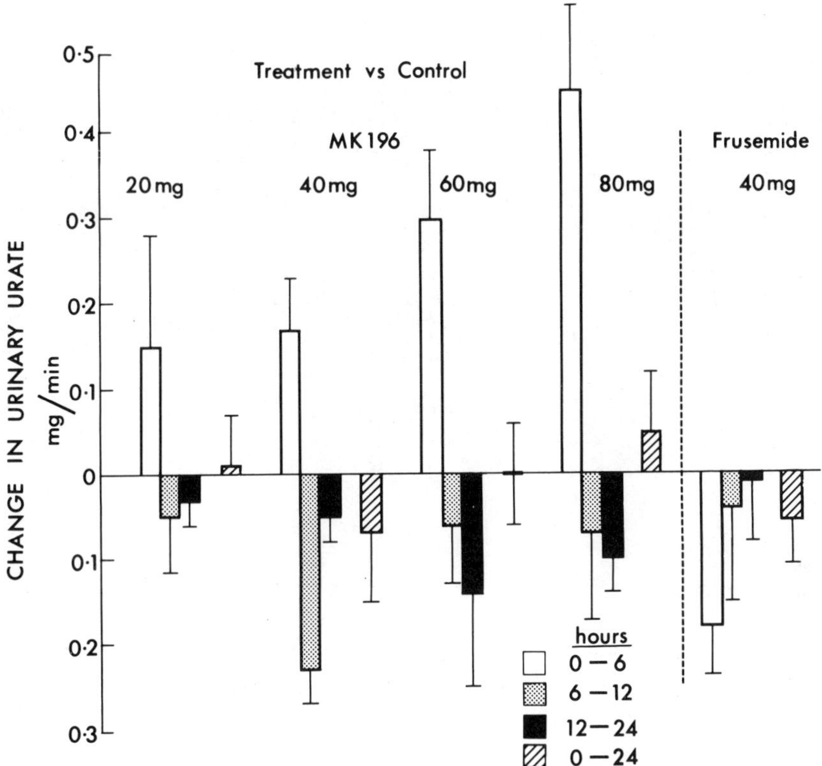

Fig. 4. Mean changes in urinary urate excretion on treatment day in comparison with control day (+ S.D.) with increasing doses of MK-196 and frusemide.

at 24 hours after administration of the drug. This was slightly more than could be accounted for by the urinary uric acid excretion, and may reflect a reduction in plasma volume. It may be noted that the urate retention of frusemide had stopped sooner than that due to MK-196 which, in view of the greater natriuresis and diuresis of MK-196, also suggests that an additional factor such as reduced plasma volume might have been involved.

Although occasional postural hypotension was seen, no consistent change in the standing blood pressure was detected with MK-196. The mean weight loss during the 24 hours following administration was 0.9 kg for the 20 mg dose, 1.4 kg for the 40 mg dose, 2.2 kg for the 60 and 80 mg dose and 0.4 kg for the 40 mg dose of frusemide. However, the pulse rate did rise, consistently reaching a peak between 12 and 24 hours after administration and

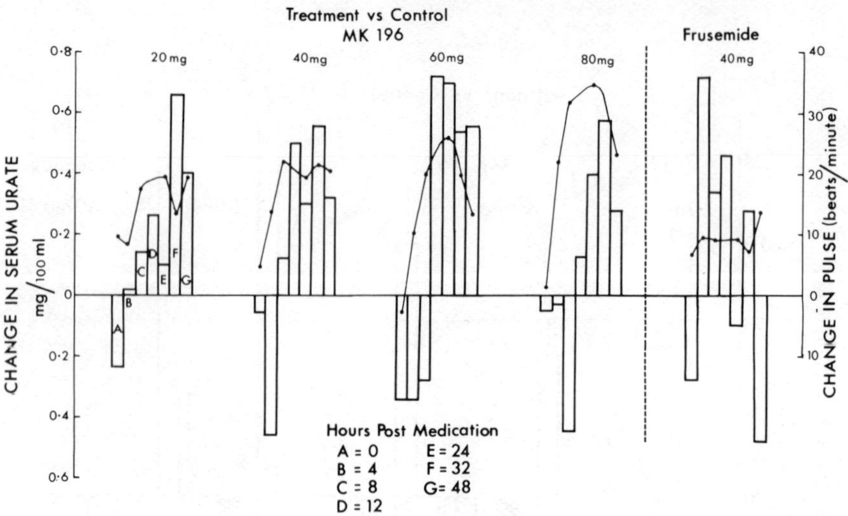

Fig. 5. Change in serum urate concentration on treatment day in comparison with control day at different times after administration of increasing doses of MK-196 and frusemide (blocks). The change in pulse rate at the different times has been superimposed and joined by a line.

tending to subside after this. It is thought that this reflects reduction in the circulating plasma volume and its degree tends to parallel the weight loss and to anticipate the extent of urate retention (Figure 5). It appears therefore that the initial uricosuric effect is limited by the urate retention associated with the volume contraction resulting from the diuresis.

DISCUSSION

There seems little doubt that this drug indanone (MK-196) does induce uricosuria, presumably by a direct effect upon the renal tubule. This uricosuric response is superimposed upon a renal tubular response to plasma volume contraction due to the diuresis. The limitation in the uricosuria may therefore be due in part to the superimposition of excessive proximal tubular reabsorption of sodium at this site in states of volume contraction (Steele and Oppenheimer, 1967). It must be remembered too that this study was undertaken in normal subjects in whom there would be a normal and active response to plasma volume contraction and definite renal retention of salt and water was observed on the recovery day. On the other hand, the uricosuria was shorter in duration than either the diuresis and natriuresis, although the rise in the pulse rate did correspond with the time of urate retention. In patients in need of diuretics, the uricosuric response may well have a different relationship with time because the plasma volume would probably be increased initially, and the diuresis would be directed towards restoring this to normal. Hence, a less significant urate retention may result from the use of this drug in oedematous patients.

It appears therefore that the indanone MK-196 has a more prolonged natriuretic than uricosuric effect. Moreover, the diuretic, natriuretic, kaluretic and uricosuric effects all increase with the increasing doses of the drug which were studied. It did not seem that the maximum response to the drug had necessarily been achieved with the doses studied and there was no basis for suggesting that the larger doses might show an increased effect on uricosuria without a corresponding effect on the diuresis. Further studies would need to involve patients requiring diuretics. A uricosuric effect more prolonged than the diuretic effect would have enhanced the usefulness of this drug, although it is encouraging to see the development of new drugs which have less tendency to cause hyperuricaemia than the previously available diuretics.

The authors are grateful to the Merck Sharp & Dohme Research Laboratories for their assistance in facilitating this study.

REFERENCES

1. FANELLI, G.M., BOHN, D.L., HORBATY, C.A., BEYER, K.H. and SCRIABINE, A. Kidney International 6:40A (1974).

2. STEELE, T.H. and OPPENHEIMER, S. Am. J. Med. 47:564-574 (1969).

THE PHARMACOLOGY OF THE HYPOURICEMIC EFFECT OF BENZBROMARONE

Irving H. Fox and David S. Sinclair

Purine Research Laboratory, University of Toronto

Rheumatic Disease Unit, Wellesley Hospital, Toronto

The pharmacological reduction of the serum uric acid in man has been commonly achieved by the use of compounds which increase the renal excretion of uric acid. Although numerous uricosuric drugs have been available, sulfinpyrazone and probenecid have gained the widest usage. Recently a new class of uricosuric compounds, the benzofurans, has been discovered and tested. One of these compounds, benzbromarone, has been found useful in the clinical management of hyperuricemia (for bibliography see 1). In addition to its uricosuric properties it has been proposed that the hypouricemic effects of benzbromarone may also result from the inhibition of uric acid synthesis. We have evaluated the mechanism by which benzbromarone could lower the serum uric acid in man (1).

Initial studies were performed to assess the effects of benzbromarone on the renal clearance of uric acid (Figure 1). Curate: Ccreatinine increased from a control value of 4.9 to a peak value of 23.1% 2 to 4 hours after benzbromarone administration. The uricosuric effect continued for 8 to 22 hours during which time Curate:Ccreatinine was 7.9% the serum uric acid decreased from a control value of 7.8 to 6.1 mg/dl after 8 hours and to 4.3 mg/dl after 24 hours.

Benzbromarone could potentially increase the renal clearance of uric acid by an elevation of the glomerular filtration rate or by an alteration in the renal tubular handling of uric acid. No evidence for an increased glomerular filtration rate was obtained since the mean creatinine clearance was 101 ml/min before and 96 ml/min after benzbromarone administration. Uricosuria may also result from the displacement of uric acid from its binding sites on plasma protein. The binding of uric acid to plasma protein was not

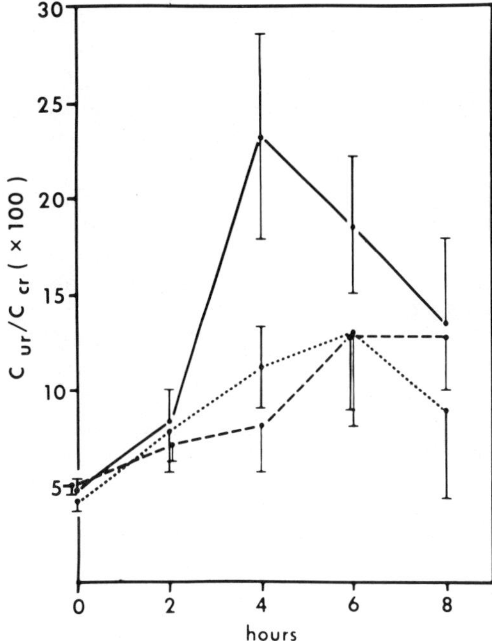

Fig. 1. Uricosuric effect of benzbromarone and sulfinpyrazone. Benzbromarone 160 mg (—), sulfinpyrazone 300 mg (...), or both drugs together (---) were given to 6, 4 or 4 patients respectively. Each point represents Curate:Ccreatinine expressed as a percentage plus or minus the standard error.

changed 3 hours following the oral administration of benzbromarone. In vitro studies of the effect of benzbromarone on uric acid binding to plasma protein showed only 22% inhibition of urate binding by benzbromarone 5 μM, a concentration transiently achieved in vivo. Since at most only 25% of uric acid is protein bound, this effect would appear to be a minor one.

The mechanism of the benzbromarone effect on the renal tubule was studied by demonstrating its interaction with sulfinpyrazone, acetylsalicylic acid and pyrazinamide, drugs that are known to modify the renal tubular handling of uric acid. The administration of sulfinpyrazone alone increased Curate:Ccreatinine from 4.4 to a peak value of 13% during the 2 to 4 hour period (Figure 1). This value was significantly less than the increase of Curate:Ccreatinine after benzbromarone. The concomittant administration of benzbromarone and sulfinpyrazone resembled the effect of sulfinpyrazone alone. Administration of both acetylsalicylic acid and benzbromarone lead to a significant diminution of Curate:Ccreatinine from 23 to 9.8% (Figure 2). Acetylsalicylic acid alone lead to a decrease of Curate:Ccreatinine. Pyrazinamide completely inhibited the uricosuric effect of benzbromarone and resulted in a response that resembled pyrazinamide administration alone (Figure 3).

Uric acid is normally handled in the kidney by glomerular filtration and bidirectional tubular transport involving reabsorption and secretion. The interaction of benzbromarone with acetylsalicylic acid, pyrazinamide and sulfinpyrazone supports a renal tubular site of activity. The conversion of the benzbromarone effect to the sulfinpyrazone effect or the complete inhibition of benzbromarone induced uricosuria by pyrazinamide implies that these compounds may be acting at the same or a related tubular site. These observations are compatible with the interpretation that benzbromarone inhibits the tubular reabsorption of uric acid.

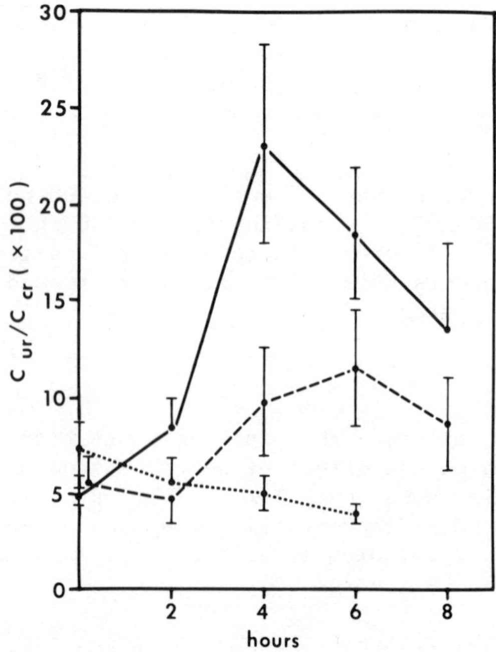

Fig. 2. The effect of acetylsalicylic acid and benzbromarone. Benzbromarone 160 mg (—), acetylsalicylic acid 600 mg (...), and both drugs together (---) were given to 6, 4 and 6 patients respectively. Format is the same as Figure 1.

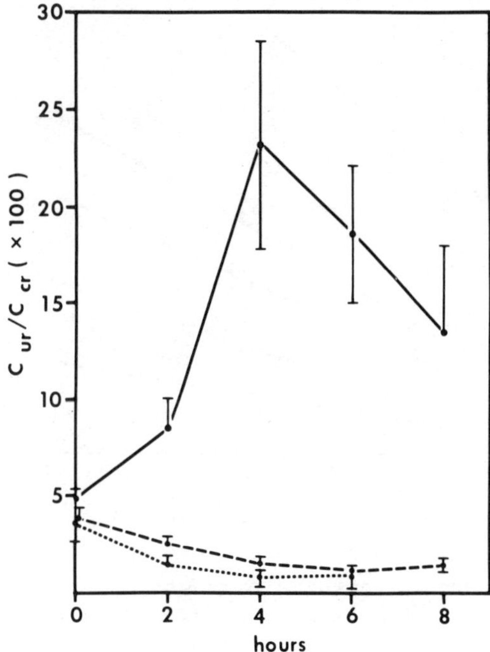

Fig. 3. Effect of pyrazinamide and benzbromarone. Benzbromarone 160 mg (—), pyrazinamide 3 g (...) or both drugs together (---) were administered to 6, 5 and 5 patients respectively. Format is the same as Figure 1.

The possibility that the hypouricemic effect of benzbromarone could originate from the inhibition of the synthesis of uric acid has been previously suggested. The effect of benzbromarone on xanthine oxidase from human liver was studied (Figure 4). Changes were observed in both the slopes and interceps, consistent with noncompetative inhibition. The secondary plot was linear and yielded a Ki slope for benzbromarone of 8.5 µM. Similar studies with allopurinol gave a Ki slope of 0.05 µM. Therefore, allopurinol, a known effective inhibitor of xanthine oxidase in vivo, was 170 times more potent than benzbromarone. As a result little or no significant inhibition of xanthine oxidase in vivo by benzbromarone would be expected, under normal circumstances. Evidence for inhibition of xanthine oxidase in man is obtained by the demonstration of an elevation of urine hypoxanthine and xanthine. These are measured as oxypurines. No increase of oxypurines was observed for up to 22 hours after benzbromarone administration, confirming the prediction of the kinetic data. It has been reported that benzbromarone may have a hypouricemic effect in renal failure and

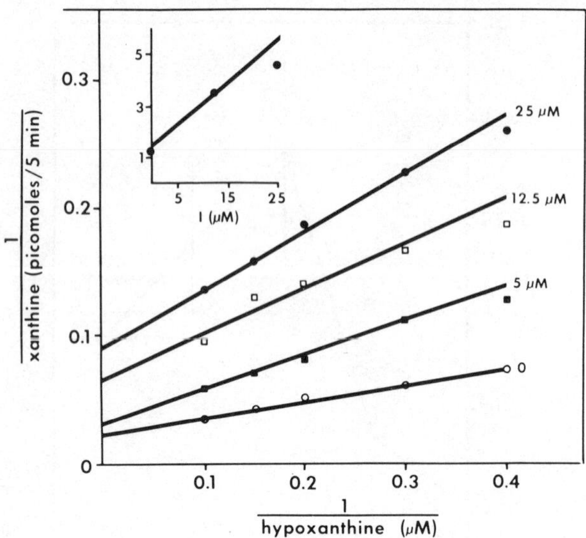

Fig. 4. Inhibition of xanthine oxidase. The effect benzbromarone 0 to 25 uM on human liver xanthine oxidase is illustrated in this double reciprocal plot. Variable hypoxanthine is plotted against the velocity. The inset shows a secondary plot of the slope against benzbromarone concentrations in micromolar.

when there are no kidneys (2,3). It is possible that higher levels of the drug occur so that more significant xanthine oxidase inhibition may be possible in renal failure.

In conclusion, these observations suggest that benzbromarone significantly increases the clearance of uric acid by a renal tubular effect. It has only a minor inhibitory effect on urate binding protein and xanthine oxidase. Only the renal tubular activity of this drug is relevant to its hypouricemic effects in man under normal circumstances. Similar conclusions have been made recently by Sorensen and Levison (4).

REFERENCES

1. Sinclair, D.S. and Fox, I.H. 1975. The Pharmacology of hypouricemic effect of benzbromarone. J. Rheumat. 2:437-445.

2. Begemann, H. and I. Neu. 1975. Die behandlung der urikopathie met benzbromaronum unter besonderer berucksichtigung der neireninsuffizienz. Therapiewoche 25:2184-2194.

3. Muller, M.M., Fuchs, H., Pischek, G. and Bresnik, W. 1975. Purinstoffwechsel und harnsaurepool bei gichtpatienten unter benzbromarontherapie. Therapiewoche 25:514-521.

4. Sorensen, L.B. and Levison, D.J. 1976. Clinical evaluation of benzbromarone. Arthritis Rheumat. 19:183-190.

THE METABOLIC EFFECTS OF TIENILIC ACID, A NEW DIURETIC WITH URICOSURIC PROPERTIES IN MAN AND DOG

Guy Lemieux, André Gougoux, Patrick Vinay, André Kiss and Gabriel Baverel
Nephrology-Metabolism Division, Hôtel-Dieu Hospital and Department of Medicine, University of Montreal
Montreal, Quebec, Canada

INTRODUCTION

Tienilic acid (2,3-dichloro-4-(2-thienyl carbonyl) phenoxyacetic acid) is a new diuretic with a chemical structure resembling that of ethacrynic acid (1). In contrast to all other diuretics in current use, tienilic acid possesses the unique property of being a potent uricosuric agent in man and mouse (1). Although its natriuretic effect is believed to take place in the cortical diluting segment of the distal nephron (2,3), the nature and site of the uricosuric effect of tienilic acid has not been established. Investigation was designed in normal man to study the metabolic effects of tienilic acid and experiments were performed in the mongrel dog, the Dalmatian coachhound, the rabbit and the chicken to clarify the nature of its uricosuric effect.

MATERIAL AND METHODS

During acute studies four healthy young volunteers took a single oral dose of 1000 mg of tienilic acid. A week later, each of the same four subjects were given pyrazinamide (PZA) as a single oral dose of 3 grams 90 minutes prior to the ingestion of tienilic acid 1000 mg. During chronic studies, six subjects were maintained on a constant diet containing protein 1.5 gram/kg, a moderately high but constant purine content, sodium 150 mEq and potassium 100 mEq/day. This diet was consumed during 11 consecutive days. After a three day control period, tienilic acid was administered during 7 days to the six subjects at a dosage of 250 mg/day during the first four days and 500 mg/day during the last three days. After a six week interval, four of the same subjects were given hydrochlorothiazide instead of tienilic acid at a dosage of 50 mg/day during the first four days and 100 mg/ day during the last three days.

All animals received tienilic acid intravenously (50 mg/kg) as a single bolus. The protocol used for the study of uric acid excretion in the mongrel dog, the Dalmatian coachhound, the rabbit and the chicken have previously been reported as well as the analytical methods (4,5).

RESULTS AND DISCUSSION

During the acute studies in man (Figure 1), it was shown that tienilic acid has a potent uricosuric effect during at least 8 hours the ratio of uric acid clearance over endogenous creatinine clearance (C_{ur}/C_{cr}) rising from a control value of 0.08 to 0.59 during the first 6 hours. The administration of pyrazinamide (PZA) prior to tienilic acid reduced C_{ur}/C_{cr} from 0.09 to 0.02. Following the administration of tienilic acid a slight depression in the uricosuric response to tienilic acid was observed C_{ur}/C_{cr} averaging 78% of the value observed with tienilic acid alone. However, with the passing of time, the difference between the two responses was accentuated. During the sixth hour C_{ur}/C_{cr} averaged only 45% of that observed with tienilic acid alone. During the last 16 hours this value was only 0.06 in contrast to the value of 0.29 with tienilic acid alone (Figure 1).

During the chronic studies, the administration of tienilic acid during 7 days led to a significant and sustained increase in uric acid excretion mean C_{ur}/C_{cr} rising from 0.11 to 0.25. At the same time plasma urate decreased from 6.0 to 3.0 mg% (Figure 2). Plasma urate rose rapidly to 5.2 mg% one day after the drug was discontinued. The administration of equivalent dosage of hydrochlorothiazide was accompanied by a modest but continued and significant fall in C_{ur}/C_{cr} from 0.09 to 0.05 while plasma urate rose from 6.3 to 7.8 mg% (Figure 2). These changes in urate excretion were not related to the natriuretic effect of tienilic acid since hydrochlorothiazide, at the dosage used, induced a greater loss of sodium. During these chronic studies it was shown that the cumulative urinary loss of potassium induced by tienilic acid was greater than with hydrochlorothiazide (139 mEq vs 45 mEq). At the same time, serum potassium fell from 4.0 to 3.4 mEq/L while no significant change was observed during hydrochlorothiazide. A moderate but significant metabolic alkalosis developed serum bicarbonate rising from 26.5 to 34.2 mEq/L after 7 days. This was accompanied by a cumulative increase in hydrogen ion excretion of 142 mEq. In contrast, hydrochlorothiazide had no significant effect on extracellular or urinary acid-base parameters. Finally, tienilic acid was shown to have a defi-

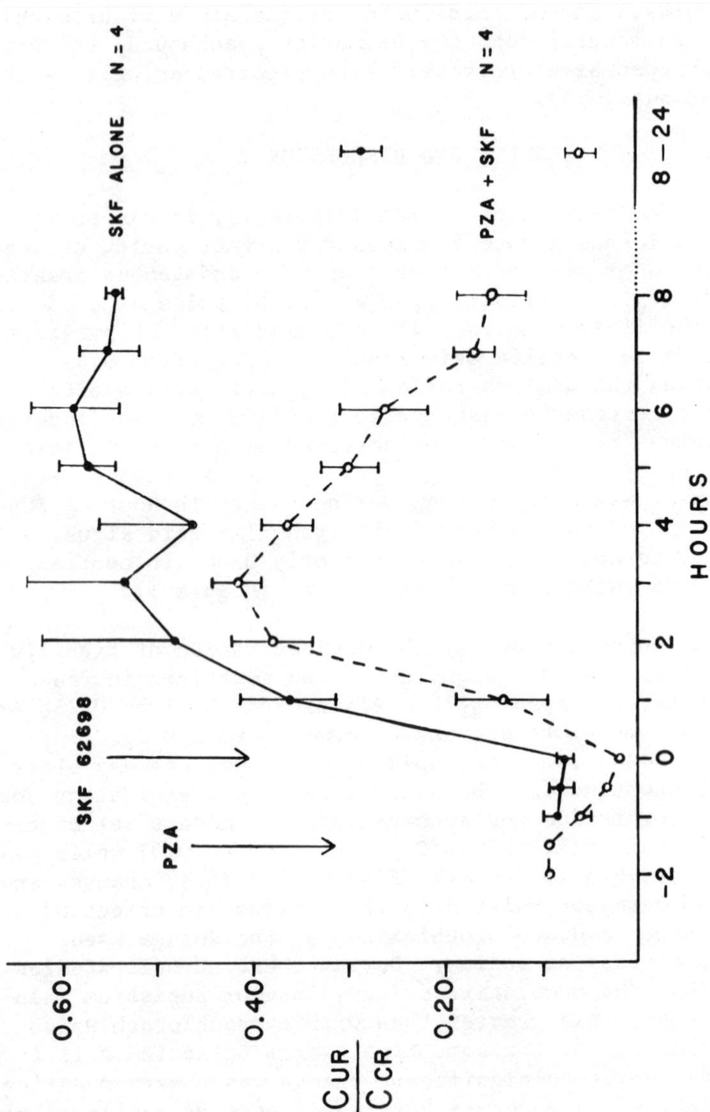

Fig. 1. Uricosuric effect of tienilic acid (SKF 62698) 1000 mg alone or following pyrazinamide (PZA) 3 grams in normal man.

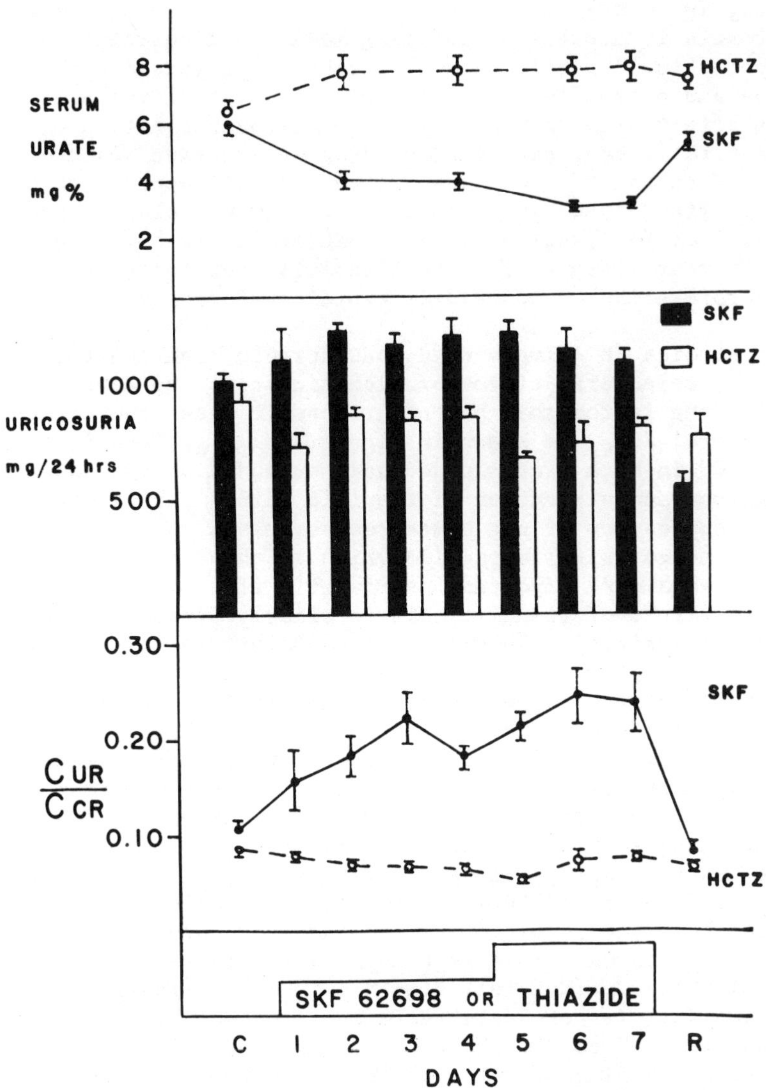

Fig. 2. Comparison of the uricosuric effect of tienilic acid (SKF 62698) with hydrochlorothiazide in normal man.

nite hypocalciuric effect. However, this effect was smaller than that observed with hydrochlorothiazide.

Thus, in normal man, in addition to its natriuretic effect, tienilic acid is capable of inducing marked and sustained uricosuria, potassium depletion and hypokalemia, increased net acid excretion and metabolic alkalosis and also hypocalciuria. The blunting effect of pyrazinamide on the uricosuric action of tienilic acid is best explained by drug interaction (6) and the duration of action of each drug. Tienilic acid appears to have a shorter effect than pyrazinamide (3,7). One could envisage a common carrier for proximal tubular secretion for both drugs having a higher affinity for tienilic acid than for pyrazinamide. Such interpretation would readily explain our results.

Our studies in animals shed considerable light on the mechanism of the uricosuric effect of tienilic acid. We have shown that this drug is the most potent uricosuric agent ever reported in the mongrel dog. As shown in Figure 3 C_{ur}/C_{cr} rose from 0.41-0.55 to 1.00 in both urate-loaded and nonloaded mongrel dogs following the administration of tienilic acid. This indicates complete suppression of net urate reabsorption. No uricosuric effect was noted with furosemide which at the dosage used (5 mg/kg) had a more important natriuretic effect. Thus the uricosuric effect of tienilic acid is clearly dissociated from its diuretic activity. In urate-loaded Dalmatian coachhounds with a control C_{ur}/C_{cr} of 1.16 the administration of tienilic acid reduced this ratio to 1.00 indicating suppression of net secretion of urate. In nonloaded Dalmatians with a C_{ur}/C_{cr} of 1.0, tienilic acid had no effect on urate excretion. In secreting urate loaded rabbits with a C_{ur}/C_{cr} of 1.5, net secretion of urate was abolished following tienilic acid administration C_{ur}/C_{cr} falling to 1.0. In unloaded rabbits with net reabsorption the control C_{ur}/C_{cr} value of 0.79 fell to 0.54. These results are interpreted to mean that tienilic acid is capable of inhibiting both the reabsorptive and the secretory fluxes of urate in the nephron. This is clearly illustrated in the chickens (Figure 4) where C_{ur}/C_{in} fell from 8.1 to 3.2 following tienilic acid, a 60% fall in net urate secretion. At the same time plasma urate rose from 25 to 38 mg%. We feel that tienilic acid inhibits the reabsorptive flux of urate which is great in the mongrel dog and the secretory flux which is predominant in the urate-loaded Dalmatian coachhound as well as in the rabbit and the chicken.

The magnitude of the uricosuric effect of tienilic acid in the mongrel dog suggests that this effect takes place in the proximal tubule. This is strongly supported by the fact that the renal extraction of para-aminohippurate (PAH) is markedly depressed following the administration of tienilic acid in the mongrel dog.

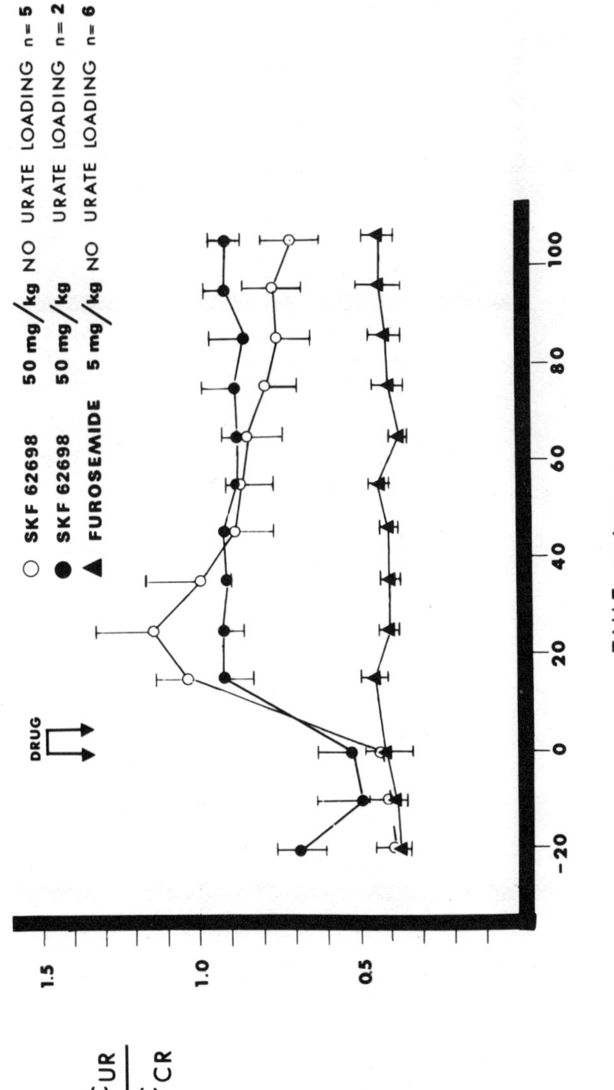

Fig. 3. Uricosuric effect of tienilic acid (SKF 62698) in urate-loaded and unloaded mongrel dogs.

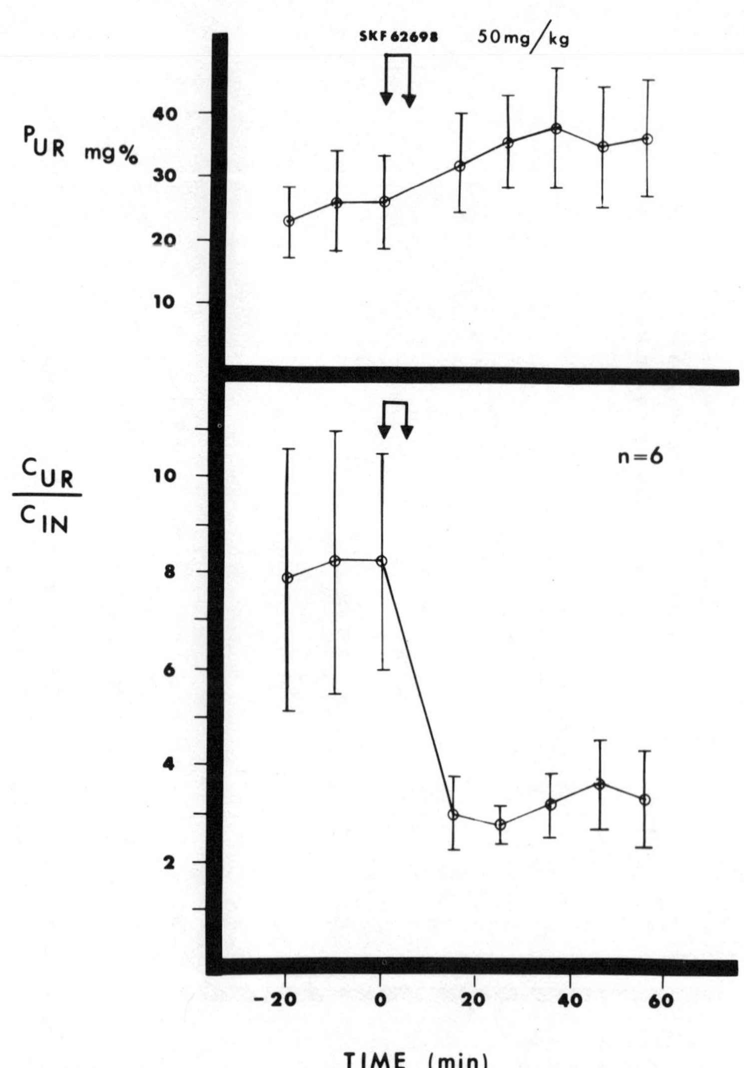

Fig. 4. Effect of tienilic acid (SKF 62698) on urate excretion and plasma urate in chickens.

Thus standard PAH clearance cannot be used to measure variations in renal plasma flow following tienilic acid. On the other hand, PAH loading (plasma PAH 30 mg%) has no effect on the uricosuric action of tienilic acid. It is most likely that tienilic acid is secreted in the proximal tubule by the common organic acid (PAH) pathway, the carrier having a higher affinity for tienilic acid than for PAH. In view of the effect of pyrazinamide on the uricosuric effect of tienilic acid in man, the presence of a second high affinity carrier between these two drugs must be postulated since pyrazinamide has no effect whatsoever on PAH transport in the nephron (8).

REFERENCES

1 - Thuillier G, Laforest J, Cariou B, Bessin P, Bonnet J, Thuillier J: Dérivés hétérocycliques d'acides phénoxy-acétiques: Synthèse et étude préliminaire de leur activité diurétique et uricosurique. Eur J Med Chem 9: 625-633, 1974.

2 - Lau K, Goldberg M, Stote R, Agus ZS: Renal effects of SKF 62698: A new uricosuric diuretic. Clin Res 24: 405, 1976 (Abstract).

3 - Stote RM, Maass AR, Cherrill DA, Beg MMA, Alexander F: Tienilic acid: A potent diuretic-uricosuric agent. J Pharmacol Clin (Special Issue): 19-27, 1976.

4 - Lemieux G, Gougoux A, Vinay P, Michaud G: Uricosuric effect of benziodarone in man and laboratory animals: A comparative study. Am J Physiol 224: 1431-1439, 1973.

5 - Lemieux G, Vinay P, Robitaille P, Plante GE, Lussier Y, Martin P: The effect of ketone bodies on renal ammoniogenesis. J Clin Invest 50: 1781-1791, 1971.

6 - Weiner IM: Mechanisms of drug absorption and excretion. The renal excretion of drugs and related compounds. Ann Rev Pharmacol 7: 39-56, 1967.

7 - Yu TF, Berger L, Stone DJ, Wolf J, Gutman AB: Effect of pyrazinamide and pyrazinoic acid on urate clearance and other discrete renal functions. Proc Soc Exp Biol Med 96: 264-267, 1957.

8 - Fanelli GM Jr, Weiner IM: Pyrazinoate excretion in the chimpanzee: Relation to urate disposition and the actions of uricosuric drugs. J Clin Invest 52: 1946-1957, 1973.

PANEL DISCUSSION: HYPERURICEMIA AS A RISK FACTOR

Dr. R. W. E. Watts (Chairman)(U.K.)
Professor W. N. Kelley (U.S.A.)
Dr. A. Rapado (Spain)
Dr. J. T. Scott (U.K.)
Professor J. E. Seegmiller (U.S.A.)
Professor A. de Vries (Israel)
Professor J. B. Wyngaarden (U.S.A.)
Professor N. Zöllner (Federal Republic of Germany)

Dr. Watts: I wish to thank Professor Kaiser and Dr. Müller for having invited us to form this panel which is to discuss hyperuricaemia as a risk factor. I wrote to my colleagues about the four main topics which we are going to discuss today. We have to consider the extent to which hyperuricaemia may be a health risk factor, or an indicator of disease; whether it ever merits treatment in the absence of known complications, and if so, when and how it should be treated. We also have to identify the risks, or the alleged risks, and the importance of known or suspected pathophysiological factors, which may modulate the serum urate and urine uric acid concentrations in particular individuals. Under this heading, we have to think of such things as genetic, ethnic and dietary factors, sex and obesity.

Having outlined my preliminary thinking on this, I would like to move to the panel for the meeting proper. I will put each of the questions that we discussed at our preliminary meeting yesterday to each member of the panel and we will then hear his view.

The first thing that we have to discuss is really what do we mean by hyperuricaemia? and I will ask Professor Wyngaarden how he defines hyperuricaemia.

Prof. Wyngaarden: I think that you might say something in both a positive and negative vein on that topic. First, the positive aspect, I think that hyperuricaemia is best defined, in physicochemical terms, as any level above the saturation value probably at 37°C. The studies that have been published indicate that the solubility of sodium urate in simulated plasma (protein free artificial plasma) is about 6.4 - 6.8 mg/100 ml. Allowing for the possibility of a very small amount of protein binding, which is of course controversial, we have rounded these figures off to give 7.0 mg/100 ml as perhaps the upper limit of normal. One needs to have information about the analytical method employed. Most of the methods in routine use clinically overestimate plasma urate by about 0.5 - 1.0 mg/100 ml when compared with, for example, analyses done with uricase. Thus, for clinical purposes something in the 7.0-7.5 mg/100 ml range represents a limit of solubility as measured with these automatic analyser methods. A second point is that the solubility of urate is very temperature-dependent. The values just given apply at 37°C, but solubility decreases to about 6.0 mg/100 ml at 30°C. The temperature in the peripheral joints is far less than 37°C. It averages about 33°C in the knee and about 29°C in the ankle and is perhaps even less in the great toe. This is a factor that we have some difficulty in evaluating, but for clinical purposes perhaps the positive answer is the solubility at 37°C adjusted for the analytical method, which brings us to 7.0 or 7.5 mg/100 ml. On the negative side, I think that it is erroneous to define hyperuricaemia by any sort of statistical approach. First of all, serum urate values are not normally distributed in the general population. The frequency distribution curve is bell-shaped in only a general way, and there is skewing towards high values. Thus, the distribution is not Gaussian, and it is not really justifiable to define a mean and two standard deviations range. Furthermore, even if one did that, the value would be different from one population to the next. It would be 1.0 mg/100 ml lower in England than in the United States, and it would be 2 or 3 mg/100 ml higher in the Pacific Islands than in the United States or Western Europe. I think that one gets into difficulty with a statistical definition in any given population whereas the physicochemical definition ought to be universally applicable although the Maoris are perhaps an exception.

Dr. Watts: Thank you very much. Professor Zöllner, would you like to add anything to that?

Prof. Zöllner: I agree completely with Professor Wyngaarden. I doubt if the term hyperuricaemia, as it is defined now and as it has been used in this Symposium repeatedly is a useful term. I think that hyperuricaemia with for example a level of 7.2 mg/100 ml has an entirely different meaning from hyperuricaemia at say 10 mg/100 ml and if we continue putting all these cases into one box we will presumably arrive at erroneous conclusions. Therefore, when we discuss hyperuricaemia we should at least say whether we are discussing the common variety or the severe cases. This is not only academic. In Germany, in the last fifteen years, as Professor Wyngaarden has pointed out for other countries, the average serum uric acid value has risen by more than 1.0 mg /100 ml and is now 6.0 mg/100 ml. Considerably more than 10 per cent of the healthy male German population have serum urate values greater than 7.0 mg/100 ml without any demonstrable disease, although they must be called hyperuricaemic in physicochemical terms. I think this is something entirely different from what we have to discuss today.

Dr. Watts: Thank you very much. Professor de Vries, what are your views?

Prof. de Vries: I agree with Professor Zöllner. Perhaps we should not talk about hyperuricaemia but about plasma urate concentrations which we need to treat if at all. Although there is not a complete consensus in our country we take the upper limit of normal as 7.5 mg/100 ml for men and 6.5 mg/100 ml for women. I do not know if this is justified.

Dr. Watts: Thank you for drawing a valuable distinction between a numerical value and an action value. Professor Seegmiller, would you like to take this subject up now?

Prof. Seegmiller: I think we run a hazard of having a great many patients referred to us with an erroneous diagnosis of gout. I think we are probably finding, at least in many areas of the world, an overdiagnosis of gout now as a result of what has been alluded to here, namely, the frequency with which serum urate determinations are performed, in many cases, on patients who have minimal if any joint symptoms. To many physicians, the presence of joint symptoms with a minimally elevated serum urate concentration has been a basis for an erroneous diagnosis of gout and a good share of the patients that are referred to me are in this category. I concur in Professor Wyngaarden's view that we should think of it in physicochemical terms rather than in terms of population groups, otherwise we are going to have separate values for women and separate values for men when I think the data that have been collected suggest that it is the same urate value which is decisive as to whether or not there is a risk of developing gout in both sexes.

Dr. Watts: Thank you. Professor Kelley?

Prof.Kelley: I certainly agree with the comments that have been made However, the degree to which urate is bound to plasma proteins may be a factor which should be considered when the significance of hyperuricaemia is being considered in a specific population. I think that is what Professor Wyngaarden was referring to when he said that the Maoris are perhaps an exception.

Dr. Watts: Thank you. Dr. Scott?

Dr. Scott: There are only one or two small additional comments. With regard to Professor Wyngaarden's points, differences attributable to differences in methodology exist but are fairly constant I think with regard, at any rate, to the Technicon autoanalyser. This gives a figure quite regularly about 0.3 mg/100 ml higher than the uricase method in serum. With this knowledge, the autoanalyser is a perfectly adequate clinical tool. This conclusion does not apply to urine where I think the results are far more fickle. I suppose that your statement that the serum levels are 1.0 mg higher in the United States than in England dates from Popert & Hewitt's survey. We have just done a survey, which I reported this morning, and you can superimpose our curves right on the Framingham distribution curves now.

Dr. Watts: Dr. Rapado, do you wish to comment?

Dr. Rapado: The levels of serum uric acid in 10,000 normal people in Spain demonstrated that 10.8 per cent of men have high levels of serum uric acid and 2.9 per cent of women.

Dr. Watts: Thank you. We now move on to the question of analytical methodology,which has been touched on. I think that we have to bear in mind that there are not only the differential U-V spectrophotometric uricase enzymatic methods,and the reductimetric methods to be considered, but also those in which one measures the amount of hydrogen peroxide produced. These methods avoid the use of a U-V spectrophotometer. Kagayama put forward such a method which I think is gaining acceptance, and a kit for doing it is available commercially. Professor Wyngaarden, do you wish to add anything on the subject of methodology?

Prof. Wyngaarden: No, I think not.

Prof. Zöllner: I think that in careful hands, all methods employing uricase give the same results but the methods utilising the determination of hydrogen peroxide as a second step are particularly tricky and liable to get out of hand if not controlled carefully.

Dr. Watts: Professor de Vries?

Prof. de Vries: No comment.

Dr. Watts: Professor Seegmiller?

Prof. Seegmiller: I think that one of the things we have seen recently is that the automated methods have greatly diminished the range of variation that different groups have reported in a normal population and in the comparison of different population groups as Dr. Scott mentioned. One of the factors that enters in, is the extent to which non-urate chromogens contribute to the values obtained by the different methods, and certainly with reductive methods, one of the largest factors has been ascorbic acid. The popularity, in the United States at any rate, of supplemental ascorbic acid, I think raises the question of the extent to which this might be altering the presumed serum urate values and perhaps even the urinary uric values. I wonder if any members of our panel have had occasion to really try to evaluate this.

Dr. Watts: Professor Kelley, would you like to take on?

Prof. Kelley: I was hoping that Dr. Watts would consider commenting more extensively since he has reviewed this recently I think.

Dr. Watts: I have no comment to make on the subject of ascorbic acid. Dr. Scott?

Dr. Scott: No comment.

Dr. Watts: Dr. Rapado?

Dr. Rapado: No comment

Dr. Watts: Thank you very much. Now I think we should turn to the question of urine analysis which,is more difficult. I think that most of the panel members would emphasise the value of mea uring the urine uric acid excretion when one first evaluates a patient. Professor Wyngaarden, can we have your views on this?

Prof. Wyngaarden: Yes, I agree that it is important to quantitate the urinary uric acid content under ideal conditions. Professor Zöllner has shown as decisively as anyone that the dietary contribution to the urinary uric acid may be very substantial,and that unless one controls the diet during the collection period the results can be only semi accurate. It is not always possible to collect urine on a purine free diet under diagnostic conditions, so we have in practice set a limit of about 800 mg per day as an alerting value. If we collect a random 24 hour urine specimen on a patient on an American diet and that value is not above 800 mg per day we feel it is unlikely that the patient has any of the special forms of gout attributable to a specific enzymatic abnormality. I think that it is useful, and far preferable from the standpoint of stone risk to establish the level of urinary uric acid output under the living conditions of the individual. The risk of urinary stone is much more closely related to the urinary uric acid excretion value than it is to the plasma urate value. Analysis on a purine free diet is preferable from the standpoint of a potential diagnosis of an enzyme deficiency. I thoroughly agree that part of the standard diagnostic work up of any moderately severely hyperuricaemi individual, certainly anyone with gout or uric acid stones, should include a careful evaluation of the urinary uric acid excretion value.

Dr. Watts: Thank you. Professor Zöllner, would you like to come in on this?

Prof. Zöllner: No, I think I agree completely.

Dr. Watts: Professor de Vries?

Prof. de Vries: Yes, I also agree, but maybe I put it a little more sharply. I think that the level of urate in blood has nothing to do with uric acid stones, but the excretion in the urine is one of the factors which determine the propensity to stone formation. It is very difficult to define an absolute value for the normal urinary uric acid excretion. The most useful information from the practical viewpoint is obtained by studying the patient under his own normal conditions. Hyperuric aciduria due to an enzyme deficiency usually exceeds 1200 mg/24 hours. One sometimes encounters patients who excrete between 1200 mg and 1500 mg/24 hours without forming stones. I think that the absolute level of uric acid excretion is relatively unimportant with respect to stone formation and the decision as to whether to treat or not to treat.

Dr. Watts: Thank you. Professor Seegmiller?

Prof. Seegmiller: I would certainly second Professor Wyngaarden's appeal to include the urinary uric acid excretion value over a 24 hour period in the work up of a gouty patient. I think there are very good reasons to do this besides the theoretical one of possibly identifying a group who may have enzyme defects that would be of value in further investigations. From the standpoint of plans for the patient's future, I feel that this is especially valuable because we are in essence committing him to a lifetime of medications, and it behoves us to make certain that we are selecting the very best medication for his particular problems. The 24 hour excretion of uric acid after equilibration on a diet virtually free from purines enables us to decide if a patient is a uric acid over-producer. Our routine is to have a six day period of purine free diet and usually a patient who has presented with an acute attack of gout is sufficiently humbled by the experience to be willing to go to the trouble of eliminating certain foods from his diet during this period immediately after the acute attack of gout. We usually try to control the acute attack of gout with colchicine so that we do not have complications from the uricosuric action of phenylbutazone during this equilibration period. We do not give uricosuric drugs or allopurinol during an acute attack of gout, so this is a very good time to arrange for the six-day period of diet. During the last 3 days, we collect 24 hour urines in a bottle containing 3 ml of toluine as a preservative and advise the patient to mark on the bottle the date and time of the beginning and end of the collection. We always do urinary creatinine determinations in order to ensure that we are getting complete 24 hour collections. If the urinary uric acid is above 600 mg/24 hours on the purine restricted diet the patient is an overproducer of uric acid and therefore deserves the benefits of allopurinol in shutting off that excessive uric acid production. If the result is in the normal range, the patient presumably has a problem in the renal elimination of uric acid so a uricosuric drug becomes the rational approach for management.

Dr. Watts: Thank you Professor Seegmiller. Professor Kelley?

Prof. Kelley: I just want to make one or two comments about some of the quantitative information that we have. As I recall the study of Gutman & Yü relating the incidence of uric acid stone formation to urinary uric acid excretion indicated that the incidence of stones in their series was 34 per cent, 35 per cent and 50 per cent in patients excreting 700 - 900 mg, 900 - 1100 mg and more than 1100 mg/24 hours respectively. I think that these data alone provide substantial evidence that a urinary uric acid excretion in an excess of 700 mg per day certainly puts the patient at a substantial risk for uric acid stone formation regardless of whether it does anything else. The two other things that I would like to re-emphasise are that in analysing the urine for uric acid it is quite clear that a uricase, or uricase modified, technique should be used. Secondly, it is also extremely important not to refrigerate the urine. We all use toluine or some other preservative in the urine and keep the sample at the ambient temperature. If we refrigerate the urine from a patient who excretes 1200 or 1300 mg of uric acid per day it is all going to be in a sediment at the bottom of the bottle, and if the technicians involved are not aware that they need to get it back into solution you will tremendously underestimate the level of uric acid excretion. While I doubt if this is a problem in any of our laboratories I am absolutely certain that it is a problem in many hospital laboratories and I think that it is a point which ought to be emphasised.

Dr. Watts: Thank you Professor Kelley. Dr. Scott?

Dr. Scott: Yes, I agree. I think that the sort of work up outlined by Professor Seegmiller is one which should be used. It is very similar to our own practice. Perhaps a few elderly patients can be let off this, but certainly in middle aged and young patients you should have this information. Of course, one has to remember even if one is only giving colchicine oneself there may have been some other physicians around who have been giving phenylbutazone or even salicylates, so one has to be very careful about inquiring what drugs have been given. Nonetheless, it is a very cumbersome procedure, and it would be nice if we could get round it. Dr. Simpkin here a couple of days ago was making a case for doing a spot morning urine uric acid and creatinine and expressing the view that the uric acid/creatinine ratio could be used in population surveys. I think that you, Professor Seegmiller, some years ago emphasised the value of the morning urate/creatinine ratio in screening for juvenile uric acid overproducers. Are you quite sure really that we cannot develop this sort of concept further?

Professor Seegmiller: We have been interested in this kind of short cut and there is no doubt that it is a very useful screening test for patients with Lesch-Nyhan syndrome and variants of the Lesch-Nyhan syndrome, and any patient who has a gross overproduction of uric acid will show up with an elevated uric acid/creatinine ratio. I should hasten to add though that this is a test that is somewhat blunted as compared to the 24 hour excretion value. Furthermore, there is a great age dependent factor here that uric acid/creatinine ratios in the new born are phenomenally high and they diminish with the years until the mid-teen age period, and then seem to be stabilised for the continuing period so one has to refer this back always to the values of the normal population. I would like to point out, however, that one of the things that we have paid very little attention to in the past has been low values for the uric acid/creatinine ratio. After this symposium it is obvious that this would be the test to use for picking up patients with purine nucleoside phosphorylase deficiency and that this would be enormously helpful as a screening test. In our own laboratory in connection with our plans for doing screening of cerebral palsy clinics and so forth for Lesch-Nyhan syndrome we have developed a simple little dip stick which is coated with plastic on one side and filter paper fairly thick on the other, and we use this as a simple way of sampling the urine for uric acid/creatinine ratios. We elute and analyse them with automated analytical equipment. We would be happy to help at least for a time with anyone who might want to set this up as a screening method in clinics for immunological defects. It is very little trouble for us to run extra samples and they can be sent in the mail.

Dr. Watts: Dr. Rapado, have you any comments?

Dr. Rapado: We also study the acid base status in our patients with renal stones.

Dr. Watts: Thank you. I think Professor Zöllner would like to come in again at this point.

Prof. Zöllner: Professor Seegmiller made a statement with which I cannot agree, namely that all patients with uric acid overproduction should be treated with allopurinol, but that patients who show a renal defect in uric acid excretion should be given a uricosuric drug. I do not think that this is really warranted. I personally believe that unless there is a particularly strong reason for using a uricosuric drug you should use allopurinol in all cases. I wonder what the other members of this panel think?

Prof. Seegmiller: I certainly agree that if the patient has evidence of renal dysfunction which we suspect might be due to urate deposits one should use allopurinol. I was thinking more in terms of a normal range of uric acid excretion in a 24 hour urine in the absence of any overt renal impairment. The other indication of course for allopurinol even in the presence of a normal uric acid excretion would be a history of frequent uric acid lithiasis.

Dr. Watts: Thank you. Professor Zöllner, do you want to add anything further?

Prof. Zöllner: Not at the moment. I think we will come back to this point. In our own clinic the incidence of some indication of renal dysfunction, be it proteinuria or something else, is rather high, and of course we are not certain that we catch all cases of early renal dysfunction. We would much rather act on the assumption that we can never exclude early renal dysfunction, and therefore prefer allopurinol in practically all cases.

Dr. Watts: Thank you. Professor de Vries?

Prof. de Vries: I would also like to take issue with Professor Seegmiller's statement, if I understood it correctly. He said that if a patient has, or has had, an attack of gout, or has had a uric acid stone and the excretion is about 800 mg/24 hours one should treat with allopurinol. I think we have all seen patients who come in with an attack of colic and who had one other attack 10 or 15 years previously. One has to ask what should have been done during these 15 years, because if you start treatment you have to continue otherwise it is illogical. I think that it is questionable whether one should treat at all except by high fluid intake and alkalinisation on the basis of one attack of stones of one attack of gout even if there is an excretion of 800 - 1000 mg of uric acid/24 hours. We start allopurinol if there are either recurrent or preferably closely recurrent episodes of urolithiasis, repeated attacks of gout or renal dysfunction. We do not treat after one acute episode unless the uric acid excretion is very high (e.g. 2000 mg/24 hours), when an enzyme defect may be present.

Dr. Watts: Thank you Professor de Vries. The next items that we have to consider are: the question of solubility, and the variability of urate levels? I would just like to see if the members of the panel wish to add anything further or whether they feel that we have explored these fields incidentally.

Prof. Wyngaarden: No, I think they have been covered.

Dr. Watts: Professor Zöllner?

Prof. Zöllner: No

Dr. Watts: Professor de Vries?

Prof. de Vries: No

Dr. Watts: Professor Seegmiller?

Prof. Seegmiller: No

Dr. Watts: Professor Kelley?

Prof. Kelley: Yes, I could comment briefly. I am sure all of you are aware that serum urate levels are somewhat variable. There is an interesting study that shows a difference 1.5 - 2.0 mg/100 ml between summer and winter. I believe that this study was in Washington D.C., and a subsequent investigation in Palo Alto, California, did not confirm these findings. However, the climate is different in Palo Alto than in Washington D.C. and this may explain the different results. One also has to be cautious about drugs that may affect uric acid excretion and thereby give a false elevation or reduction. The one series of drugs that I think one is most likely to ignore are the oral gall bladder contrast agents, and yet they are potent uricosurics and will give a transient very high uric acid excretion level in the urine.

Dr. Watts: Thank you very much. Dr. Scott?

Dr. Scott: I agree with that. You are often called to see a patient after they have had an intravenous pylegram or cholocystogram and the levels are quite bizarre.

Dr. Watts: Dr. Rapado?

Dr. Rapado: No comments.

Dr. Watts: I think we now come to what is perhaps a really central part of this meeting, the question of the possible correlations between hyperuricaemia or gout, or both, and ischaemic heart disease. I will now ask the members of the panel to comment and give us their views and experiences in this area.

Professor Wyngaarden: I think it is a little difficult for me to say much on the basis of personal experience, and I suspect that unless someone has a very large experience and analyses it very carefully one can be impressed by specific events in specific patients. There has, of course, for a long time, been a discussion, perhaps an argument, as to whether there is a more rapid rate of development of atherosclerotic disease in the gouty population. A more frequent occurrence of coronary oclusions in gouty patients has been reported in the literature for at least 50 years. There have recently been some very important epidemiological studies, and these have tended to emphasise that hyperuricaemia is only one of many factors that may influence the rate of development of hypertension or atherosclerotic disease. I think that the following statements can be made on the basis of the Framingham and Tecumseh Michigan studies, and a large investigation of 10,000 Israeli civil service workers. In the 13 or 14 year period of the Framingham study, there were twice as many coronary heart disease events in the hyperuricaemic population as in the normo uricaemic population. However, when one analysed the events, they were entirely in those patients with clinical gout and there was no excess of coronary events in the asymptomatic hyperuricaemic group. The Tecumseh Michigan study gave a similar result. In the Israeli investigation, a large number of variables was analysed in the population and looked at by multiple regression analyses as potential contributing factors to hypertension. Such things as body weight and pulse rate were important factors, whereas when all of the other factors were normalised only 1 per cent of the influence on the development of hypertension could be assigned to hyperuricaemia. If one looks at some of the studies that have come out of the Kaiser Permanente Unit in California, there have been some papers there relating the development of hypertension to hyperuricaemia, but almost all of the patients who developed hypertension in the short term were markedly obese. I think that Dr. Scott has emphasised this point and will perhaps want to elaborate on it. I came to the conclusion in my own recent review of this,and as a more or less outside observer, that the emphasis on the correlation of hyperuricaemia with hypertension or vascular disease had been too narrow, and that when one corrects for weight and for age and other anthropomorphic factors there is very little evidence to implicate hyperuricaemia itself as a risk factor in vascular disease.

Dr. Watts: Thank you very much. Professor Zöllner?

Prof. Zöllner: It is our experience that if patients with hyper- or hyperuricaemia plus gout are studied for long enough they show a high incidence of hypertension. We wonder if hyperuricaemia is only a risk factor, or risk indicator, in gouty hyperuricaemic patients with renal damage, due to stones or other causes, which have produced the hypertension.

Dr. Watts: Thank you. Professor de Vries?

Prof. de Vries: No, I have no comment except from literature.

Dr. Watts: Professor Seegmiller? Professor Kelley? Dr. Scott?

Dr. Scott: Could I go on to a rather more general point? I think that when we start to talk about these risk factors it is very important to distinguish between the associations of hyperuricaemia and the effects of hyperuricaemia. These associations of hyperuricaemia with obesity, hypertriglyceridaemia, ischaemic heart disease, hypertension etc. are very complex. Of course, not all such associations are necessarily harmful. Some people believe that there is an association with intelligence, and Professor Kellgren in Manchester, was very worried about lowering the serum uric acid concentration even in gouty patients let alone in asymptomatic hyperuricaemia because he thought it might "turn them off a little bit". I do not think that he still believes this, but he really did make the point. In general, these are associations and there is no evidence really, despite the paper we heard this morning by Dr. Dudley Jacobs from South Africa, that they are influenced by changing the serum uric acid concentration. On the other hand you have got effects, and as I see it, there are two effects of hyperuricaemia, one is joint disease and the other is certain forms of kidney disease. Now the joint disease does not matter very much because you can wait for your attack of gout and then treat it along the lines with which we are all familiar. I would very much like to hear the views of other members of the panel on the more difficult problem of what to do with a patient who has substantial hyperuricaemia, a raised output of uric acid in the urine and who must be at some risk. What is the risk, and what do we do about it? I do not know the answer.

Dr. Watts: Thank you. We shall come to this a little later, of course, but specifically on the subject of ischaemic heart disease, would Dr. Rapado have any comments to make from Spain?

Dr. Rapado: The frequency of cardiac complications in the gouty population is not higher than in the normal population in my country.

Dr. Watts: Thank you. Dr. Scott has emphasised the importance of distinguishing clearly between associations and cause and effect. Another association suggested has been with diabetes. I wonder if any of the panel would like to comment on that? Professor Wyngaarden?

Prof. Wyngaarden: It is confusing in the hospitalised populations, and there does seem to be an association between diabetes and gout in a number of studies. However, there is a negative association between hyperuricaemia and hyperglycaemia in population studies. It has been emphasised recently of course that perhaps the association of gout and diabetes in some patients is really an association with one or more other variables such as obesity and hypertriglyceridaemia. The emphasis is once again placed upon the obesity and perhaps on the alcoholism, which is also often present. I think that the distinction which Dr. Scott drew between associations with hyperuricaemia and the effects of hyperuricaemia merits emphasis. Diabetes mellitus is more related to obesity than to any direct biochemical interaction between purine and carbohydrate metabolism.

Dr. Watts: Thank you very much. Professor Zöllner?

Prof. Zöllner: We do not find any significant correlation. As Professor Wyngaarden has said, if you study these cases carefully you will find impaired glucose tolerance, but that can entirely be attributed to obesity and overfeeding.

Dr. Watts: Professor de Vries?

Prof. de Vries: I do not have the data available here, but I think that the incidence of diabetes mellitus was higher in our series of gouty patients than in the non gouty population.

Dr. Watts: Professor Seegmiller?

Prof. Seegmiller: Dr. Fessels of the Kaiser Permanente Hospital in San Francisco found that hyperuricaemic patients have an increased tendency to develop diabetes. I do not think that we have any evidence of a cause and effect relationship here now. However, at the chemical level I have always been intrigued with the fact that uric acid is a very close chemical relative of alloxan. In fact, alloxan is produced under acid conditions during the oxidative degradation of uric acid. To my knowledge, no one has investigated the possibility that human cells can produce an agent such as this and perhaps, if we got a biochemical handle on this aspect of uric acid metabolism, possibly as part of the degradation of uric acid by leukocytes it would help us to get more definitive information about a possible cause and effect relationship. At the present time, I concur that most of the increased incidence of diabetes seems to be related to obesity.

Dr. Watts: Professor Kelley?

Prof. Kelley: Only to say that I think Dr. Fessels' study is a good example of a study of obesity and not of hyperuricaemia.

Dr. Watts: Dr. Scott?

Dr. Scott: I think that a lot of the associations of diabetes have not taken body weight into account as the Glasgow group (Boyle and others) showed a few years ago, but even when weight is taken into account, and Herbert Diamond studied this very carefully a year or two ago, there does appear to be some abnormality of glucose tolerance in hyperuricaemic people. It seems to be a very subtle sort of association and probably of not much clinical importance.

Dr. Watts: I take it then that you are somewhat in disagreement with the previous speaker on this Dr. Scott?

Dr. Scott: Yes, I think that I am mildly so.

Dr. Watts: Yes, you are in mild disagreement. Dr. Rapado?

Dr. Rapado: In our normal population the presence of diabetes was 6 per cent and in the gouty population 5.7 per cent so there does not appear to be any correlation.

Dr. Watts: Thank you. The next topic I have on my list is genetic and ethnic factors. I wonder again if I could take the panel in the same order. Professor Wyngaarden? Have you anything to add on that?

Prof. Wyngaarden: I think a number of points were made throughout this week that there are some populations in which hyperuricaemia is much more frequent than for example in Western populations, particularly in the Pacific Islands among the Micronesians and the Polynesians. Some of this may be intensified by the adoption of a Western diet. However, isolates have been studied where there has not been much admixture of Western blood, and where the indigenous population are neither obese nor alcoholic, but where, nevertheless, hyperuricaemia is also frequent. Thus, there does seem to be a very important genetic factor at play. I think that is perhaps all I have to say at the moment.

Dr. Watts: Thank you. Professor Zöllner?

Prof. Zöllner: No comment.

Dr. Watts: Professor de Vries?

Prof. de Vries: In our country where we have had waves of immigration, we originally thought about genetic predisposition. For instance, the immigrants from the Yemen were all thin people, and we never saw gout among them. In recent years when they eat much more, some of them have become obese and gouty. I do not think that there is now a real difference between them and the rest of the Israeli population.

Dr. Watts: Thank you. Professor Seegmiller?

Prof. Seegmiller: No comment.

Dr. Watts: Dr. Scott?

Dr. Scott: No comment

Dr. Watts: Professor Kelley?

Prof. Kelley: No comment.

Dr. Watts: I think that one other point we might bring in here is the proposals that have been made for using some unconventional sort of food stuffs, namely, protein derived from algae, bacteria and other microorganisms. Protein preparations from these sources contain a large amount of nucleic acid and if these are fed to a population,which is genetically predisposed to hyperuricaemia and hyperuric aciduria,we could markedly increase these values unwittingly. I also wish to consider other possible modulating factors. Professor Wyngaarden? Do you feel that we have dealt with everything?

Prof. Wyngaarden: I have nothing to add this time.

Dr. Watts: Professor Zöllner?

Prof. Zöllner: I think that the purines are the most important dietary contributor to uric acid. However, we do not now have sufficient evidence to exclude the possible influence of total protein intake, or the type of protein, and we cannot say for sure how far the influence of dietary alcohol goes. Many people are unaware of the fact that the average dietary consumption of alcohol in Europe amounts to between 10 and 15 per cent of the total calorie intake and of course if you assume that only every second person really drinks sizable amounts of alcohol, then the consumption in these persons must supply between 10 and 30 per cent of their total energy requirements. There is not enough experimental nutritional investigation to exclude the possibility that these amounts of alcohol play a significant role in the production of hyperuricaemia.

Dr. Watts: Thank you. Professor de Vries?

Prof. de Vries: I have no real statistics but the gouty patients whom we see are not alcoholics or alcohol consumers. There is very little alcoholism in our country, so what the other factor is I do not know, but it is certainly not alcohol.

Dr. Watts: Professor Seegmiller?

Prof. Seegmiller: In the San Diego area, we have been concerned about possible lead poisoning arising from improperly fired pottery imported from Mexico. This has so far not been a serious matter but I think Professor Kelley has had some more detailed experiece with what is a more common problem, at least in the southern part of the United States.

Dr. Watts: Thank you. Professor Kelley?

Prof. Kelley: May I just comment briefly? I am sure all of you are aware that chronic lead intoxication from "moonshine" is a cause of gout, and this seems to be a serious problem in the South Eastern part of the United States. For example, in 37 of 43 consecutive inpatients with gout at the Birmingham Alabama,Veteran's Administration Hospital over the period of 1 year, the disease was apparently due to "moonshine" ingestion and consequent chronic lead intoxication. In our series at Duke University Hospital in Durham, North Carolina, of 100 patients with crystal-documented gout, 65 of those patients had a history of heavy "moonshine" drinking. I should, however, point out that we do not have firm evidence that these patients actually have chronic lead intoxication now, but we certainly have a strong history of heavy "moonshine" ingestion. They drink a lot, but it is a special kind of home made brew.

Dr. Watts: Thank you. I am sure that is not the problem in Bavaria! Dr. Scott?

Dr. Scott: It is just interesting to go back to Professor de Vries' point that gouty patients in Israel do not consume much alcohol, because I think this is different from the experience in most other parts of the world. There is little doubt that alcohol ingestion is very closely tied up with the hypertriglyceridaemia problem, and the body weight problem in many countries.

Dr. Watts: Dr. Rapado?

Dr. Rapado: I have seen patients who change drastically from high protein intake to a vegetarian diet, and their serum urate concentration is dramatically lower for years.

Dr. Watts: We now come to the problem of treatment. Not treatment of gout because that is not what we were asked to talk about, but the treatment of hyperuricaemia itself. Should we ever treat it or not? and if so, when? Most of us now work in hospitals where a large amount of biochemical information comes to us in a block on the initial evaluation of the patient. We have all got to learn how best to use this new routine information, and I wish to discuss this with respect to uric acid. I will begin again by asking Prof. Wyngaarden to give us his views as to what one should do about the chance finding of a high serum urate concentration.

Prof. Wyngaarden: I think there are really two parts to this, one is the high serum uric acid. Do you want to talk about the urine uric acid later or combine now?

Dr. Watts: I think we might take them both together.

Prof. Wyngaarden: The data that we have found most useful in terms of potential risk of the development of articular gout in hyperuricaemia has been the data of the Framingham study and then a more recent French study. Perhaps we could see the first slide at this point. On this slide, the prevalence of gout in males is related to the serum urate concentration and the age of the patient, or really the mean age of the population. You see the serum urine concentration range on the left and the per cent of patients in that range having gout at mean age 49 years in the centre column and at mean age 58 years on the right. The centre column is a study of Zallocar of France, and the righthand column is the study in Framingham which was published by Dr. Arthur Hall. You will notice that the prevalence of gout increases as the height of the serum urate concentration increases. The data are not really very much different if one uses the first value, the average value or the highest value of serum urate recorded. The likelihood of gout developing increases with the height of the serum urate, but it is also very much age dependent, and even at values of 10 mg/100 ml or above at mean age 49 years only one half of the males will have gout, whereas at mean age 58 years, by a slightly different method, values of 9 mg/100 ml or above, 90 per cent will have gout. This may be rather a soft figure. It is 9 out of only 10 patients, but there they are, those are the best data that I am acquainted with. Based on such figures it seems to me that there is probably almost never a need to treat a patient with a serum urate value under 9 mg/100 ml and in the 9-10 mg/100 ml range it is probably not necessary until a patient has an acute attack or until they are perhaps more than 50 years old. I personally take a fairly conservative view about the treatment of asymptomatic hyperuricaemia from the standpoint of the possibility of articular gout because it seems to me that this is not an emergency situation. One can delay treatment until the first attack occurs in most instances, and I think Dr. Scott's point about renal disease is much more important. I also think that it becomes particularly important to know if the hyperuricaemic patients are also hypertensive. The progression of renal disability in the gouty population correlates much better with the degree of control of hypertension than it does with the degree of control of hyperuricaemia, therefore, we ought to be alerted by such values but hit the hypertension hard and worry less about the hyperuricaemia.

Dr. Watts: Thank you very much. Professor Zöllner?

Prof. Zöllner: I agree, but I would add possibly two points. First I would like to stress that patients with rather high uric acid values in the serum should not only be watched but be subjected to certain diagnostic procedures. We have found two cases of otherwise unsuspected myelo-proliferative disorders which were detected by the incidental recognition of a high serum urate concentration. A minimum renal check up should, in my opinion, also be done. I would like to turn to the people who have moderate hyperuricaemia, let us say between 7 and 8 mg/100 ml. I think that in most of these cases, it indicates something wrong with the diet, one should take a careful dietary history and possibly only weigh them. If there is something wrong with the diet they should be advised about this. Thus, moderate hyperuricaemia can be a reason for medical intervention although certainly not by drugs.

Dr. Watts: Thank you. Professor de Vries?

Prof. de Vries: I do not know how justified it is but I wish to tell you about our policy. In Dr. Klinenberg's series of hyperuricaemic patients, who were not particularly gouty, there is a very large percentage of those who have what he calls subtle kidney function disturbances in urine concentrating ability, and BSP excretion. Our experiences are less impressive from this point of view but we feel that if somebody comes in or is screened, and he has a serum urate concentration greater than 9.0 mg/100 ml this is impressive enough to start to treat him from the point of view of preventing renal impairment, particularly if he has had a stone with the further complication of infection or obstruction. The occurrence of gout is not a factor in this decision. Hypertension should, of course, also be treated.

Dr. Watts: I think that what the audience would most like to hear my colleagues say is a firm statement on the level, if any, of serum uric acid which they treat in the absence of any manifestations of gout. It remains to be seen if they will get that information.

Prof. Seegmiller: I wish that we had more reliable data to go on, but I think this is in the realm where one does find differences of opinion among rheumatologists. On the basis of the data that Professor Wyngaarden just gave, many of us have been feeling very uncomfortable about patients with serum urate concentrations of 9.0 mg/100 ml and above, in whom there is no overt cause such as myelo-proliferative disease or renal impairment. We have discussed this on panels with Dr. Gutman when he was with us, and we generally feel very uncomfortable about not treating these patients in order to bring them into a more favourable category as far as the development of gouty arthritis is concerned. In general we tell them that they are likely to develop gout, and encourage them to drink large amounts of fluids. They learn the high purine foods when their 24 hour urine uric acid excretion is being assessed, although I do not say that we must restrict high purine foods indefinitely in these people. I think that the situation is much better handled with the medications that are available, so in general if the serum urate concentration is more than 9.0 mg/100 ml, I treat. If it is between 7.0 and 9.0 mg/100 ml, I advise the patients that they have a greater risk than average for developing kidney stones and possibly gouty arthritis. I advise them to take large amounts of fluid and perhaps to go easy on the alcohol, and if they are eating large amounts of meat, I advise them to cut down a bit on that.

Dr. Watts: Thank you. Professor Kelley?

Prof. Kelley: I have just one comment. The preamble is to say that I think the question of symptomatic versus asymptomatic is not quite as easy to determine. I might give you several examples. First of all, I think all of us would agree that the patient with acute gouty arthritis where we demonstrate urate crystals, or the patient with uric acid stones, or the patient with tophi, obviously should be treated, and the patient who has hyperuricaemia in a modest degree and just aches and pains, or low back pain or other ill defined symptoms, probably should not be treated. There are several grey zones. For example, the patient with hyperuricaemia and calcium oxalate stones, is that asymptomatic hyperuricaemia? I am not sure. I think there is some recent data which suggests that treatment of hyperuricaemia in that setting may be helpful, and thereby relieve symptoms. Another area which I think is a grey zone is the patient with aseptic necrosis of the femoral head and hyperuricaemia. The incidence there has shown that, in at least a substantial number of patients, aseptic necrosis of the femoral head may be due to hyperuricaemia. To get to the specific question of the treatment of asymptomatic hyperuricaemia, I do not have any more data than anyone else, and of course, we are all of us in this room asked this question and all of us respond I am sure with just as much hesitation. Since Dr. Watts is forcing us to put a number here, I would say first of all that I think we should make an effort to determine the urinary uric acid excretion. If it exceeds 700 mg/24 hours then I think that it should be treated. In terms of hyperuricaemia, I personally do not recommend treating patients with secondary hyperuricaemia unless I think they are going to be hyperuricaemic for 20 years or more, because that would seem to put them into the same category as primary hyperuricaemia physiologically. If they have primary hyperuricaemia and are not overproducers, in other words if they excrete less than 700 mg of uric acid per 24 hours, I usually pick a number of 9.0 mg/100 ml for the same reason that the rest of us do. However, I think we should reserve a special place in our hearts for those patients that come from families where there is a significant incidence of renal disease associated with hyperuricaemia, and for patients with a single kidney and other similar special circumstances.

Dr. Watts: Thank you very much Professor Kelley for those nice straight answers.

Dr. Scott: I could not add to that. If you want figures, all right, 9.0 mg/100 ml in the plasma and 800 mg/24 hours in the urine. I think one has to consider these two things together, and then of course you are dealing with a very very small group of people. We are not talking about mild hyperuricaemia (7.5 or 8.0 mg/100 ml) which we see every day. We are now talking about a very special and small group of patients. I only have one such patient, he has HGPRT deficiency, his brother presented with gout, and he is a young boy, a gross overproducer who I am sure will be in for trouble. I am therefore treating him.

Dr. Watts: Thank you. Dr. Rapado?

Dr. Rapado: I think that the hyperuricaemic patient must be kept carefully controlled, but not actively treated until some symptomatic effects appear.

Dr. Watts: Thank you very much. I think that one point has not been emphasised, namely, the importance of weight reduction in gouty patients. We have not considered the possible side effects of the drugs that we use. I would also like to know from other members of the panel how many of their patients actually turn out to be over-excretors because it is my impression that, even in Britain, it is less than the 25 per cent that one has traditionally thought about. Could I ask the members of the panel if they would like to comment on the value of weight reduction, whether we should take note of the side effect of drugs when deciding to treat otherwise asymptomatic patients, though they be rare, and what proportion of their patients they find to be over-excretors. Prof. Wyngaarden?

Prof. Wyngaarden: Perhaps in reverse order, I too thought for a long time that the data Dr. Gutman published of I think 22 or 25 per cent of patients being overproducers or overexcretors within the New York population that he served was higher than perhaps elsewhere. Professor Kelley may have more precise data on the patients in North Carolina. It is my impression that the figure is about 5 per cent of our population. We do have more lead gout, which of course is not a high excretor type of gout, but we have certainly not found it easy to locate overproducers for studies when we wanted them. On the point of weight reduction, a few years ago Dr. Emmerson published a paper and described a patient I think, as a very important example, of what may be achieved by weight reduction alone. This patient was gouty by many criteria: clinical gout, a large urate pool, hyperuricaemia, increased glycine incorporation into urinary uric acid and a rather low renal urate clearance. The patient was treated by weight reduction for about two years and then thoroughly restudied. By that time the serum urate, urate pool-size and turnover rate, glycine incorporation value and the urinary urate clearance were normal. The patient's hypertension was also improved during that time. I think that this was a very important study. Siegfried Hyden has also shown that the serum urate concentrations in a series of very obese patients were reduced, often into the normal range, when they reduced to their normal body weights. I think most of us are now a little embarrassed, that, with the flush of success of the uricosuric agents and allopurinol, attention to diet has wained and almost disappeared. Now gradually we are much impressed with the factor of obesity and there is a renewed interest in dietary control. I think it is well placed.

Dr. Watts: Thank you Professor Wyngaarden. Professor Zöllner?

Prof. Zöllner: With respect to weight reduction, it is important to differentiate clearly between the period of reduction and the evaluation of the patient who has obtained a new lower weight. It is well known that reducing in itself may increase keto-acidosis and inhibit the urinary excretion of uric acid, so that during reduction you may well have increased plasma levels and once the lower weight is obtained the plasma concentrations are usually lower than before. With respect to drugs, I think that drugs should be considered in the evaluation of all cases of hyperuricaemia. Regarding the over-excretors, we had one case many years ago, who came from Yugoslavia and at that time, when enzyme determinations were very new we could not find any evidence of a causative enzyme deficiency or other abnormality. Since Dr. Wolfgang Gröbner's return from Durham, North Carolina, we have studied every single case and found only two over-excretors in more than one hundred. However, when we took them on to the metabolic ward and gave them a reliable diet their over-excretion disappeared. So, as of today, many more than a hundred cases and not a single over-excretor in Munich.

Dr. Watts: Thank you very much. Professor de Vries?

Prof. de Vries: I think that Dr. Sperling had the same experience. When you put them on a strict diet some of them disappear from the over-excretor group. However, this raises the question of whether they are over-excretors in their natural conditions, and we cannot evalutate this any more because the peripheral and non-hospital physicians have access to automation. All citizens ask for complete check-ups, and many of them are now treated with allopurinol outside so that we cannot really have good statistics any more. We see much less stone disease, and much less gout because the patients are treated elsewhere. In my view, we have failed in the treatment of obesity, except in the most severe cases, and particularly in the general population. I think that most of us are overweight, and the average overweight citizen loses weight if you put him on a regime, but he gets it back anyway. Therefore, weight reduction is not a good way of preventing hyperuricaemia.

Dr. Watts: Thank you. Professor Seegmiller?

Prof. Seegmiller: The presence of hyperuricaemia is an additional factor that I present to obese patients as a further incentive to try to get them to lose weight. The matter of losing weight is a difficult problem and I think in my experience is best done if you can persuade the patient to come in just for a weighing-in period and work with the wife or other member of the family who is concerned with making the food. I have obtained the best results when the wife was fully convinced that it was in the interest of her husband's health to lose weight, and I have seen some very dramatic responses of the serum urate concentration to just weight reduction. As to the proportion of the gouty patients who are overproducers, I think our values at the National Institutes of Health, and also probably Dr. Gutman's, reflect perhaps a bias of patients who are more severely affected having found their way to our respective clinics. We find in San Diego, and I have heard in many other places, that the incidence of overproduction of uric acid as measured by the 24 hour excretion of uric acid, is substantially less. In our experience in San Diego it is perhaps between 10 and 12 per cent.

Dr. Watts: Professor Kelley?

Prof. Kelley: In answer to your 3 questions: yes, yes and 5 per cent.

Dr. Watts: Thank you very much. Dr. Scott?

Dr. Scott: I am very interested and pleased to hear Professor Seegmiller say what he has just done because it has worried us for a long time. Our figures for overexcretors in England were far lower than the standard published figures, being in the region of between 5 and 10 per cent, and it is interesting to hear you say what I have long thought that probably you had selected patients sent to you. Yes, no more on weight reduction. On average, if you lose 8 kg in weight your serum urate comes down by 0.8 mg/100 ml.

Dr. Watts: Nothing from Dr. Rapado? Do any members of the panel wish to add anything at this point? Silence from my colleagues. So I think that, if I might summarise for them, there are perhaps five headings under which we would ask you to take away a message. Firstly, that one should analyse urine as well as blood. Secondly, that in recent years we have underestimated the importance of dietary factors and alcohol intake in the production of excessive uric acid levels in the blood and urine. Thirdly, the treatment of the asymptomatic patient: bearing in mind Professor Kelley's strictures both before the meeting and since on how we define asymptomatic, the members of the panel would certainly not treat patients with asymptomatic hyperuricaemia, by which I mean patients who lack the specific concomitants of hyperuricaemia or hyperuric aciduria, namely arthritis or renal complications, unless the urate or uric acid levels were very high. I think that none of the members of the panel would treat below 9.0 mg/100 ml and the figure of 10 mg/100 ml was in fact mentioned by one of my colleagues. Fourthly, the members of the panel are not of the opinion that there is any significant risk of ischaemic heart disease to patients with asymptomatic hyperuricaemia. Fifthly, we consider that patients who are obese should have their obesity treated regardless of whether they are hyperuricaemic or not.

That concludes the panel meeting, and leaves me, as the Chairman of the last scientific session, with one additional duty or pleasure. It is to thank Professor Kaiser, Dr. Müller and their many colleagues from the University who are unknown to us, but who must have made a very large contribution to the success of this meeting. I must say a special word of thanks to our projectionist who coped so ably with the slides going forwards and backwards, and with, as far as I could see, not a mistake until today when he was probably given a box of shuffled slides anyway. So from all the people who have attended the Congress, may we give our heartfelt thanks to Professor Kaiser, Dr. Müller and their colleagues and we hope that we will see them at our own respective Institutions in the not too far distant future.

AUTHOR INDEX

SUBJECT INDEX